From Smart Grid to Internet of Energy

From Smart Grid to Internet of Energy

Ersan Kabalci
Electrical and Electronics Engineering, Faculty of Engineering and Architecture, Nevşehir Hacı Bektaş Veli University, Nevsehir, Turkey

Yasin Kabalci
Electrical and Electronics Engineering, Faculty of Engineering, Niğde Ömer Halisdemir University, Nigde, Turkey

Academic Press is an imprint of Elsevier
125 London Wall, London EC2Y 5AS, United Kingdom
525 B Street, Suite 1650, San Diego, CA 92101, United States
50 Hampshire Street, 5th Floor, Cambridge, MA 02139, United States
The Boulevard, Langford Lane, Kidlington, Oxford OX5 1GB, United Kingdom

Library of Congress Cataloging-in-Publication Data
A catalog record for this book is available from the Library of Congress

British Library Cataloguing-in-Publication Data
A catalogue record for this book is available from the British Library

ISBN 978-0-12-819710-3

For information on all Academic Press publications visit our website at https://www.elsevier.com/books-and-journals

Publisher: Joe Hayton
Acquisition Editor: Lisa Reading
Editorial Project Manager: Ali Afzal-Khan
Production Project Manager: Anitha Sivaraj
Cover Designer: Miles Hitchen

Typeset by SPi Global, India

Dedication

This book would not be succeeded without endless support and love of our family. We dedicate this book to our families and to junior Bilge, Ersan's daughter and Yasin's niece, welcome to us.

Contents

Chapter 1

Introduction to smart grid and internet of energy systems

Chapter outline

1.1 Introduction

The energy demand is intensively increased since a few decades due to improving industrial consumption and residential usages. On the contrary, the environmental concerns and regulations are playing pivotal role on energy generation and consumption policies. The conventional power grid is composed of widely used large scale power plants, aging transmission lines, traditional substation management systems, and consumers that have no chance to act in generation cycle. The traditional power grid either had to be comprehensively rehabilitated which could cost billions of dollars or would be revised to a new and improved grid structure that has been enriched by intelligent solutions.

Under these circumstances, the *Smart Grid* term has been first appeared late 1990s and early 2000s as a concept making the existing power grid smarter. This improvement was aiming to tackle power quality, reliability, resiliency, and flexibility problems of power grid against increasing energy demand and control requirements. The bulk generation should meet this power demand itself, and aged transmission and distribution systems were expected to handle this heavy duty while consumers were requesting high quality of service.

From Smart Grid to Internet of Energy. https://doi.org/10.1016/B978-0-12-819710-3.00001-6

This big picture has promoted improvement and enhancement of Smart Grid by two brilliant idea: two-way flow of power and communication signals. The Smart Grid can be summarized by these two contributions if we want to explain in the widest sense. However, to obtain a power grid with two-way power flow and two-way communication infrastructure is much more complicated than summarizing with a few words. The Smart Grid architecture requires widespread technologies for each infrastructure of generation, transmission, distribution, and consumers. The distributed generation and microgrid concepts are the most prominent concepts supporting to accomplish two-way power flow task. Therefore, consumers can be converted to prosumers that do not only consume the energy but also participate to power generation by constructing their own distributed generation plants with several micro sources such as renewable energy sources (RESs), energy storage systems (ESSs), and conventional generation plants and so on. Inevitably, communication technologies are involved for measurement, monitoring, and control aims in such a scenario where the prosumers either consume or generate. These improvements are assumed as the most important contributions of intelligent systems to existing power grid to comprise the Smart Grid concept [1].

In the early times of Smart Grid researches, it has announced with several names such as intelligent grid, future grid, intelligrid, and intergrid by different research groups [2]. However, Smart Grid has been widely accepted and assumes as a standard definition of this new power grid technology [1]. It will be useful to remember some perspectives of conventional grid before describing Smart Grid infrastructure.

The traditional power grid is comprised by four main infrastructures which are generation, transmission, distribution and consumption layers. This infrastructure is survived by large generation plants that are installed at several MW power levels, and it is connected to a transmission substation where the transmission system is get started [3]. The transmission line is responsible for delivering required power to distribution substations to feed the generated power to several distribution networks. Consequently, the connection between generation and consumption layers is managed by the intermediate layers. The typical transmission systems are designed to operate in large voltage operating voltage levels ranging from 150 to 765 kV while the distribution networks operate at 11–110 kV voltage ranges. The traditional power grid is unidirectional in terms of power flow since distributed generation (DG) sources have not been allowed to participate in this system. Moreover, the electricity price is determined by utility system operators where customers had no chance to choose their electricity tariff [4].

On the contrary to increasing demand and consumption rates, the central generation approach used in traditional power grid was not sufficient to meet the requirements of growing societies. In addition to generation, the control process of traditional grid was not appropriate for the aging power networks since it has been installed according to manual monitoring and manual restoration approaches due to limited control ability [2]. The schematic diagram of traditional power grid has been illustrated in Fig. 1.1 to visualize generation, transmission, distribution, and consumption levels. This hierarchical architecture is installed in a unidirectional structure allowing power flow from large and central generators to consumers over transmission and distribution lines. The centralized generation sources are mostly based on conventional and fossil fuel-based plants such as thermal, diesel or combined heat and power (CHP) cogeneration plants, nuclear power plants, hydro plants or similar generators. The generated power is firstly increased to extra high voltage levels to prevent transmission line losses to seriously decrease the power level.

The transmission line voltages can be at 765, 500, 345, 230, and 138 kV depending to distances and grid codes of utilization. While the transmission substations are required to increase the voltage levels to carry high power, the distribution substations and transformers decrease the voltage level. The voltage levels of distribution network vary according to load and consumer types where substation consumers are fed by 69 and 33 kV, primary consumers at 11 and 4 kV level, and secondary or namely residential and industrial loads are supplied with 0.4 kV voltage level. The architecture of traditional power grid is mostly assumed in a vertical structure to describe unidirectional power flow. However, we describe the traditional power network horizontal since the Smart Grid is vertical due to its multilayer architecture comprised by information and communication technologies (ICT) layer and control and management layer [3]. The comparison prominent grid features are compared for traditional and Smart Grids in Table 1.1 [5]. The limited control and monitoring features of traditional grid have forced independent system operators (ISOs) and regional transmission operators (RTOs) to improve communication capabilities of existing power network to obtain more flexible system. In the 1980s, advanced metering requirements have been improved to provide averaging on power prices and the limitations of tariff selection has been removed. In the late 1990s, environmental concerns have increased to prevent fossil fuel-based generation that was one of the milestone to improve distributed generation, demand side management, and decentralized control and monitoring operations. Thus, a new grid concept researches have been intensively started.

The distributed generation and decentralized control were main drivers of RES usage in power generation. Besides, the microgrid term has been improved in early 2000s that aided to increase capacity and resiliency of existing power grid. Therefore, it was possible to mention about two-way power flow and two-way communication options in the improved power grid infrastructure that is named as Smart Grid. The improvements have also provided self-healing and widespread control capabilities to this new power grid due to high number of sensor usage almost at each node and line of whole grid. Once a failure occurred in any section of network, the sensors detect the failure and protection system manages power flow by comprising new relaying paths for power. This control capability is provided three major technical infrastructures of Smart Grid that are smart infrastructure, smart management and smart control systems. The smart infrastructure definition stands for power and ICT interfaces

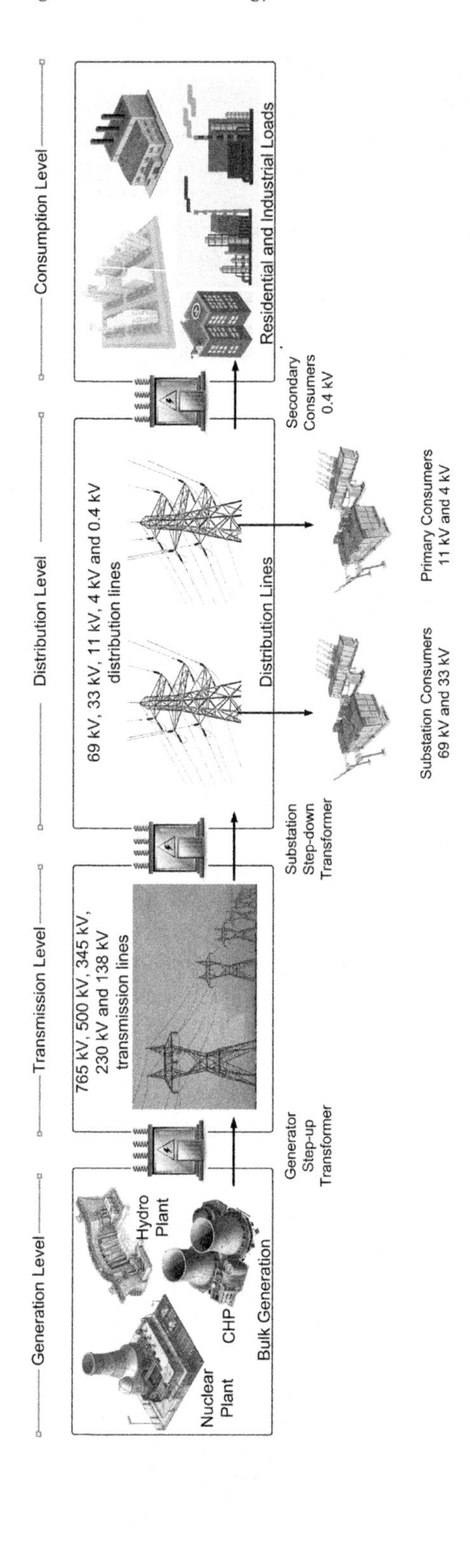

FIG. 1.1 Overview of traditional power grid.

TABLE 1.1 Comparison of traditional and smart grid

Feature	Traditional grid	Smart grid
Control method	Electromechanical	Digital
Communication	One-way	Two-way
Generation type	Centralized	Distributed
Sensing method	Limited sensors	Widespread sensor usage
Monitoring	Manual	Self and remote
Restoration	Manual	Self-healing
Control options	Limited	Widespread
Customer options	Limited	Various

that are responsible for advanced generation, distribution and consumption. On the other hand, the advanced metering infrastructure (AMI), monitoring, management, and communication technologies also comprise the smart infrastructure. The smart management system is related with monitoring, management, and decision-making subsystems along the power network. The smart protection system provides failure detection and protection, security and privacy protection, services, and system analysis during the operation of power network [1, 2].

The tailored architecture of Smart Grid is provided by intensive control and communication systems that make it able to react to any change in any section of grid. It improves reliability and resiliency of degraded generation sections by allowing to penetration of distributed generation sources and microgrid applications. Thus, security, reliability, efficiency, and sustainability of entire power network are ensured. This chapter presents an overview of Smart Grid evolution with its architectural structure in general, fundamental components of new grid structure, applications and requirements, and an introduction to internet of energy concept. The evolution of Internet of Energy (IoE) is presented in an overview section that we describe this new concept as Smart Grid 2.0.

1.2 Overview of smart grid evolution

It is assumed that the first definition and description of Smart Grid has been regulated by Energy Independence and Security Act of US Government in 2007 [6]. The general perspective and characteristic features have been defined in 10 items as,

(1) Increased use of digital information and controls technology to improve reliability, security, and efficiency of the electric grid.

(2) Dynamic optimization of grid operations and resources, with full cyber-security.
(3) Deployment and integration of distributed resources and generation, including renewable resources.
(4) Development and incorporation of demand response, demand-side resources, and energy-efficiency resources.
(5) Deployment of "smart" technologies (real-time, automated, interactive technologies that optimize the physical operation of appliances and consumer devices) for metering, communications concerning grid operations and status, and distribution automation.
(6) Integration of "smart" appliances and consumer devices.
(7) Deployment and integration of advanced electricity storage and peak-shaving technologies, including plug-in electric and hybrid electric vehicles, and thermal-storage air conditioning.
(8) Provision to consumers of timely information and control options.
(9) Development of standards for communication and interoperability of appliances and equipment connected to the electric grid, including the infrastructure serving the grid.
(10) Identification and lowering of unreasonable or unnecessary barriers to adoption of Smart Grid technologies, practices, and services.

It has been declared as a policy of US Government to modernize the electricity transmission and distribution system to maintain a reliable and secure electricity infrastructure that can meet future demand growth and to achieve each of the following, which together characterize a Smart Grid. These items summarize the technical merits of Smart Grid infrastructure in terms of characteristic features. The digital control and information technologies are widely used in Smart Grid applications to increase reliability, security, and efficiency of power grid. It was well-known that security of this cyber-physical system (CPS) is depended to dynamic optimization of grid operations and sources. Therefore, demand response (DR), demand side management (DSM), deployment of distributed source in generation and deployment of smart technologies such as remote monitoring, advanced metering, and distribution automation control have been described as crucial characteristics of a Smart Grid system. In addition to contributions in generation, transmission, and distribution systems; the consumption level is also considered in Energy Independence and Security Act of 2007. The consumer devices and appliances should be converted to smart ones while improving the existing power grid.

The smart appliances, smart applications, and smart devices are targeted to convert consumers to prosumers that plays active role in Smart Grid environment. The prosumers can install their microgrid plants with RES and distributed micro-sources, and thus they can participate to generation and increasing the grid efficiency with their plug-in electric vehicles (PEVs), smart home management systems, AMIs, smart meters, and smart appliances. The advanced metering systems that are improved technologies of regular automated meter reading (AMR) and automatic meter management (AMM) systems provide increased accessibility for DSM operations in distribution network operators (DNOs). A robust communication infrastructure is involved to accomplish these tasks while converting the conventional power network to Smart Grid. The advanced and sophisticated communication systems are operated either in wireline or wireless mediums in Smart Grid infrastructure. The wireline communication technologies include power line communication, fiber optics, ethernet or digital subscriber lines (DSLs) while the wireless communication is provided by wireless personal area network (WPAN), wireless local area networks (WLAN), IEEE 802.22 protocol wireless regional area network (WRAN), worldwide interoperability for microwave access (WiMax), cellular and satellite based or several other IEEE 802.15 based technologies.

IEEE 2030-2011 standard describes communication architecture for Smart Grid in a hierarchical arrangement of applications, services, and infrastructures. It is required to create a consensus of numerous technologies and to refine them into some limited ICT infrastructures. Thus, the Smart Grid communication systems have been described with three subnetworks where the first one includes building area networks (BAN), home area network (HAN), and industrial area networks (IAN). These area networks are defined to be used in consumption level while neighborhood area network (NAN) and field area network (FAN) comprises the local area network (LAN) that is utilized in distribution level of the power network. The last and widest network type is described as wide area network (WAN) which is used for communicating in several km distances and includes a number of LANs, virtual private networks (VPNs) and data management systems (DMSs) for transmission level operation [1, 7].

The operation of Smart Grid is depended to several intelligent electronic devices (IEDs), smart transformers, smart power converters, phasor measurement units (PMUs), and remote terminal units (RTUs) that all are equipped with sensors to provide required measurement and monitoring information to control center. Thus, the entire power network is converted to an adaptive, flexible, predictive, interactive, secure and self-healing system. The Smart Grid is sometimes defined as system of systems due to these features. The smart systems support all types of generation and energy storage options to perform two-way power flow in this new grid architecture. Moreover, it also provides resilient operation against CPS attacks by predicting possible faults and failures along the grid, and rapidly reacts to these changes or attacks to overcome faults occurred at any level.

This section presents architecture and conceptual models of Smart Grid, introduction to distributed generation and microgrid structure, smart devices used in transmission and distribution networks, energy storage systems, and control, resiliency, and flexibility features in terms of Smart Grid evolution.

1.2.1 Architecture of smart grid

European Committee for Standardization (CEN), European Committee for Electrotechnical Standardization (CENELEC), and European Telecommunications Standard Institute (ETSI) are requested to develop a standard framework for Smart Grid, and they have provided one of the most prominent reference architecture for Smart Grid by the Smart Grid Coordination Group (SG-CG) to accomplish European Commission requirements on new grid infrastructure. The Smart Grid Reference Document of CENELEC presents a detailed conceptual model and reference architecture principles, Smart Grid architecture model framework, and reference architecture elements to improve a standardized research infrastructure for any shareholder of power grid [8].

Another important reference guide has been presented by National Institute of Standards and Technology (NIST) of US because of Energy Independence and Security Act (EISA) of 2007 in 2010. The main contribution of NIST report was related to proposing roadmap for interoperability and standards of Smart Grid. In this context, a conceptual reference model has been proposed and it has been used as a standard architecture to define Smart Grid infrastructure [9]. The conceptual reference model of NIST has proposed a interconnected network where each shareholder and section has been defined as a domain and whole Smart Grid has been divided into seven domains as presented in Table 1.2. The main goal of conceptual model is to provide support for planning and organization of domains and their sub domains including shareholders, participants, and applications. The shareholders that are defined as actors in NIST

TABLE 1.2 Domains and shareholders in NIST smart grid conceptual model

Domains	Shareholders in the domain
Bulk generation	Centralized or distributed generation units with optional energy storage systems for massive generation
Transmission	Carriers of generated energy in bulk quantities to long distances and transmission substations
Distribution	The electricity distributors to consumers and also connections of prosumer side generation with energy storage options
Consumption	Consumers and end-users of generated electricity. They can also generate, store, and manage their energy sub-system
Market	Operators and participators in electricity market
Operations	The managing authority of generated and transmitted electricity
Service provider	The shareholders and operators to provide generated electricity to consumers and utilities

model describe the infrastructure including devices, systems and operation software to manage the whole system. On the other hand, the operations are performed by one or more shareholders in any domain division. The remote monitoring, smart metering, energy generation with distributed sources and RESs, energy storage and management are assumed as the operations and applications in the conceptual model. The shareholders and actors in the same domain systems are responsible for similar objectives. The bulk generation, transmission, distribution, and consumer domains are connected with power and communication networks while the other domains are interconnected by using only communication flow that is based on two-way signal flow [9].

The domain infrastructure and interaction of domains is illustrated in Fig. 1.2 regarding to NIST conceptual reference model. The power flow of conceptual reference model is depicted in lower horizontal plans starting from bulk generation to consumer domains. In any Smart Grid architecture, theses domains include central generators in generation domain, substation transformers in Transmission Domain, substations and distributed generation plants as microgrid integration in distribution domain, and micro-sources and ESSs in consumer domain. The Distribution Domain also includes distribution management systems, ISOs and RTOs which are mainly located in Markets and Operations domains. On the other hand, the Consumer Domain includes smart meters and smart energy management systems.

The proposed conceptual reference model of NIST is not only a framework to define domains and shareholders, but also an efficient roadmap to comprise a Smart Grid architecture complying with regulations, standards, and interoperability requirements along the grid codes. The upper layer of conceptual reference model shown in Fig. 1.2 is mostly related with communication and operation bases. The Markets Domain is equipped by ISO and RTO participants that are in connection with ISO and RTOs comprising Operations Domain.

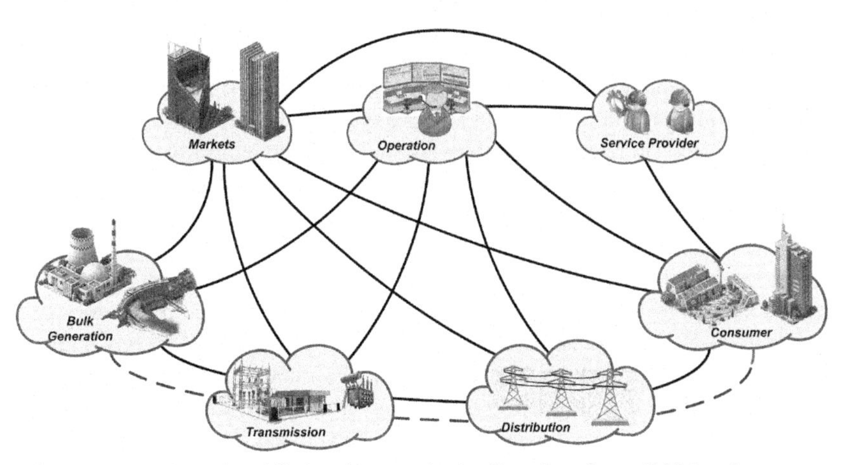

FIG. 1.2 Power flow (dotted line) and communication flow along Smart Grid domains.

The business objectives, investment and installation plans and modernizing the existing grid infrastructure are decided with the interaction of Markets and Operation Domains. The transmission and distribution operations are also located in Operation Domain in addition to ISO and RTO operations. The transmission operations are comprised by energy management systems and WAMs that interact with Transmission Domain over enterprise bus and transmission Supervisory Control and Data Acquisition (SCADA). The distribution operation in Operation Domain includes DMS, metering data management system (MDMS), DR and asset management systems. These systems interact with lower horizontal domains over metering system and distribution SCADA infrastructures using WAN and FAN connections [9].

The Service Provider Domain includes utility provider and third-party provider applications both including customer information service (CIS) and billing services. The utility provider applications are comprised by these services while the third-party provider applications include additional home/building management, aggregation and other related services. Service Provider Domain interacts with Operation Domain over enterprise communication bus or internet-based connections. According to the NIST conceptual reference model, the communication applications along domains are performed by gateway actors, information networks, and communication paths. The gateway actors are responsible to interaction between different domains by using several communication protocols and services. The implemented networks are called information networks which is a collection of communication devices, ICT systems, and network servers. Moreover, communication paths describe the secure communication infrastructures and data exchange between domains and ICT devices [9].

The European Commission has requested from CENELEC to develop a Smart Grid reference model as a standard for interoperability as like NIST conceptual model. CENELEC has comprised a research group and put into action this reference model by handling interoperability concerns at first. The interoperability categories have been defined into three branches as technical, informational and organizational divisions. The technical interoperability category comprises fundamental physical and logical connections by three subcategories that are described as basic connectivity, networks interoperability, and syntactic interoperability. These three subcategories out of eight establish physical connection, exchange message control, and implementing data structures along exchanged messages. The informational interoperability category provides semantic understanding and business context by using message data structures presented from technical interoperability section. The organizational interoperability level as the highest layer among others includes business procedures, business objectives, and economic policies. The interoperability sections are illustrated in Fig. 1.3 where the conceptual framework model of GridWise Architecture Council reference have been developed regarding to refined three different interoperability layers as seen on the left-hand side of figure.

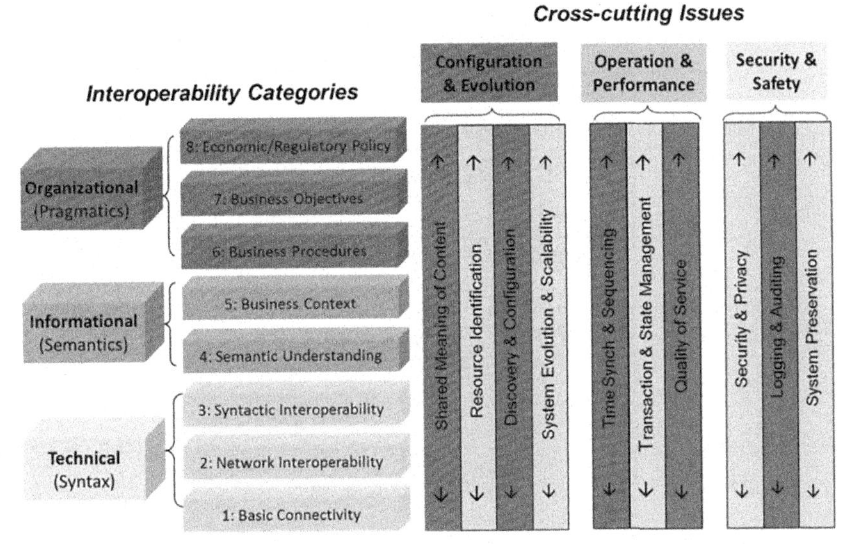

FIG. 1.3 Interoperability framework of GridWise Architecture Council [10].

The summarized sections are titled as technical, informational, and organizational layer that are interfacing eight component and communication layers. Each layer covers any smart grid plane related to architecture model in terms of power flow or communication signal flow.

The component layer is physical structure of Smart Grid context similarly the lower horizontal domains of NIST model presented in Fig. 1.2. This layer includes bulk generation, transmission, distribution, and consumer domains with their required and related devices used for metering, monitoring, control, and protection objectives. Moreover, the communication connections comprising CPS are also located at this layer. The communication layer hosts protocols and interoperability mechanisms for ICT devices and related services, data management interfaces and components are located at this layer. The information layer defines data models for information exchange between functions, services, and devices. Thus, semantic understanding and business context are managed and interoperability of information exchange is performed at this layer. The function layer includes required functions and services to comprise an architectural viewpoint as its name implies. The business layer meets the markets and operation domain of NIST conceptual model where it is used to determine paths for policies, business models, involved products and services, and decision-making actions [8].

On the contrary of NIST architecture model, European Smart Grid architecture is described in a three-dimensional structure due to interoperability layers and zone definitions. The basic smart grid plane is comprised by domains and zones as shown in Fig. 1.4. The domains include SGAM component,

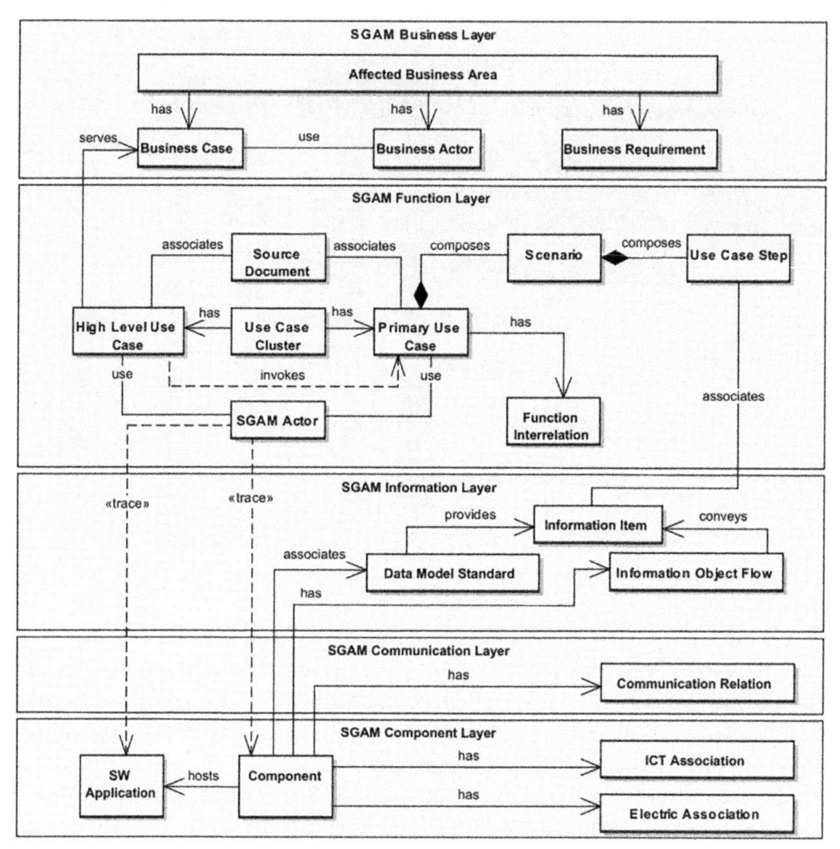

FIG. 1.4 SGAM model and its application in INTEGRA Meta-model [11].

communication, information, function, and business layers. The domains except DER is like NIST model while DER domain is comprised by small-scale power generation plants up to 10 MW generation capacity. The customer premises may also include micro-sources in terms of generation that enables customers to participate in generation.

The contribution of CENELEC architecture model represents zones that are describing hierarchical levels of power grid management at process, field, station, operation, enterprise, and market levels are modeled in a project titled INTEGRA and layers are operated as seen in Fig. 1.4. [8, 11]. The Process Zone includes energy generation methodologies by using chemical, fuel based, renewable sources. The power generation and management equipments such as generators, substations, circuit breakers, overhead lines, distribution transformers, sensors, and measurement devices are also located in Process Zone. The Field Zone of CENELEC model is comprised by protection, control, and monitoring devices such as relays, IEDs, data processors, data acquisition devices, and aggregators. The Substation Zone represents area aggregation based on field level acquisition obtained from data concentrators, substation automation, SCADA systems, and plant supervisors. The Operation Zone is comprised by high level control systems such as DMS, energy management systems (EMSs), microgrid management systems, DER managements and other related management systems. The Enterprise Zone interacts between Operation and Market Zones by providing commercial and organizational processes, services, and applications while Market Zone is dedicated to energy conversion operations such as trading, retail market and mass management.

In addition to NIST and CENELEC conceptual architecture models, IEEE have proposed a guide for Smart Grid interoperability, end-use applications, and loads based on IEEE Standard 2030. This standard a Smart Grid interoperability reference model (SGIRM) that is a reference guide for describing terminology, characteristic features, performance evaluation criteria, and application principles for end-user applications and loads [12]. IEEE Standard 2030 handles a Smart Grid system as a system of systems including power networks, communication technologies, and information technologies. Thus, the power and signal flow along power network are defined with the classification and characteristic features in terms of interoperability. Another aspect of system of systems description is related with *smart infrastructure systems*, *smart management systems*, and *smart protection systems* definitions given in [1]. The smart infrastructure system is comprised by ICT and power networks at each domain of Smart Grid system while the smart management system includes control and protection subsystems that are required to ensure power quality and resiliency of power network. The smart protection system includes particular security and privacy protections for whole power grid at hardware and software domains [1].

It is obvious that the communication infrastructure has crucial role in Smart Grid applications and services that are involved to coordinate each domain and zones. It has been previously discussed that the services and applications of Smart Grid installs a connection base with each domain by using area networks. Regardless of any grid type integrated to predefined Smart Grid architecture, IEEE Standard 2030 accommodates a synchronous operation among different frameworks and architectures in the context of interoperability requirements. It suggests a three-layer vertical organization scheme as shown in Fig. 1.5 where IEEE Standard 2030 Smart Grid Interoperability Guidance is in the middle of conceptual reference model and Smart Grid Applications. The conceptual reference models of NIST, IEC, IEEE, CENELEC and so on should comply with IEEE Standard 2030 interoperability guidance for communication, power network and ICT segments. The architectural principles of Smart Grid are described in 12 terms by IEEE Standard 2030 as standardization, openness, interoperability, security, extensibility, scalability, manageability, upgradeability, share ability, ubiquity, integrity, and ease of use [12].

The standardization implies for the definitions of elements and methods used in the Smart Grid infrastructure to provide clearness while openness is

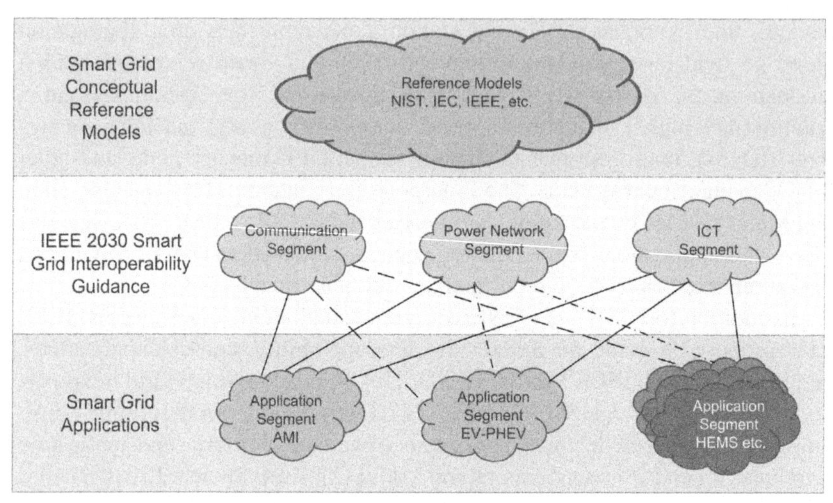

FIG. 1.5 IEEE Standard 2030-2011 Smart Grid interoperability.

required for generating a standard technology which is available for each stakeholder. The interoperability describes the standardization of interfaces and applications without depending to any geographical, preferable or methodological difference in the use of system. The security is required to prevent unauthorized access and interference to the system operation to improve information privacy and security policies. The extensibility of Smart Grid architecture is requested to enable the infrastructure to adopt newly discovered and developed applications while scalability implies for expansion capability of infrastructure against integration of new power plants to existing network.

The manageability conditions of Smart Grid architecture include configuration and management capability of each component throughout the infrastructure, fault isolation and remote management abilities. The interoperability guidance layer accommodates connection to any Smart Grid application including AMI, PEVs, and home energy management system (HEMS). Thus, interoperability brings the ability of any networks, applications, services, devices, and systems to communicate by using the ICT in a secure and reliable way. The Smart Grid interoperability infrastructure is expected to incorporate with hardware and software systems, several data transmission systems, and data exchange networks. The ICT interaction of interoperability is improved regarding to Open Systems Interconnect (OSI) reference model where functions are placed into seven layers, and layers are connected with service interfaces as done in the internet infrastructure [1, 12].

The communication networks providing connection along applications and conceptual reference models of Smart Grid architecture are illustrated in Fig. 1.6. The communication infrastructure includes core network, WAN, and private networks in component layer that is comprised by generation,

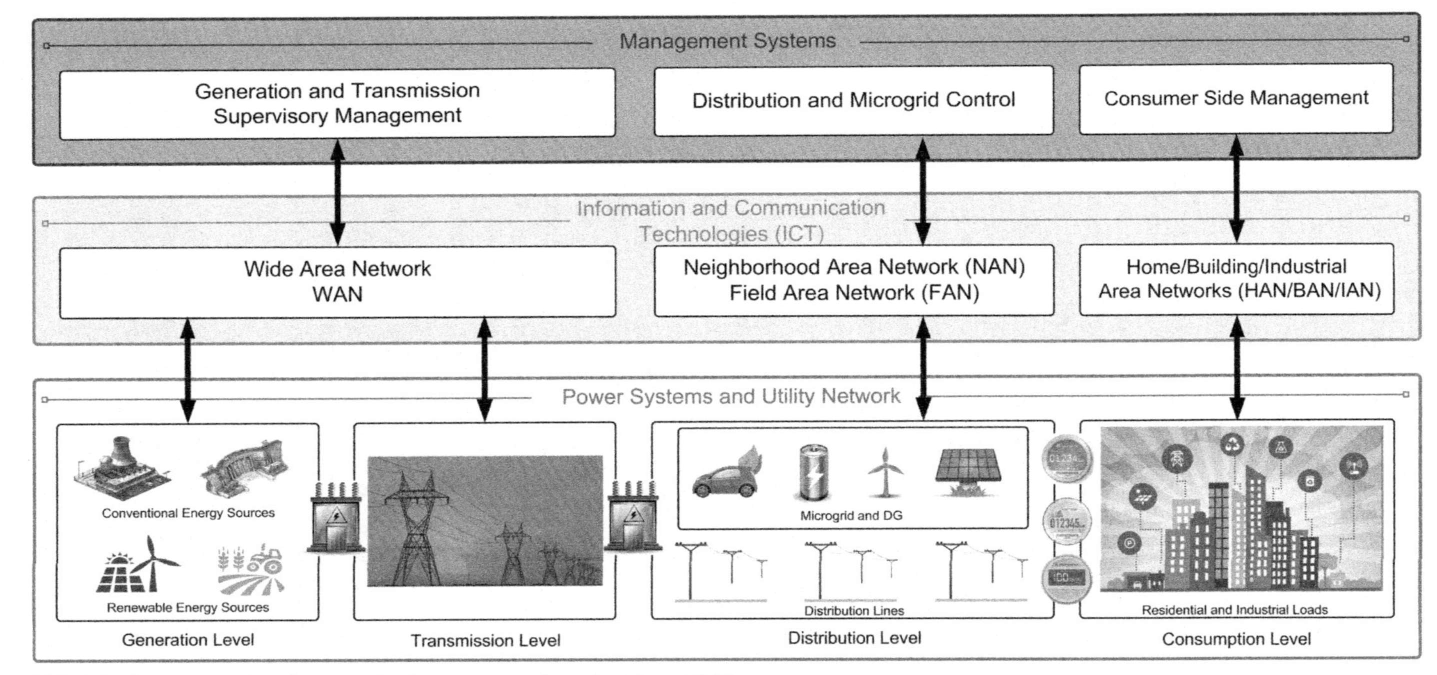

FIG. 1.6 Power network and communication ecosystem throughout Smart Grid.

transmission, distribution, and consumption levels. The ICT backbone network is composed of DSL and fiber optic wirelines connecting control and management systems of substations and plants. The particular control and management of applications are performed by using WAN, LAN, NAN, FAN, BAN, HAN and IAN type networks from bulk generation to consumption domains respectively [1, 13]. The generation system of Smart Grid includes DER integration in addition to conventional bulk generation and connects to transmission level by a step-up transformer as illustrated in Fig. 1.6. The ICT interface of generation and transmission level is accomplished by using WAN networks that provides the connection between domains and supervisory management system. The distribution and microgrid management systems are connected to distribution level and substations over LAN, NAN, and FAN networks. The distribution level includes microgrid and substation distribution that enables two-way power flow. The residential, industrial, and substation loads comprise the consumption level where communication interface to consumer side management systems are HAN for residential loads and BAN and IAN for industrial and substation loads. The main contribution of WAN is to install connection between substations, DERs, ESSs, feeders, transformers and other bulk equipments. The bandwidth of WAN is the highest among others and enables long distance data and control signal transmission with very low latency. NAN is essential to manage AMI networks to transmit measured consumer data such as demand rate, consumption level, and power quality while FAN is used to install a communication environment between backhaul and distribution networks services.

Thus, data transmission between management system and distribution substations, feeder points, and services are accomplished by FAN networks. Another AMI network throughout Smart Grid ecosystem is located at consumption level where data transmissions such as smart meter, HEMS, EV and PHEV consumption, and microgrid generation rates of customers are acquired by using HAN networks. Table 1.3 presents a detailed list of wireline and wireless communication technologies, standards that rely on, data rates, transmission distances, network types that they are used, and comparisons in terms of advantages and disadvantages. The wireline communication technologies include power line communication (PLC), fiber optic, and DSL while wireless technologies are WPAN, Wi-Fi, WiMAX, GSM and satellite.

The smart metering and smart monitoring systems based on these communication technologies are presented in Chapter 2, and smart grid network architectures are particularly introduced and presented in Chapter 3. However, a brief description of a standard on interoperability, IEC 61850, is essential to be presented in this section. The IEC 61850 is improved as an international standard for substation and feeder equipment automation including IEDs, EMS, SCADA, distribution control, and information exchange between these systems.

The first edition of IEC 61850 has been introduced in 2004 as a communication standard, and then the next one developed in 2011 has provided

TABLE 1.3 Wireline and wireless communication technologies in Smart Grid architecture [1]

Tech.	Standard	Data rate	Distance	Network	Advantage	Disadvantage
Wireline technologies						
PLC	• NB-PLC: ISO/IEC 14908–314,543–3-5, CEA-600.31, IEC61334–3-1, IEC 61334–5 (FSK) • BB-PLC: TIA-1113 (HomePlug 1.0), IEEE 1901, ITU-T G.hn (G.9960/G.9961) • BB-PLC: HomePlug AV/Ext., HomePlug Green PHY, HD-PLC	• NB-PLC: 1–10 Kbps for low data rate PHYs, 10–500 Kbps for high data-rate PHYs • BB-PLC: 1–10 Mbps (up to 200 Mbps on very short distance)	• NB-PLC: 150 km or more • BB-PLC: about 1.5 km	• NB-PLC: NAN, FAN, WAN, large scale • BB-PLC: HAN, BAN, IAN, small scale AMI	• Large-scale communication infrastructure is already established • Physical separation from other networks • Low operational costs	• High signal attenuation and channel distortion • Disruptive interference from electric appliances and other electromagnetic sources • Difficult to support high bit rates • Complex routing
Fiber optic	• AON (IEEE 802.3ah) • BPON (ITU-T G.983) • GPON (ITU-T G.984) • EPON (IEEE 802.3ah)	• AON:100 Mbps up/down • BPON:155–622 Mbps • GPON: 155–2448 Mbps up, 1.244–2.448 Gbps down • EPON: 1 Gbps up/down	• AON: up to 10 Km • BPON: up to 20–60 Km • EPON: up to 20 Km	• WAN	• Long-distance communications • Ultra-high bandwidth • Robustness against electromagnetic and radio interference	• Higher installing costs (PONs are lower than AONs • High cost of terminal equipment • Not suitable for upgrading • Not suitable for metering applications

Continued

TABLE 1.3 Wireline and wireless communication technologies in Smart Grid architecture [1]—cont'd

Tech.	Standard	Data rate	Distance	Network	Advantage	Disadvantage
DSL	• ITU G.991.1 (HDSL) • ITU G.992.1 (ADSL), ITU G.992.3 (ADSL2), ITU G.992.5 (ADSL2+) • ITU G.993.1 (VDSL), ITU G.993.1 (VDSL2)	• ADSL: 8 Mbps down/1.3 Mbps up • ADSL2: 12 Mbps down/3.5 Mbps up • ADSL2+: 24 Mbps down/3.3 Mbps up • VDSL: 52–85 Mbps down/16–85 Mbps up • VDSL2: up to 200 Mbps down/up	• ADSL: up to 5 km • ADSL2: up to 7 km • ADSL2+: up to 7 km • VDSL: up to 1.2 km • VDSL2: 300 m–1.5 Km (50 Mbps)	• AMI, NAN, FAN	• Large-scale communication infrastructure is already established • Most commonly deployed broadband	• Telco operators can charge utilities high prices to use their networks • Not suitable for network backhaul (long distances)
Wireless technologies						
WPAN	• IEEE 802.15.4 • ZigBee, ZigBee Pro, ISA 100.11a (IEEE 802.15.4)	• IEEE 802.15.4: 256 Kbps	• ZigBee: Up to 100 m • ZigBee Pro: Up to 1600 m	• HAN, BAN, IAN, NAN, FAN, AMI	• Very low power consumption • Low cost deployment • Easy to develop • Fully compatible with IPv6-based networks	• Low bandwidth • Limitations to build large networks

Wi-Fi	• IEEE 802.11e • IEEE 802.11n (ultra-high network) • IEEE 802.11s (mesh networking) • IEEE 802.11p (WAVE—wireless access in vehicular environments)	• IEEE 802.11e/s: up to 54 Mbps • IEEE 802.11n: up to 600 Mbps	• IEEE 802.11e/s/n: up to 300 m (outdoors) • IEEE 802.11p: up to 1 Km	• HAN, BAN, IAN, NAN, FAN, AMI	• Low-cost network deployments • Cheaper equipment • High flexibility, suitable for different use cases	• High interference spectrum • Too high-power consumption for many smart grid devices • Simple QoS support
WiMax	• IEEE 802.16 (fixed and mobile broadband wireless access) • IEEE 802.16j (multi-hop relay) • IEEE 802.16m (air interface)	• 802.16: 128 Mbps down/28 Mbps up • 802.16m: 100 Mbps for mobile, 1 Gbps for fixed users	• IEEE 802.16: 0–10 km • IEEE 802.16m: 0–5 (opt.), 5–30 acceptable, 30–100 km low performance	• NAN, FAN, WAN, AMI	• Supports huge groups of simultaneous users • Longer distances than Wi-Fi • A connection-oriented control of the channel bandwidth • More sophisticated QoS mechanisms than 802.11e	• Complex network management is • High cost of terminal equipment • Licensed spectrum requirement
GSM	• 2G TDM, IS95 • 2.5G HSCSD, GPRS • 3G UMTS (HSPA, HSPA+)	• 2G: 14.4 kbps • 2.5G: 144 kbps • HSPA: 14.4 Mbps down/5.75 Mbps up • HSPA+: 84 Mbps down/22 Mbps up	• HSPA+: 0–5 km • LTE-Advanced: optimum 0–5 km, acceptable 5–30, 30–100 km	• HAN, BAN, IAN, NAN, FAN, AMI	• Supports millions of devices • Low power consumption of terminal equipment	• High prices to use service provider networks • Increased costs since the

Continued

TABLE 1.3 Wireline and wireless communication technologies in Smart Grid architecture [1]—cont'd

Tech.	Standard	Data rate	Distance	Network	Advantage	Disadvantage
	• 3.5G HSPA, CDMA EVDO • 4G LTE, LTE-Advanced	• LTE: 326 Mbps down/86 Mbps up • LTE-Advanced: 1 Gbps down/ 500 Mbps up	(reduced performance)		• High flexibility, suitable for different use cases • Licensed spectrum reducing interference • Open industry standards	licensed spectrum
Satellite	• LEO: Iridium, Globalstar, • MEO: New ICO • GEO: Inmarsat, BGAN, Swift, MPDS	• Iridium: 2.4 to 28 Kbps • Inmarsat-B: 9.6 up to 128 Kbps • BGAN: 380 up to 1000 Kbps	• 100–6000 km	• WAN, AMI	• Long distance • Highly reliable	• High cost of terminal equipment • High latency

substation automation in addition to communication standards. Although IEEE 1815 is one of the most widely adopted standard in substation automation, it only includes communication standards for low bandwidth monitoring and control operations. On the contrary, IEC 61850 provides real time protection and control applications in high bandwidth in addition to others provided by IEEE 1815 standard. Thus, IEC 61850 is the most convenient and widespread protocol used for substation automation and communication systems. IEC 61850 standard describes a wide variety of IED types as breaker, switches, merging units, protection and control devices in addition to generic IED definitions. The protection and control IED devices accomplish supervising the control and protection operation throughout their service bay unit while breaker and switches are required to monitor the actual situation of circuit breakers, to communicate with protection and control IEDs for information signals and manages trip or close operations regarding to receive protection signals. The merging unit type IEDs are responsible to provide sampled status signals to protection and control IEDs corresponding to acquired measurement signals from field sensors such as current and voltage transducers that generate sampled measured values (SMV). The acquired analog values are converted to digital signals and transmitted to protection and control system to provide instant status information. The communication architecture of IEC 61850 standard has been constructed in a hierarchical structure that is comprised by station, bay, and process levels as illustrated in Fig. 1.7 [13]. The Process Level includes field

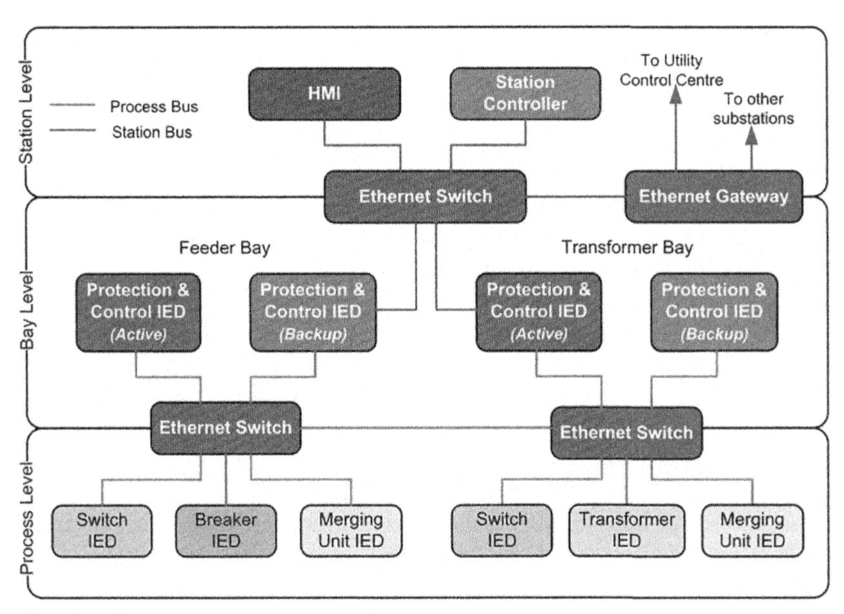

FIG. 1.7 Hierarchical diagram of an IEC 61850 based substation automation [13].

devices such as current transformers, power transformers, data processing devices, sensors, transducers, breakers, switches, and actuators. All the field devices are interfaced with Bay Level over ethernet switches that are separated for feeder and transformer bays in Bay Level. Each bay is managed by two dedicated protection and control IEDs where one is used as active IED while the other one is reserved as backup IED. The protection and control IEDs in Bay and Station Levels are managed by computerized interfaces that are known as human machine interface (HMI), station computers and supervisory computers. The communication buses along the hierarchical architecture have been separated as process bus and station bus where the process buses are responsible to manage communication with latency between protection and control devices and field devices while the station buses are dedicated for instant communication through separated bays and station controllers in addition to remote communication and area networks [2, 13].

The preliminary protocols of early IEC 61850 versions were describing data transmission methods over communication lines but they were lacking on data management and data organizations. This caused manual configuration of objects and system variables regarding to register definitions, input/output modules, and device configurations. However, recent versions of IEC 61850 have brought exact definitions on data organization in addition to data transmission methods over communication lines. In this definition, a physical field device is connected to communication networks and it is addressed by network operator with a unique identify. IEC 61850 allows logical devices to connect a physical device to use it as a gateway or proxy server since each physical device includes one or more logical devices. Thus, the data organization is organized in this architecture. In this structure, each logical device includes one or more logical nodes to organize one or more data element. The physical and logical device interaction provides a communication structure transmitting SMVs as logical data. The mapping process approach provides a standard method to describe the power system devices regarding to abstract data and object models.

There are several communication services have been defined in IEC 61850 to describe priority and differences of applications in the power system. IEC 61850 functionalities and associated communication profiles are grouped into five types as SMV multicast (SMV), Generic Object Oriented Substation Event (GOOSE), Time Synchronization (TS), Generic Substation Status Event (GSSE), and Abstract Communication Service Interface (ACSI). The SMV and GOOSE applications are utilized to map into ethernet data frames and to eliminate the middle layer processes. Manufacturing Message Specification (MMS) protocols provide operation ability on TCP/IP layers and GSSE enables communications on ISO architecture. The ACSI models are designed to bring a set of services for IEDs to adopt to the network structure. This abstract model is essential to provide interoperability between operation levels. Therefore, the service sets require additional protocols to operate in practical systems and to integrate to the power system network [2, 13].

The Smart Grid is accepted as the integration of distributed system automation and data transmission network by contributions of IEC 61850 standard. Therefore, the matured power grid can be converted to Smart Grid infrastructure with automation and communication systems from generation to consumption layers. The required comprehensive automation and communication infrastructure are ensured by IEC 61850 services and protocols. Smart Grid enables distributed automation for distribution generation, microgrid aggregation, transmission and distribution network management, ESSs, and management concepts such as control, resiliency, and flexibility depending to IEC 61850 standards. These infrastructures and concepts are discussed in the following sub headings as enablers of Smart Grid evolution.

1.2.2 Distributed generation and microgrid

The conventional electricity generation has been depended to primary energy sources such as coal, natural gas, oil, and other fossil fuels. It is noted that the annual global power generation is around 20,250 TWh in 2012, and it is foreseen to reach up to 25,500 TWh in 2020 [14]. The conventional power plants are based on thermal, hydraulic, and chemical systems. However, development of power networks has promoted innovative researches on alternative energy sources such as wind, solar, wave, geothermal, and biomass. The alternative sources are defined as DERs which play crucial role in the improvement of Smart Grid in the context of power system. The DER integration to conventional power grid enabled transforming the centralized generation to distributed generation (DG) to facilitate Smart Grid evolution in power generation layer. The integration of RESs to centralized generation cycle improved a new concept called the decentralized generation where the consumers are also able to generate by using micro-sources. Besides, high level penetration of wind turbines, solar plants, geothermal plants, hydro-plants and various other sources at multi MW level provide flexibility to generation initiatives.

The integration of DERs to existing power grid has brought a new concept, microgrid, which is comprised by DERs, manageable loads, ESSs, and centralized or decentralized controllers. The generation capacity of a microgrid can be a few kilowatts to multi megawatts regarding to its installation layer throughout the power grid. The regular operation of a microgrid is described as grid-connected (or grid-tied) to provide two-way power flow either generating or consuming cycles of operation. The second operation mode of a microgrid is performed when a fault or maintenance requirement occurred in main grid, and thus microgrid shifts to islanded operation mode [15]. On the contrary of its advantages, the penetration of large DERs cause to several challenging issues for distribution grid and main grid. The intermittent structure of RESs may pose fluctuations on injected power to the distribution grid and it can damage overall stability of power grid at high integration levels of different power plant types. Moreover, DERs are expected to tackle with extraordinary operation conditions such as off-shore wind turbines, increasing temperature of PV plants and so on. Therefore, the transmission system operators (TSOs) and distributed system operators (DSOs) requires strict interconnection grid codes to prevent unexpected operation and maintenance requirements. While the improved power electronics devices play crucial role in development of DG; the resource-based challenges make it difficult to improve the reliability, resiliency, flexibility, efficiency and scalability of DG integration to power grid. Therefore, the researches and implementations are focused on improving highly reliable and resilient DG plant integration to main grid. The resiliency of a DG system requires some features such as anticipating the potential events, rapidly recovering from detected events, and adapting to prevent future events. When a sudden fault is detected in any section of the system, DG should rapidly respond to the disturbances due to its control and operation management systems. The response time, recovery duration and self-healing features are main indicators of DG system resiliency [16].

Although DG is assumed as small-scale generation system located at the consumer side to meet power demand of customer loads, large scale DG plants are also integrated to transmission and distribution level of main grid. The integration levels and installed capacities of DG plants are rapidly increasing day by day due to electricity power requirement all over the world. The decentralized generation option provided by DG plants have brought several technical, economic and environmental benefits. The technical advantages provided by DG plants are seen in reliability improvement, voltage quality improvement, decrements on the line losses, enhancements on security issues, and operational benefits. The DG integration increases power system reliability by decreasing peak power demand and capacity releases. On the other hand, it improves generation diversity enabling integration of a wide variety of power resources to the main grid. The voltage quality of main grid is improved by reducing flickers and providing better voltage regulation operations. Moreover, the active and reactive power controls are improved and line losses are gradually reduced. The resource diversity and increased integration ensure the security of critical loads and increases power utility security by decreasing blackout risks. In addition to this, CPS security is ensured by preventing vulnerabilities to intentional cyber and physical attacks [16, 17].

The DG technologies are assumed as the initial applications that have improved development of microgrid systems. DG plants have become essential backup systems for traditional power system to prevent blackout and curtailments due to their flexible integration features to the main grid. The DERs used in a DG system are classified into two categories as RESs and conventional resource-based plants. The conventional power plants are mostly comprised by traditional power plants including synchronous generator such as gas turbines, CHPs, steam turbines, micro turbines, and hydro plants. On the other hand, RES based DG systems may include some synchronous generator-based resources such as wind turbines, micro hydro plants, or geothermal sources while other types that are not including synchronous generator are PV plants, biomass plants, fuel cells, and batteries. The expressed generator structures require some featured control systems to ensure the stability and quality of voltage, frequency, active and reactive power of microgrid system. These are directly related to microgrid operation, control, and protection issues that will be introduced in detail in the following chapters.

The generated power of DERs throughout a DG system is defined as input power as illustrated in Fig. 1.8. The generated power is converted to electricity regarding to operation structure of aforementioned DERs. The microgrid can be operated in islanded mode to supply the local loads or in grid-tied mode to integrate to the utility grid and perform power exchange along microgrid and main grid. One of the most important issue in this structure is obviously control operation. The required control operations are performed in two sections where one is related to source side namely input power control, and the other one is grid side control. The input or source side controllers target to maximize the achievable power rate by using several algorithms called maximum power point

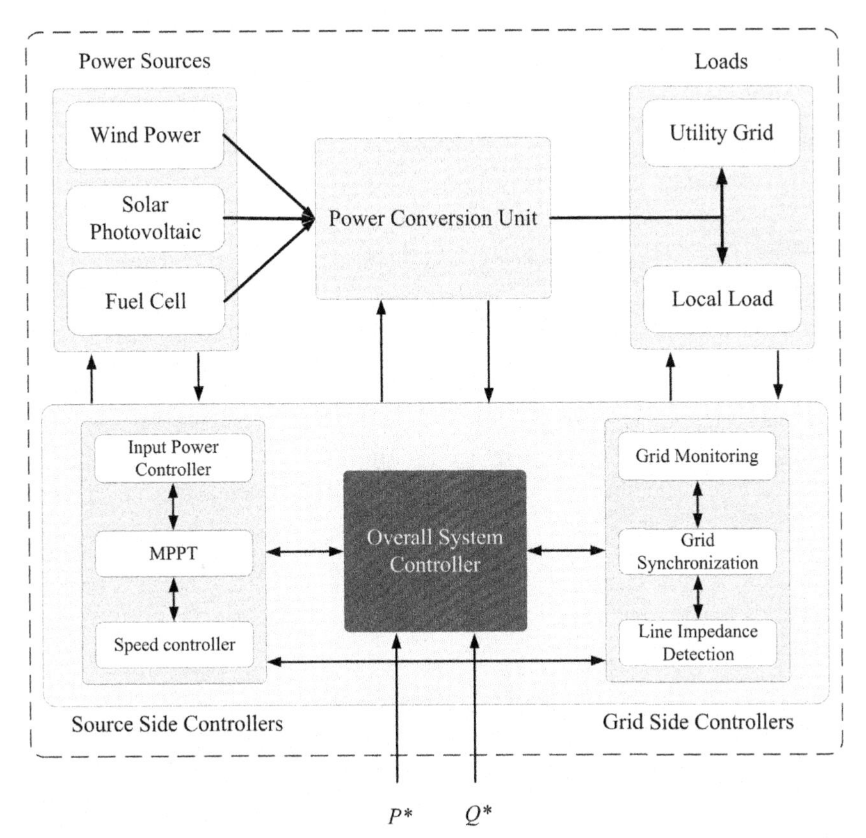

FIG. 1.8 Block diagram of a general DG system with controllers.

tracking (MPPT) algorithms. These algorithms are based on physical and electronic control of DERs while an MPPT algorithm is calculating input power and controls the power electronics in a PV system, another controller in a wind turbine can additionally adjust the generator speed.

The grid side controllers are responsible to perform sophisticated controls on generated and supplied active power to the utility grid, reactive power exchange between DG and grid, dc link voltage rates, supplied power quality, and grid synchronization. These requirements comprise the fundamental features of a power converter used in DG system. In addition to fundamental requirements, accompanying control services such as voltage and frequency regulation, harmonic and other disturbance compensations, and filtering issues are also required [18].

The microgrid that are operating in an integrated structure with ESSs is one of the most important concepts of DG where the generation section of Fig. 1.9 illustrates this concept. In addition to be an improved infrastructure to tackle integration problem of DERs to utility grid, the microgrid plays a vital role with participation of consumers to generation, energy storage, control and management issues in the grid. This development transforms consumers to prosumers and enables them to play active role in the grid as a part. The microgrids can be

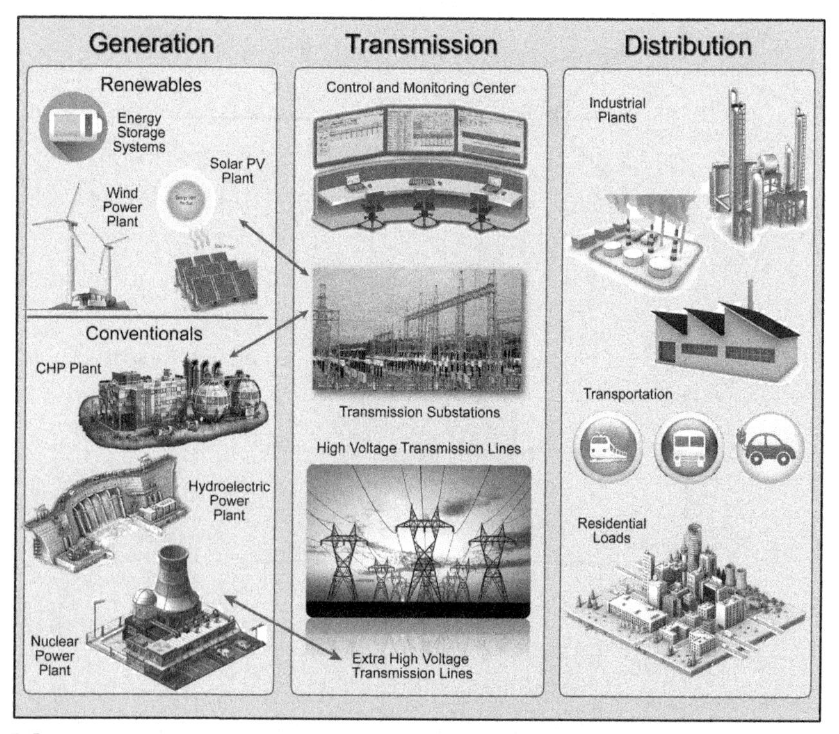

FIG. 1.9 Key enabler technologies for Smart Grid evolution.

improved either with DC sources or with AC sources. Moreover, hybrid microgrids including DC and AC DERs are recently being improved. The microgrid integration to existing distribution system has transformed it from passive structure to an active distribution system.

Besides its DER types, the microgrids are classified into several groups regarding to their characteristics, features, operation and control modes. The operation characteristics and feeder architectures have pave the way to define microgrids as urban, rural, and off-grid microgrids [19]. The urban microgrids are installed in concentrated industrial and residential areas as its name implies. The feeders are in densely populated areas and thus, the distribution lines are relatively short. One of the main indicators of an urban microgrid is its short circuit ratio that is determined by the ratio of utility grid short circuit capacity at the point of common coupling (PCC) to the total DER generation capacity of the microgrid. Therefore, the short circuit ratio of an urban microgrid should be over than 25 to manage voltage and frequency by the utility grid during grid-tied operation. Besides, the transient disturbances are easily tackled in a such integration. The rural microgrid architectures are widely seen in rarely populated areas and thus the distribution lines are relatively longer than urban microgrids. On the other hand, short circuit ratio is not strictly considered but it causes to several voltage and frequency fluctuations along the microgrid. The microgrid is much more vulnerable to reliability and stability of DERs in rural architecture. The last prominent architecture is based on off-grid structure that has no connection to the utility grid and is utilized to supply local loads in geographically spanned areas [19].

The microgrid integration of RESs and DERs should comply with IEEE P1547-2003 which is a standard for interconnecting distributed resources with electric power systems. IEEE P1547 standard series provide several scope and purpose to microgrid control and management operations. A microgrid central controller (MGCC) ensures interconnection and coordination of DERs in a microgrid to utility grid. A MGCC mainly controls the load demand and generation capacity of plants to manage energy balance between generation and demand. The MGCC is capable to load shedding to sustain critical load supply during excessive load demands where the generation has been limited. Therefore, the non-critical loads are shaded when the demand excesses generation. Block diagram of microgrid architecture with DERs, MGCC and distributed controllers have been shown in Fig. 1.10. There are different types of loads and resources exist in a microgrid and almost all of them are integrated with utility grid. Therefore, a robust and reliable controller infrastructure is required to meet operational requirements of microgrid. The MGCC accomplish this task by monitoring the electrical parameters and making decisions to synchronize, to connect to grid or disconnect from grid since it is located between utility grid and microgrid. Once the MGCC decides to disconnect microgrid from utility grid, then it monitors resynchronization periods to detect restoration of utility grid and connects the microgrid to restored utility grid in a properly

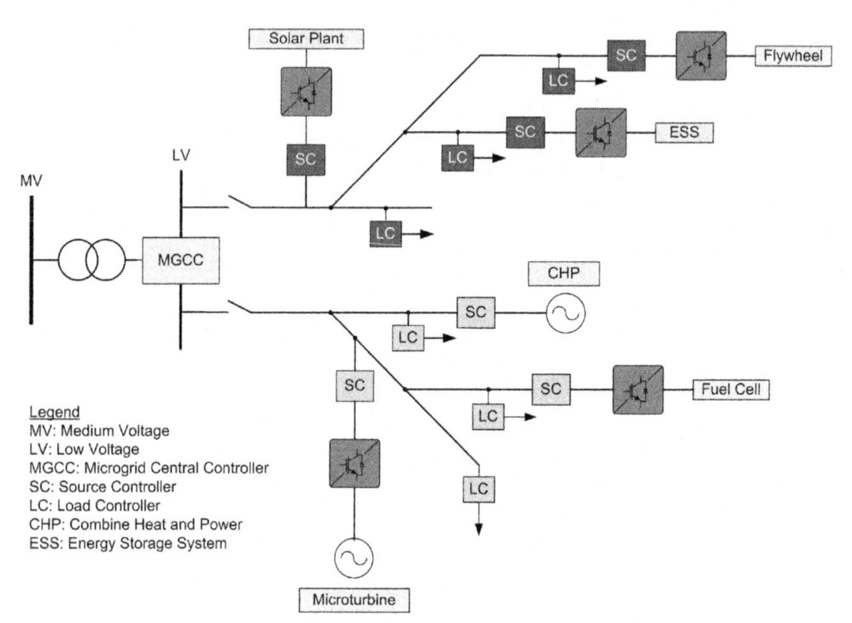

FIG. 1.10 Block diagram of a microgrid architecture [20].

synchronized interval. This operation is called island detection and resynchronization process [21]. The centralized microgrid control includes hierarchical control levels that are performed by interaction of MGCC with source controllers (SC) and load controllers (LC) as shown in Fig. 1.10. The hierarchical control of microgrid is generally performed in three levels regarding to SC and LCs, centralized MGCC control, and distribution management system at the last level. The hierarchical control optimizes efficiency and reliability of microgrid by preventing curtailments and blackouts. In addition, it is noted that MGCC can provide around 22% savings for daily operation in high price regime, and 2% decrement on operation and maintenance cost per year [21].

The MGCC is utilized to interface microgrid and distribution management system (DMS) which is used to detect blackouts and other fault types during the operation of utility grid. The MGCCs are featured as AC and DC central controllers depending to microgrid types that will be used. In an AC microgrid the load and sources operate in AC waveforms and thus the DERs should be connected to microgrid by inverters. In addition to voltage and frequency control performed by power inverters, several featured power controllers such as unified power quality conditioners (UPQCs) and P-Q controllers are required in AC microgrid and MGCC. Therefore, the hierarchical control scheme of AC microgrid is more complicated and more sophisticated controllers are required to ensure appropriate connection to utility grid. on the other hand, DC microgrid is operated much more easily comparing to AC microgrid since it does not require frequency and phase control. Since the most of RESs as well as ESSs operate in DC, DERs can be easily integrated to microgrid and utility grid with light controllers. The DC microgrid eliminates installation cost and line losses due to a smaller number of power converters are required.

The microgrid controllers should comply with IEEE P2030.7 *Standard for the Specification of Microgrid Controllers* and IEEE P2030.8 *Standard for the Testing of Microgrid Controllers* () standards in addition to IEEE P1547-2003 *Standard for Interconnection and Interoperability of Distributed Energy Resources with Associated Electric Power Systems Interfaces.* IEEE P2030.7 standard describes fundamental control operations and functions for transmission and dispatching conditions of microgrid. Thus, a microgrid can be capable to autonomously operate and administer interconnection requirements itself. The communication protocols and infrastructure are also highly required as well as power sources in microgrid. The MGCC provides a standard and manageable communication interface such as SCADA systems for supervisory control and energy management systems [20].

1.2.3 Transmission and distribution networks

Transmission systems are used to carry multi MW rated power of high-voltages at the output of bulk generation plants. Therefore, smart grid transformation of conventional power grids should consider to improve transmission and distribution networks as well. The integration of microgrids and large-scale RESs makes existing transmission and distribution networks more complicated and interconnected to serve power delivery. In addition to conventional AC transmission networks, high-voltage direct current (HVDC) networks are intensively being improved to enhance efficiency by decreasing line losses in long distance transmissions. However, the high installation and operation costs of HVDC networks limit the widespread utilization. The highly complex infrastructure of transmission and distribution networks due to integration of distributed generation requires sophisticated monitoring and control systems to ensure reliability, sustainability and efficiency of smart grid deployment. Moreover, monitoring and control technologies are required to prevent system faults caused by excessive power demand, resource failures or unexpected system faults. Some of smart grid systems used for transmission networks perform preventing actions for automatic control of network by a centralized monitoring and control center while others are based on decentralized control infrastructures where TSO and DSOs take action against occurred faults [4, 22].

In the conventional power grid, generated high-voltage is transmitted to substations that are closely located to consumers and the distribution networks are used to adjust and deliver low-voltage from substations to residential and industrial consumers. Besides substation technologies in smart grid transformation of conventional power grid, the grid-tie inverters play a crucial role in integration of DERs to transmission substations and distribution networks. The grid-tie inverters are essential devices to integrate DER and RES based microgrids to utility grid due to its synchronization and power control capabilities while interconnecting to grid. The efficient integration of microgrids and RESs to high-voltage transmission networks and low-voltage distribution networks are dependent to efficiency of grid-tie inverters. Furthermore, these inverter types provide presuming options to low voltage consumers by feeding generated power of their own microgrid. One of the major contributions of grid-tie inverters to transmission and distribution networks are their synchronization and phasor measurement features that ensure synchronized operation with grid or safely disconnecting from grid to operate the microgrid in island mode. Thus, it provides self-healing and restoration capability to power network during fault conditions. These capabilities of inverters are involved with monitoring and EMS integration to control power flow throughout the smart grid.

EMS is a software-based control system that provides management, monitoring, optimization and operating generation, transmission and distribution networks. A widely known component of EMS is SCADA system that is based on centralized control to monitor and to control numerous generation plants and transmission networks in long distances. The integration of EMS and SCADA enables smart communication and CPS installation along the smart grid with automatic control processes. Outage management system (OMS) is another significant component of EMS that are developed to detect and to solve outages occurred in the power network. Most of transmission and distribution networks include special OMS to identify outages instantly, and to store historical data of outages in databases. The improvement of smart grid communication systems has developed traditional technologies of transmission network such as flexible AC transmission systems (FACTS) and static var compensation (SVC). FACTS is essential to stable voltage and reactive power of networks while voltage and current are out of phase. On the other hand, SVC is required to compensate reactive power in an AC power network to facilitate stabilizing grid voltage and frequency. Thus, grid voltage deficiencies such as sags, swell, fluctuation, dips, and curtailments are efficiently prevented. The substation automation is another important contribution of smart grid to transmission and distribution networks to resolve synchronization and fault situations [4, 22].

A comprehensive list of devices and technologies used in smart grid transformation of transmission and distribution networks has been presented in Table 1.4. The substation automation is required in transmission and distribution networks to provide several automation applications for voltage control, synchronization, load transfer, curtailment prevention and fault detection controls. The relays and circuit breakers are major protection devices brought by smart grid transformation. Although these devices also exist in conventional power grid, the communication infrastructure enabled by smart grid architecture facilitates the integration of more sophisticated relays and breakers to be used in transmission and distribution networks.

The relays are controlled by CTs to indicate any fault occurred along the system and trip signals are transmitted by the controller to prevent damages

TABLE 1.4 Smart grid technologies and their contribution to transmission and distribution networks

Grid level	Device and technology	Description
Transmission	FACTS and HVDC	AC or DC (HV or LV) power transmission from generation plants to distribution network
	Power transmission analysis systems	This system is comprised by a package consisting utilities for configuring, customizing, and managing the power transmission network models like the real-world transmission systems
	Inverters and rectifier	Utilized for AC to DC and DC to AC conversion
	Static var compensator	Utilized for static VAR compensation as lines have mutual inductance consuming reactive power
Transmission and distribution	Substation automation	Automation applications, voltage control, synchronization, load transfer, curtailment prevention, fault detection
	Relays and breakers	Relays are controlled by CTs to detect faults in power network and trip signals are provided to open the breaker to disconnect circuit and avoid equipment damages
	Fault locator for distribution system	Fault locators are devices and software typically installed at a substation to identify fault event and fault types, and calculate the distance from a monitored to the identified fault source in a distribution system
	SCADA	Computerized supervisory control systems for monitoring and controlling industrial, infrastructural, or facility-based processes
	Remote terminal unit (RTU)	The RTU operates as a field device at the remote location wherever a SCADA system requires equipment monitoring or control process
	Energy management system (EMS)	An EMS is used by TSO and DSO to monitor, control, and optimize the performance of the generation and/or transmission or distribution system

Continued

TABLE 1.4 Smart grid technologies and their contribution to transmission and distribution networks—cont'd

Grid level	Device and technology	Description
Distribution	Advanced metering infrastructure (AMI) meters	AMI involves two-way communications with smart meters and other energy management devices. Allows DSO to detect problems and to communicate to define real-time tariffs
	Advanced substation gateway	Namely symmetric multi-processing (SMP) gateway—an advanced computing platform serving as a single point of all IEDs in the distribution network
	Distribution automation	Consists line equipment, communication infrastructures, and ICT that are used to acquire data of distribution network
	Metering data management system (MDMS)	Utilized to acquire metering data from multiple meter technologies. Evaluates the quality of acquired data and generates estimations on where errors and gaps exist

in the whole network. The trip signals provided by central or decentralized controllers open breakers and thus, the fault location is isolated from power network. Fault passage indicators (FPIs) and fault locators are software assisted devices to detect and define the fault type occurred in transmission and distribution network. The FPI also detects the location and distance of identified fault to generation resource in distribution network.

The RTUs are used as field devices at remote locations to provide device monitoring and control processes. Besides, LV feeders need RTUs as smart meter applications to monitor and control the system conditions to detect faults and disturbances. In case of any fault or unbalance is detected along the system, RTU provides location and magnitude of disturbance and provides resolving command to central controller. The AMI can be assumed as the developed version of AMR infrastructure with its improved communication, metering, monitoring, and operation capabilities. It provides two-way communication with smart meters and EMS devices to perform instant data transmission. Besides, it provides required data to DSOs to detect faults or problems in the distribution network and to determine real time electricity tariffs to manage DR. The development of AMI allowed to precise control of demand management due to its area networks based on HAN and NAN, and any DSO would be capable to plan and optimize the utility grid in much more complicated manner [23].

The advanced substation gateway is also known as symmetric multiprocessing (SMP) gateway that is an improved computerized system merging the acquired data from IEDs. The IEDs are also associated with circuit breakers to perform several protection measurements, fault detection, monitoring and control processes.

In this context, the distribution automation incorporates with line equipment, ICT infrastructures, and data acquisition system along the distribution network. Another major system is MDMS that is utilized to associate measurement data acquired from several smart meters and diverse metering technologies [4, 23]. A wide variety of smart grid technologies are available to improve efficiency and resiliency of transmission and distribution networks. In addition to microgrid and substation automation technologies, other major smart grid applications are distribution network automation and fault detection, isolation, and recovery (FDIR) technologies. All this smart grid enabled transmission and distribution automation technologies provide several benefits such as decreased operational and maintenance costs, enhanced grid security, and economic earnings.

1.2.4 Energy storage systems

The energy storage is an old application that has been applied since several years. Moreover, ESSs are important systems for DR, electric vehicles, PV energy systems, uninterrupted power supply (UPS), and remote energy plants. The widespread usage of ESS is seen in pumped hydro plants and solar energy plants. The storage applications are essential when operating intermittent RESs and islanded and grid-connected operation modes shift regularly. The intermittent structures of RESs are forced by weather changes and thus energy storage provides a buffer in these conditions. The ESS selection is based on several parameters such as each unit capacity, storage bank capacity, on-demand capacity, self-discharge and deep discharge cycle, efficiency, life cycle of units, reliability, and cost. The storage capacity means the available energy rate after a full charge while self-discharge is the rate of the time required to reach to the deep discharge capacity.

The fundamental ESS technologies are pumped hydro plants for long term usage and advanced batteries, compressed air energy storage (CAES), superconducting magnetic storage systems, super capacitors and flywheels [24]. It should be noted that ESSs are mostly categorized according to rated power capacity and time of discharge. Three use conditions of ESSs that are defined depending to classification categories are power quality, power recovery, and energy management. The power quality of an ESS is its capability to prevent the voltage disturbances such as sags, flickers or supply interruptions during supplying the load. The power demand from an ESS occurs in very short intervals as seconds and can be rated from a few kW to MW rates. Therefore, ESSs should be comprised by high power capacity flywheels, super capacitors or several types of batteries to rapidly respond to energy demand. The energy management capability of an ESS is required to meet load demands such as power balancing, peak shaving and smoothing, and energy storage requirements during low price tariff periods while participating to load supply in high price tariff periods. This feature is one of the key contribution of ESSs to Smart Grid evolution [14].

1.2.5 Control, interoperability and flexibility

The interoperability, flexibility and security issues of a smart grid infrastructure require communication and control system to operate together. One of major concern in smart grid flexibility is DSM to meet power requirement of load shifts and load changes. Therefore, an ICT infrastructure with communication and widespread control options is required to sustain reliability and flexibility of grid. The control requirements are focused on generation, distribution, storage and interoperability flexibilities where generation and storage control play vital role. Moreover, TSO and DSOs can control DSM by load shedding and managing the smart load by disconnection some consumers to balance generation and load demand. The DSM is performed by applying particular DR policies by shifting tariffs and incentive based programs for consumers to sustain grid reliability [7].

The microgrid and distributed generation infrastructure enables distributed control opportunity along smart grid infrastructure. The local microgrid controllers include load controllers, energy storage controllers, and micro source controllers. These controllers provide system reliability by detecting any disturbance or fault in time and response to changed conditions to ensure reliability of interoperability served by diverse sources and loads. Local controllers employ sensors, relays, breakers and IEDs to facilitate remote monitoring and measurement to detect probable disturbances. A microgrid management system is used to optimize different microgrid controls in a smart grid infrastructure. It is based on remote metering of various node points over smart meters and transmitting the measurement data to central management and control system over DMSs. The smart metering system provides additional data for real time estimation and load forecast applications in addition to regular metering data. The DMS system provides concentrated measurement data to MDMS system that includes CIS, OMS, geographical information system (GIS), and distribution management system. These infrastructures comprise AMI interface of a Smart Grid infrastructure. OMS is essential to detect power quality disturbances and faults at the customer side of system. In case of any fault or disturbance detected, OMS transmit a control signal to MDMS to response the abnormal operation of related customer. The customer information, consumption rates, billing data and consumer identification data are provided by GIS and CIS infrastructures to MDMS. The distribution management system acts as a supervisory controller of whole grid infrastructure by observing power quality, load demand amounts, and performing estimation algorithms in the system [1, 4].

The control and interoperability of Smart Grid can be performed by a wide variety of communication system but complying with IEC 61850 standard that is an open system standard regardless to type of vendor or infrastructure. The standard communication interface enables interaction between MDMS and different field devices such as IEDs, power plants, feeders, substations, and power system utilities installed at any DER. PEVs and ESSs are also integrated to utility grid in addition to power plants installed at any microgrid along a Smart Grid infrastructure. Therefore, any device or micro resource penetrated to Smart Gris is handled in a "plug and play" structure due to IEC 61850 standard [4].

Another major control strategy used in distributed system automation and control is agent-based control approach. In this strategy, each DER has a local controller agent (LCA) that is responsible to monitor and to control distributed source in remote areas. Each LCA communicates with other LCAs throughout microgrid and comprises a global control agent (GCA) infrastructure.

When any DER faces with any fault during operation, it is disconnected from grid and LCA transmits an emergency signal to other LCAs and GCA to express that its DER has been disconnected. Afterwards, multiagent based system rearranges new operation conditions regarding to instant generation capacity and load demand. The GCA runs the control algorithm procedures to response the fault and recovers the disconnected DER in a short while. In this step, GCA uses offline calculation and tries to find an optimized solution to operate self-healing procedure of microgrid structure [25].

1.3 Fundamental components of smart grids

The Smart Grid infrastructure is defined as system of systems and includes many analog and digital technologies altogether due to its complex structure. The fundamental components of Smart Grid infrastructure are new and advanced grid sections, integrated communication systems, advanced control systems, smart devices and smart metering, and decision support environment with HMI. These components are presented in following topics as smart sensors and sensor networks, phasor measurement units, smart meters, and wireless sensor networks. The smart metering, management, and control infrastructures require smart devices that are equipped with sensors and sensor networks. Sensors can be used at any location throughout Smart Grid infrastructure from generation level to customer sides and play crucial role for monitoring and data acquisition. The sensor infrastructure provides required measurement data to control generation and consumption rates instantly to transmit actual situation data of utility grid. Thus, several management processes such as DSM, demand response and generation planning can be performed. The developed sensor networks facilitate management issues of TSO and DSOs to detect faults and disturbances or provide instant information to manage real time pricing and detecting losses and fraud usage if applicable.

The power flow is unidirectional in conventional grid where the power generated at power plants are supplied to consumers over transmission and distribution networks. The deployment and distribution of power is managed by using distribution substations where medium and low voltage levels are obtained. The improvements on distributed generation and penetration of RESs forced the conventional grid infrastructure to provide two-way power flow. Therefore, efficient and reliable operation of a such grid transformation brought some monitoring and control requirements. The transformed grid infrastructure requires secure, interoperable and cost-effective use of sources by the contribution of information and two-way communication technologies [26]. The fundamental structure of a power network transformation to smart grid has been shown in Fig. 1.9 with source and load types. Although the conventional generation is based on centralized power plants as CHP, large scale hydroelectric power plants, and nuclear power plants where most of them use fossil-fuels, smart grid infrastructure allows decentralized generation and microgrid power plants. In a such scenario, all the sources should be integrated to widespread monitoring and control system. Thus, generation and consumption balance are effectively managed and DSM requirements are met in a reliable way.

The power control and management systems include demand forecasting (DF), automatic generation control (AGC), automatic voltage regulator (AVR), and load frequency control (LFC). The challenges and requirements for modernizing the conventional power network are summarized by four titles as environmental challenges, consumer requirements, infrastructure challenges, and innovative technologies in [27]. On the other hand, the proposed solutions to address these challenges are listed as digitalization, intelligence, flexibility, resiliency, sustainability, and customization. A collection of smart grid applications has been presented in Table 1.5 according to power system level, measurement and control applications, ICT technologies and used device types. The critical applications are real time monitoring, power plant control, RES and alternative energy sources integration, distributed generation monitoring and quality of service (QoS) detection in generation level of smart grid. These applications can be performed by using Metropolitan Area Network (MAN), WAN, LAN and SCADA networks to provide power control and DSM measurements. The most important devices used to ensure these tasks are AVRs, PMUs, IEDs and smart transformers in generation level.

The transmission level applications include substation automation, transmission line control and monitoring, and power monitoring applications based on power loss and leakage measurements, fault localization and prevention, phasor measurements, network topology analyses, and security monitoring. These operations are performed by use of PMU, FPI, RTU, and GPS systems with WAN and LAN communications or EMS and SCADA systems. The substation automation is required in addition to smart transformer control, DLC, AMI and AMR applications in distribution level.

TABLE 1.5 A collection of critical applications and ICT in SG

Power system level		Critical applications	Measurement and controls	Communication technology and applications	Devices used
Utility	Generation	Real-time monitoring	• Energy efficiency • Demand profile • Voltage, current, frequency, phase, and power control • Droop control, • Active/reactive power • DSM	• Metropolitan area network (MAN) • Wide area network (WAN) • Local area network (LAN) • Backbone and core networks • SCADA	• Automatic voltage restorer (AVR) • PMU • IED • Smart transformers
		Power plant control			
		Alternative energy sources			
		Distributed generation monitoring			
		Quality of services (QoS)			
	Transmission	Substation automation	• Power loss and leakage • Localization, fault prevention • Phasor measurement • Network topology analysis • On-line power flow • Security monitoring	• Wide area network (WAN) • Local area network (LAN) • Energy management systems (EMS) • SCADA	• PMU • Smart fault passage indicator (FPI) • Smart RTUs • Global positioning system (GPS)
		Transmission line control and monitoring			
		Power monitoring			
		Quality of services (QoS)			
	Distribution	Substations automation	• Cost and pricing • Phasor measurement • Optimization periods • Machine learning and control services • Demand profile	• Extended area network (EAN) • Neighborhood area network (NAN) • Field area network (FAN)	• IED • PMU • Wide area monitoring, protection and control devices (WAMPAC) • Smart MV/LV controller • Smart RTUs
		Smart transformer control			
		Direct load control (DLC)			

Continued

TABLE 1.5 A collection of critical applications and ICT in SG—cont'd

Power system level		Critical applications	Measurement and controls	Communication technology and applications	Devices used
		Advanced metering infrastructure (AMI)	• Demand side management and control • Distribution management system controller (DMSC)	• Wireless regional area network (WRAN) • Wireless personal area network (WPAN) • MDMS	
		Automatic meter reading (AMR)			
Consumer	Consumption	Home (residential) energy management	• Consumption control • Microsource generation control • Monitoring and automation • Home automation • Security control	• Home area network (HAN) • Industrial area network (IAN) • Building area network (BAN) • Digital subscriber lines (DSL) • Advanced metering infrastructure (AMI)	• Smart appliances • Smart meters • HomePlug • Sensors (noise, temperature, humidity etc.) • Smart plugs • Ethernet
		Microgrid management			
		Prediction and forecast			
		Electric vehicle integration			

The main control and measurement procedures are required for billing and pricing, phasor measurement, optimization, DSM, distribution management system controller (DMSC). A wide variety of area networks and control systems including Extended Area Network (EAN), NAN, FAN, WRAN, WPAN, and MDMS are used to meet requirements of smart grid applications. The consumer side applications can be listed as residential or HEMS, microgrid management, source prediction and forecasting, and PEV integration to distribution grid. The communication requirements of consumption level applications are coped with limited area networks such as HAN, BAN, IAN or DSL connections for consumption control, microgrid control, home automation and security control. The devices that are used in consumer side include smart appliances, smart meters, HomePlug, sensor types, smart plugs and ethernet based devices. The technical details and discussions on these applications and requirements of smart grid will be presented in the following chapters of this book. Therefore, it is preferred to present a brief introduction in this section.

The smart meters that are located at customer sites also provide important contribution to grid management by detecting power consumption and quality parameters. The smart meters provide actual consumption data by local or remote metering features. Moreover, smart meters with remote control capability are used to load shading and remote closing of any customer if it is required. Therefore, smart meters enable DSOs to plan a detailed and more accurate DSM program for customers or load sites. The PMUs are essential components of Smart Grid infrastructure to monitor power quality, voltage and frequency values of any transmission line and overhead transmission systems. They provide instant, low latency and highly accurate measurement signals to monitoring center [2].

The fundamental components of Smart Grid infrastructure are introduced as smart sensors, sensor networks, PMUs, smart meters, and wireless sensor networks in the following sub headings to provide further understanding on communication and monitoring features.

1.3.1 Smart sensors and sensor networks

Remote monitoring and control is one of the major functions required in Smart Grid infrastructure. The ICT enables this requirement by use of integrated communication systems, sensor and transducer interfaces, advanced measurement systems, enhanced control functions and decision-making algorithms in AMI environment of Smart Grid. Accordingly, communication systems integrated with sensors and transducers comprise sensor networks to build an interactive and dynamic power system corresponding with all upstream and downstream components of utility grid. Therefore, an integrated communication system encompasses the entire power network from generation to consumption sites. The data acquisition and transmission operations are performed to meet required functionality to connect widespread sensors and actuators for meter reading, real time data reading and measurements to obtain fault or disturbance indicator signals along the utility grid. Thus, it enables to create a reliable, optimized, resilient and flexible power grid with enhanced control options and services. The sensor and transducers integrated with communication features allows meeting metering and controlling requirements such as penetration of RESs, two-way power flow infrastructure, fault detection and isolation, DSM, and high-quality power supply of developing power grid. The sensor networks are involved to implement data acquisition interface to transform inherited data to processable information to sustain reliability and integrity of Smart Grid. For instance, a smart meter can be handled as a sensor node to measure power consumption, to monitor customer conditions, and to improve DSM program [28].

The Smart Grid resiliency and flexibility are highly dependent to accurate and instant data acquisition from each nodes, substations, and customer locations. The smart sensors play vital role in advanced data acquisition and association with widespread control systems. Since geographical area of Smart Grid infrastructure from generation to consumption levels can be extended up to several kilometers and a huge amount of data is collected from many nodes, smart sensor nodes are required to be installed for comprising a distributed data processing and management operations. The distributed and smart sensor networks are the most recent approaches to tackle these difficulties. In addition to smart sensor networks; smart meters, communication technologies, and decision-making algorithms are required to transmit collected data to operation centers either in centralized or in decentralized structure. Thus, operation center considers Smart Grid infrastructure as a smart sensor network in terms of ICT interface. A distributed communication and control system as shown in Fig. 1.11 acts like a data server exchanging data between distributed management systems. Therefore, heavy load operation of a centralized communication and control center structure is distributed to different local sensor networks and eliminates high capacity central server requirements [29].

The use of sensor networks is generally identified into four areas such as generation, transmission and distribution, consumption, and energy storage. The data acquisition requirements of generation level are mostly focused on generation plant monitoring, power quality of grid, source and generation capacity, fault detection, and load demand balance between utility grid and DERs. In addition to utility scale, it is needed to monitor generation conditions of distributed generation sources and RESs along the microgrids interacting with grid. The sensor networks are required for data acquisition from substations, underground and overhead lines, and distribution transformers in transmission and distribution levels. The smart sensors are used to meter instant and average voltage, current, phase, frequency, and THD rates of utility grid. The load and generation analyses provide around 10–15% capacity increment due to metered data processing. Furthermore, sensing infrastructure ensures instant detection of faults and losses in transmission and distribution lines,

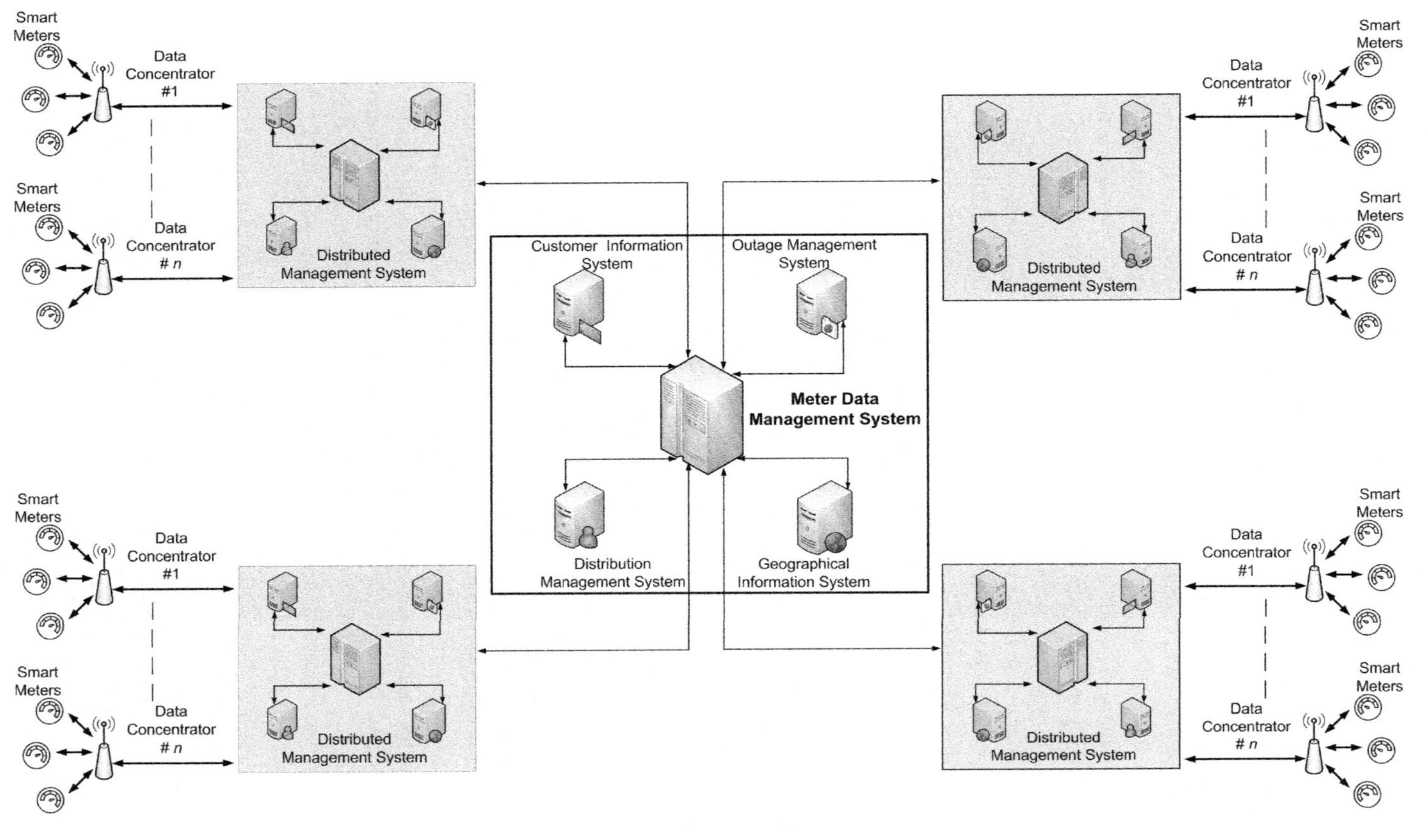

FIG. 1.11 Block diagram of distributed communication and control system in smart grid.

and overall efficiency can be increased by preventing detected disturbances. The overhead transmission lines are mostly monitored by using wireless sensor networks and fiber optic current and voltage sensors while smart current and voltage sensors with communication interface are used in LV and MV distribution systems such as substations and feeders. There are several high voltage (HV) sensors with wireless communication capability used for synchrophasor monitoring in transmission lines to detect line voltages in high accurate magnitudes. The PMUs are also classified in this group of wireless smart sensor networks with global positioning system (GPS) and coordinated universal time (UCT) units. The advanced additional units of PMUs provide synchronized and highly reliable data transmission with support of satellite or cellular communication technologies [28].

The consumer side sensing and metering processes offer a wide range of applications to both customer and DSOs in term of DSM and demand response programs. Moreover, it provides real time power consumption for billing and control option to limit or stop the power consumption automatically for defined DSM requirements. The smart meters and AMI are most critical sensor networks that are used to detect time of use (TOU) and demand rates at consumer side. Another important use of smart sensors and sensor networks are applicable for ESSs and batteries that are used to comprise a backup for RESs and DERs along distributed generation plants. The significant parameters considered for ESSs are state of charge (SoC), state of health (SoH), voltage, current, and temperature of batteries. The SoC is one of the most important parameter to be continuously monitored to detect serious disturbances on acid level and acid density of lead acid batteries. In addition to ESSs, energy storage applications are run across EVs and PHEVs where the sensor networks are widely used to monitor SoC and SoH during their autonomous operation and grid interaction.

The smart sensors and sensor networks are depended on several wireline and wireless communication technologies such as listed in Table 1.3. Furthermore, cognitive radio networks and internet based recent communication systems are adopted to Smart Grid infrastructure to perform data transmission. There are several technical challenges such as transmission losses, fading effects, channel noises and bit error rate (BER) affect selection of proper communication method in smart sensor networking. Other critical parameters of communication method selection are reliability, security, accuracy, and latency. In order to ensure to select and use of reliable communication interface, a number of standards have been defined for smart sensor networks used in Smart Grid infrastructure [28]. A list of featured active standards and short descriptions of these standard types are presented in Table 1.6. One of the most significant standard on data transmission along Smart Grid is IEEE 1159.3. Any system located in the grid provides its own control, communication, and transmission method to interact with other systems. Under these circumstances, each system and owner may face with problems while exchanging measurement and monitoring data between other systems. Therefore, IEEE has developed a standard titled IEEE

TABLE 1.6 Standards for smart sensor networks

Standard type	Short description
IEEE Standard 1159.3 *Recommended practice and applications for the transfer of power quality data in the grid*	The recommended file format utilizes a highly compressed storage scheme to minimize disk space and transmission times. The utilization of globally unique identifiers (GUID) to represent each element in the file permits the format to be extensible without the need for a central registration authority
IEEE Standard 1379-2000 *IEEE recommended practice for data comm. between RTUs and IEDs in a substation*	A uniform set of guidelines for communications and interoperations of RTUs and IEDs in an electric utility substation is provided. A mechanism for adding data elements and message structures to this recommended practice is described
IEEE 1451 *Standard for smart transducers*	A family of smart transducer interface standards for sensors and actuators; describes a set of open, common, network-independent communication interfaces for connecting transducers (sensors or actuators) to microprocessors, instrumentation systems, and control networks
IEEE 1547 *Standard for interconnecting distributed resources with the electric power systems*	This standard defines physical and electrical interconnections between utility and distributed generation (DG) and storage. It is intended to facilitate the interoperability of distributed resources (DR) and help DR project stakeholders implement monitoring, information exchange, and control (MIC) to support the technical and business operations of DR and transactions among the stakeholders
IEEE 1588 *Standard for a precision clock synch. protocol for networked measurement and control systems*	This standard defines a protocol enabling precise synchronization of clocks in measurement and control systems implemented with technologies such as network communication, local computing and distributed objects
IEEE C37.111-1999 *Standard common format for transient data exchange (COMTRADE) for power systems*	A common format for data files and exchange medium used for the interchange of various types of fault, test, or simulation data for electrical power systems is defined. Issues of sampling

Continued

TABLE 1.6 Standards for smart sensor networks—cont'd

Standard type	Short description
	rates, filters, and sample rate conversions for transient data being exchanged are discussed. Files for data exchange are specified, as is the organization of the data. A sample file is given
IEEE C37.118 *Standard for synchrophasor measurements for power systems*	This standard defines synchrophasors, frequency, and rate of change of frequency (ROCOF) measurement under all operating conditions. It specifies methods for evaluating these measurements and requirements for compliance with the standard. Time tag and synchronization requirements are included in the standards while the performance requirements are confirmed with a reference model and provided in detail
IEEE C37.2-2008 *Standard electrical power system device function numbers, acronyms, and contact designations*	This standard applies to the definition and application of function numbers and acronyms for devices and functions used in electrical substations and generating plants and in installations of power utilization and conversion apparatus. These numbers and acronyms may also be used to represent individual functions within multi-function devices or software programs, and that may contain both protection- and non-protection-oriented functions

Standard 1159.3 to provide a recommended practice for power quality data interchange format (PQDIF). Thus, any system can interact with any other one regardless of vendor or data format due to unique PQDIF documents and files generated by various platforms or PQ meters. The data acquisition and sharing applications are performed in more efficient, accurate and reliable ways by the arrangements of IEEE 1159.3 standard [30].

Another outstanding standard is improved to present arrangements data communications between RTUs and IEDs in a substation that is known as IEEE 1379-2000 standards. It also targets interoperability issues such as interfacing any equipment with other equipment and systems along the power grid. The purpose of recommended practice is to provide further understanding on RTU and IED communications as well as SCADA system and its central station for implementers. The standard recommendation presents two different protocols in terms of many aspects. Both protocols are used to meet communication requirements of RTU to IED interaction and are completely compliant with other to provide flexibility to users. One of the protocols is distributed network protocol (DNP3) that is a three-layer protocol meeting 1st, 2nd, and 7th layers of open system interconnection (OSI) reference model. DNP3 protocol is particularly improved for data acquisition and control operations that are essential in utility data transmission. The other protocol handled in IEEE 1379-2000 standard is IEC 60870-5 that is implemented for frame formats and services in three-layer reference model like DNP3. It aims to increase efficiency of RTUs, smart meters, relays and related equipments. Moreover, it uses the layered structure complying with OSI reference model for synchronization and file transfer applications [31].

IEEE 1451 standard purposes to provide general application requirements for smart transducer interfaces (STIs) in the context of intelligent systems. It is implemented to increase advantages of smart sensors and transducers used for real-time monitoring, metering, control and interfacing applications. The standard proposes network independent communication infrastructure to connect several sensors to microprocessors and control interfaces to comprise a sensor network [28, 32]. IEEE 1547 describes recommendations for electric power systems interconnection with DERs such as wind turbines, PV plants, fuel cells, micro turbines, hydro plants and other generators. The standard presents technical specifications for test and interconnection conditions related to performance, maintenance, reliability, and operation. Moreover, general requirements for islanding of DERs, power quality, design standards, generation, and fault response are included in the context of IEEE 1547 standard [33, 34]. IEEE 1588 standard defines common environments for precision time protocol (PTP) use in power systems for protection, control, automation, and data exchange services using Ethernet based communication architectures. Although it is implemented for Ethernet communications, it is not limited to Ethernet and can be used in LAN supporting multicast messaging systems. Therefore, the standard specifies PTP protocol to operate heterogeneous systems together [28, 35].

IEEE C37.111 standard is titled as Common format for transient data exchange (COMTRADE) for power systems and specifies a format for data files including waveforms and event data acquired from power system via sensors and sensor networks. The standard defines file format to convert databases for easily exchangeable along various media types such as external hard disk drives, USB and flash drives or similar disk medias. IEEE C37.118 standard describes PMU device which can be a stand-alone physical unit or a functional unit within another physical unit. This standard does not specify hardware, software, or a method for computing phasor, frequency, or ROCOF, but extends dynamic performance requirements for synchrophasors and ROCOF measurements. This description of standard includes basic synchrophasors measurements and the requirements for synchronization and data rates in the transmission of measurements. IEEE C37.2 standard defines electrical power system device function numbers, acronyms, and contact designations. The functions of each device used in electrical equipments are defined with a device function number including a prefix and a suffix for identification. The designated numbers are used to imply devices in schematics, connection diagrams, specifications and manuals [36–38].

1.3.2 Phasor measurement units

The PMU which is also defined as synchrophasor is fundamental component of a Wide Area Monitoring Protection and Control (WAMPAC) system. A PMU acquires current and voltage waveforms of the power system by using transformers and generates their phasor by sampling measured data. The generated phasor of current and voltage are complex representation of magnitude and phase angles of sampled waveform. A phasor sampling for cosine wave $v(t) = V_m \cos(\omega t + \varphi)$ can be exemplified as shown in Fig. 1.12 where the signal waveform is illustrated in Fig. 1.12A with its maximum and root-mean-square (rms) values while Fig. 1.12B represents magnitude of waveform and phase degree with φ. The V_{mag} denotes rms magnitude of measured and sampled waveform.

The synchrophasor terms describes a phasor that has been measured and calculated at any interval tagged with a time stamp. In order to generate accurate time stamps to describe each measured synchrophasor, a synchronization cycle should be arranged and each measurement should be tagged with exact timings. The timing tags are generated by GPS satellites and transmitted over wide area communication networks to synchronize the phasor information of power system. The accurate timing requirements are described by IEEE 37.118 standard that defines recommendations for PMU measurements in terms of the

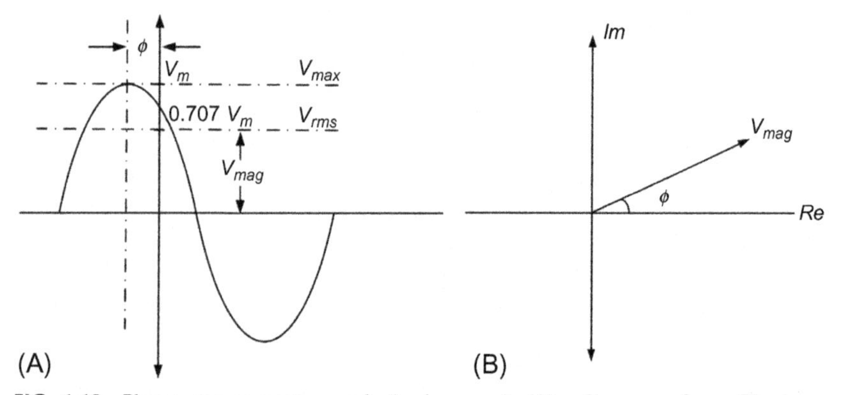

FIG. 1.12 Phasor measurement example (cosine wave), (A) voltage waveform, (B) phasor representation.

steady-state performance evaluation quantities such as total vector error, frequency error, and rate of change of frequency error as discussed earlier. The featured technology of PMU differs from conventional SCADA system in terms of timing, synchronization, and reliability where PMU can provide more than 60 samples at each cycle while SCADA can only provide 4 samples utmost for each cycle. Moreover, PMU samples frequency and ROCOF data in addition to voltage and current phasors [4, 7].

It is noted that the first PMU was developed by Virginia Tech complying with initial standard that have been approved in 1992. The recent PMUs are more accurate and can acquire different type of phasors from power network and substations. Some commercial PMUs are integrated with protection and detection equipments such as relays, circuit breakers, fault detection devices and timers. The block diagram of a typical commercial PMU system is illustrated in Fig. 1.13 where data acquisition and measurements have been shown as analog inputs. The data acquisition equipments that are used in analog inputs are generally fundamental current and voltage transformers to provide attenuated current and voltage waveforms that are inherited from transmission and distribution networks in MV and kV rates. The attenuated measurement signals are fed to data acquisition interface where the sampling process is carried out with synchronization data provided by phase locked loop (PLL). The accurate and real time synchronization signals are generated by PLL that is associated GPS receiver subsection with GPS antenna. The synchronized phasor measurement process is completed by a high-capacity microprocessor that is used to transmit generated synchrophasor data over a modem. The high sampling rates

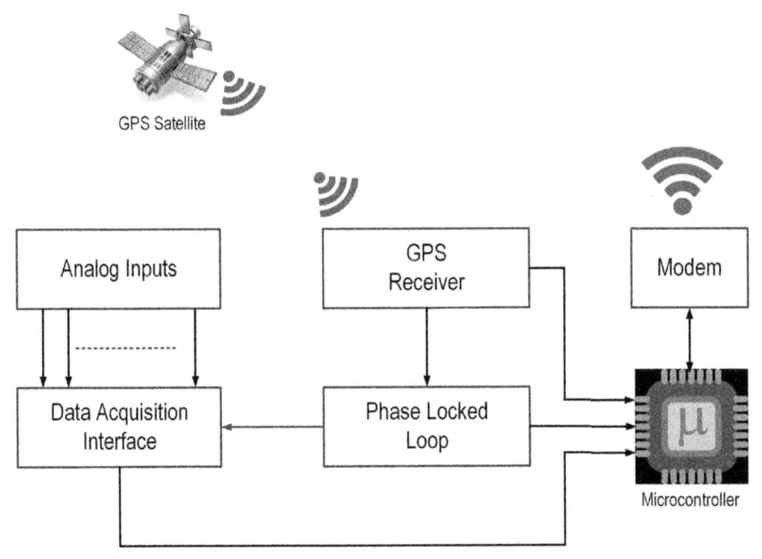

FIG. 1.13 Block diagram of a commercial PMU system.

and rapid frequency response capabilities of PMU provides accurate detection of power network disturbances. The application areas of PMU include advanced network protection and advanced control besides real time monitoring of power networks [4, 7, 28].

The PMUs play crucial role in wide area monitoring of power systems due to its rapid and accurate measurement capability. PMUs associated with IEDs are used for fault detection, fault recording and advanced protection controls. The acquired and processed measurements are transmitted to phasor data concentrators (PDCs) with tagged time stamps in a high-speed transmission technology where maximum delay should be lower than 20 ms and each measurement is repeated between 20 and 100 ms intervals. PDCs are located at monitoring and control centers and proceed power network monitoring depending to analysis of different measurement data acquired at independent nodes and IEDs. The analyzed data of PMU and PDC provide real time monitoring of power network, detection of any disturbances along power network, immediate power flow control, accurate and instant detection of faults and fault locations, and utilization monitoring [14].

The PDC is used to combine several PMU or PDC data together and assigns time stamps for each data set to transmit combined data stack to microprocessor. PDCs uses data buffer to cope with time delays occurred during transmission from sensors to PMUs and data packet latencies. The data streams are stored in data buffer in order to wait compilation of each PMU data receive and then aggregated data sets are transmitted to microprocessor at one interval. The PDC provides data checksum control, validity check, integrity and completeness controls to detect if there is a faulty data exists in dataset. PMUs operate according to particular standards such as IEEE 1344, IEEE C37.111 and IEEE C37.118 for data format, data rates and data acquisition operations, and several communication protocols as TCP, User Datagram Protocol (UDP) and Internet Protocol (IP) for data stream process. There are three types of PDCs as substation PDC, control center PDC, and super PDC are used in WAMPAC applications. The substation PDCs are also known as local PDCs that are located in substations to collect datasets from various PMUs and perform communication along the control center.

The latency of substation PDC is as low as possible since it is very close to PMU sources in substations, and it is used as local substation controllers. The substation PDC stores acquired data in very short intervals to prevent data loss against any communication failure. Therefore, it collects measurement data and immediately transmits. The control center PDC is associated with PMU of any field device, substation PDCs, and other neighbor control center PDCs. It is capable to collect and transmit very long data sets generated from multiple devices and applications in the field. It is compliant with plug-and-play type device integration to utility grid, with new protocol and data formats, and new applications. The super PDC is on the top of hierarchical organization comprised by local PDCs, and control center PDCs. It is used to combine wide

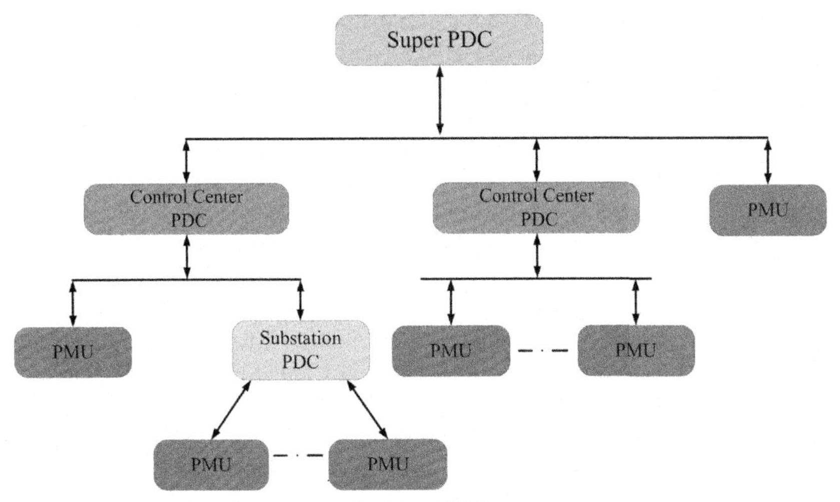

FIG. 1.14 Hierarchical organization of PMU and PDCs.

datasets generated by many PMUs and lots of substation and control center PDCs. The super PDCs perform PMU measurement in a large application group including wide area monitoring system (WAMS), monitoring, EMS, and SCADA applications. It can store collected data for long times due to its large data storage capability [4]. The hierarchical organization of PDCs and PMUs is shown in Fig. 1.14 where they are configured in star topology. The substation PDCs cannot provide technically expected efficiency when PMUs in a substation exceed 100 [2].

The collected data from substations are transmitted to control center to perform several decision-making and operation processes. The featured applications include monitoring power system conditions and faults, analyzing vulnerability of utility grid to demand response and DSM situations, and automatic gain control. The PMU based WAMPAC system combined advantages of SCADA, EMS and WAMS.

1.3.3 Smart meters

The smart metering is one of the essential operations in smart grid infrastructure. Smart meters are improved versions of conventional power meters that are developed after AMR and AMI improvements. The smart meters are equipped with advanced ICT interfaces making them quite sophisticated and detailed. In addition to metering features, smart meters are capable to calculate several parameters as power factor or THD and to predict power consumption at particular intervals. Besides its remote-control features used by MDMS, it also allows users to remote monitoring and remote control for their home energy management systems. The smart meters are also defined as smart socket due to its distribution ability of residential grid to houses. The AMI is used to define a smart meter based infrastructure along smart grid applications since it provides measurement of consumed energy, power demand rates, and power quality of entire grid. The typical features of a smart metering applications are listed as two-way communication between grid and smart meter, data recording capability at intervals of 10–60 min, at least daily data transmission to monitoring center, integrated remote disconnection switch, HAN interface, data storage capability for blackouts, voltage and current values, and secure data communication infrastructure [1, 4].

Nowadays, several millions of smart meters have been deployed all over the world, and it is noted that over than 140 million of smart meters have been installed where the most prominent usages is in China [39]. It is expected that the number of installed smart meters will exceed 400 million by 2020. Although all the installed smart meters provide previously mentioned features, almost all of them provides AMR functions such as detailed consumption storage. Due to close relation of smart meter and AMI, several functions required by MDMS and OMS are carried out. Moreover, improved remote monitoring and control capabilities of smart meters ensure two-way communication, instant data acquisition, and remote billing requirements.

The latest smart metering researches have been dealt with power quality issues such as automatic voltage restore, frequency and voltage control, active and reactive power control, DSM, decentralized generation in the context of microgrid and cyber-secure communication systems. The smart metering interface can be illustrated as seen in Fig. 1.15 where smart meter interfaces consumption and distribution grids in terms of metering and communication substructures. The metering section is comprised by three major subsections that are AMR, DMS and TOU pricing while the communication system includes control infrastructure and network connection interfaces such as HAN and WAN comprising the HEMS. The communication interface may

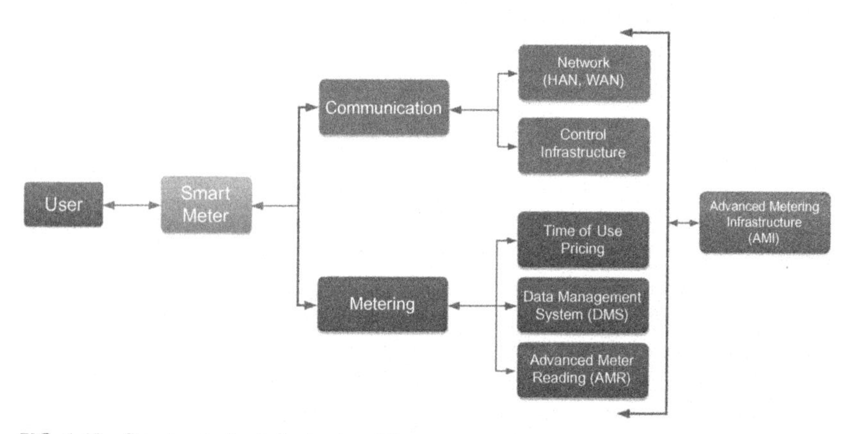

FIG. 1.15 Smart metering infrastructure [1].

provide wireline and/or wireless communication methods that allows two-way data stream between user and DSO. In addition to two main sections of a smart meter, it may also include auxiliary modules such as power supply, controller, metering and data acquisition interface, timer, protection devices, data logging module and encoding/decoding modules. The data logging module that is essential for smart meters provides data storage capability for consumer information such as identification, energy consumption logs, time stamp and outage records. The metering section comprises analog interface that interacts with grid to connect residential wiring to distribution network and is equipped with voltage and current transducers to install metering interface. The billing module is associated with timing module to generate TOU pricing data with timestamp [1].

Besides remote monitoring and smart metering applications, smart meters provide several adjunct applications and services such as DSM, energy theft protection, and CPS security. The DSM and DR programs are improved to meet consumers energy demand while protecting the balance between generation and consumption. Moreover, DSM and DR programs enable DSOs to effectively manage energy generation demands by preventing mass generation in unnecessary times. One of the most important aim of DSM is load shifting by controlling peak clipping, load shedding, valley filling, and peak shifting methods. The balance is determined according to load sufficiency and demand rates to operate these loads shifting methods where peak clipping allows reducing energy consumption during peak time while valley filling is performed by increasing energy consumption during valley times. The peak shifting which refers to composition of peak clipping and valley filling facilitates reducing the peak loads while responding to energy demand of base loads. Due to these requirements of DSM and DR programs, the instant and precise metering should be addressed by using smart meters. The DR programs are based on price-based or incentive-based approaches where most common price-based programs are TOU, critical peak pricing (CPP), critical peak rebate (CPR), and real-time pricing (RTP) [7, 40].

1.3.4 Wireless sensor networks

The wireless sensor network (WSN) implies for a network comprised by wireless sensors cooperatively operating to sense physical and electrical parameters, to control changes along their operation environment, and to interact with its CPS by machine-to-machine (M2M) communication interfaces. The recent WSNs are comprised by nodes of sensor, transducers, actuators and gateways. A wide variety of sensor nodes are deployed around the monitoring and control area to install a sensor network that each sensor is connected to each other by hopping methods. The acquired data is collected, processed and transmitted to other nodes by single or multihop routing until it reaches to management node. The WSN is configured by users or services to get required management processes. The improved WSN technologies such as IEEE 802.15.4 based ZigBee,

ZigBee Pro, WirelessHART, Wireless network for Industrial Automation – Process Automation (WIA-PA), and ISA100.11a have gradually decreased costs and leveraged widespread use of WSNs in industrial automation, residential applications, and related remote monitoring applications. The main parts of a WSN are comprised by power management module, the sensor, a microcontroller for data processing, and a wireless transceiver. The sustainability of sensors along WSN is ensured by power management module that is mostly desired to be capable for energy harvesting by solar, kinetic or thermal energy sources. The sensors are responsible for acquiring analog or physical signals such as irradiation, vibrations, pressure, humidity or electrical signals and to transform them into electrical signals to be processed by microprocessor. Then the microcontroller processes inherited signals to generate transmission signals such as modulated waveforms and additional recording or data storage applications may be realized at this stage. The modulated waveforms are supplied to transceiver module that is a wireless RF transmitter and receiver in one module. Thus, communication interface is obtained and wireless transmission medium is used to communicate with nearest sensor node or management node [41].

The organization and transmission process of WSNs is shown in Fig. 1.16 where the sensor nodes firstly communicate with each other by deploying their status and receiving others. Afterwards, sensors are organized into a networks structure as selecting one of star, mesh, linear or tree topologies and then data transmission paths are calculated to transmit data over suitable routes. The transmission distance of WSNs are relatively shorter around 1000 m since the power of sensors is generally provided by batteries. The sensor networks are operated in multihop networks that is achieved by using sensors in

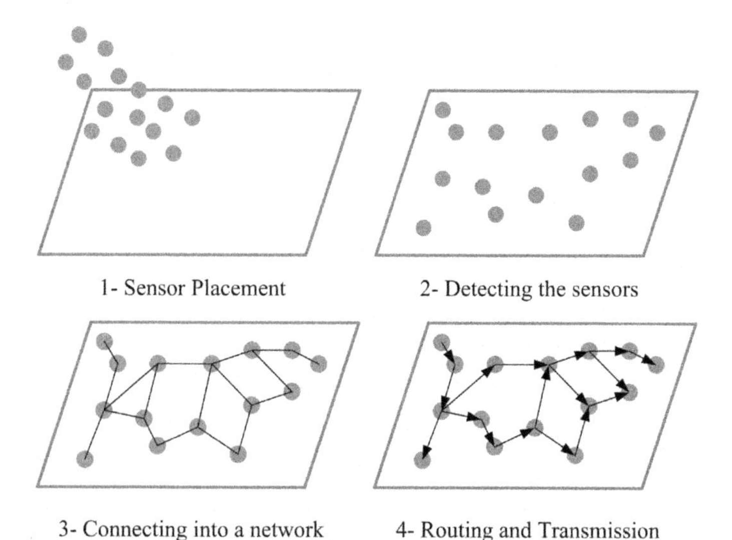

FIG. 1.16 Organization and transmission of WSNs.

transceiver mode as transmitter and receiver to increase transmission distances. Thus, each sensor receives transmitted data from previous one and forwards to next sensor until data reaches to gateway. The WSNs ensure this operation by their self-organization, self-adaption, limited node energy, and unstable transmission links as expressed in [41]. The placements of nodes along a WSN are random and each node may be replaced to another location, protected or may be interfered by any noise source. The network topology also affects the efficiency and reliability while mesh topology ensures more flexible and reliable operation among others. The self-organization of network nodes provide better management and smart mesh networking structure where nodes initially monitor the neighbor nodes and detect signal strength to select the most suitable node for communicating. The time synchronization is established throughout nodes to gateway by this way. When the gateway receives request of nodes, it assigns network resources for nodes and two or more transmission paths are assigned to ensure network reliability.

The recent advances in WSN have promoted its widespread use in smart grid applications including power generation, transmission, distribution and power consumption levels for monitoring and control processes in efficient and reliable ways. The WSNs are suitable for smart grid applications such as AMI, DR and DSM, dynamic pricing, fault detection and prevention, distributed generation control, and remote detection. Since the CPS of smart grid is completed by communication networks, selecting the proper communication method is primarily effective on the performance of WSN associated smart grid applications. The requirements of each application are depending to its data transmission and latency rates, i.e. distribution feeder automations require higher data rates with low latency communications since they should immediately detect the faults and protect the system by isolating. The smart metering and monitoring applications are not sensitive to latencies since they transmit measurement data to MDMS in predefined intervals. It is noted by several studies that the improvements of WSN will improve efficiency and reliability of smart grid applications by the enhancement of distributed automation, dynamic load and source control, advanced management systems and smart sensors with energy harvesting capabilities. The WSN applications can be classified into three levels of smart grid as given in Table 1.7 as generation, transmission and distribution, and consumption levels [42].

The generation level applications include remote monitoring of conventional and RESs such as wind and solar plants and even ESSs. The related measurement and monitoring data may be wind speed, wind direction, temperature, irradiation and humidity for RESs while others are electrical such as voltage, current, power and power factor of generation plant. The distributed generation is another aspect of generation level since small scale and micro plants can be considered in the context of generation level. The security and reliability of generation level is directly affected by the reliability of distributed generation and microgrid plant. Therefore, grid-connected and island mode operations should

TABLE 1.7 WSN application in smart grid levels

Power network level	WSN applications
Generation level	Remote monitoring of power plants
	Power quality analysis
	Distributed generation
	Substation monitoring
Transmission and distribution	Transmission line monitoring
	Outage detection
	Overhead and underground cable monitoring
	Fault detection
	Distribution automation
	Distribution substation control and monitoring
	Protection control
Consumption	AMR and wireless AMI
	Home automation
	Microgrid and Nano-grid monitoring
	Demand side management
	Outage and fault control
	Remote monitoring and control

be fairly monitored by using sensors and WSNs. The communication infrastructure should also ensure reliable and secure monitoring process with low-cost features. The power quality analysis and substation monitoring are closely associated to install a resilient generation level due to WSN and smart grid integration. The measured voltage and current values are transmitted to monitoring center to detect power quality.

The sustainability of smart grid infrastructure is mostly based on security of transmission and distribution level applications since they play vital role on entire system functionality and performance. The prominent WSN applications include transmission and distribution line monitoring such as overhead and underground cables, outage and fault detection, distributed automation, distribution substation control and monitoring, and protection control for fault isolation in transmission and distribution levels. The outage detection is required for financial and social reasons since it is directly related to losses of capital and daily life comfort. The WSNs are appropriate detection devices for faults, outages, and protection systems along transmission and distribution systems. The equipment fault detection and prevention are also coped with WSN that is based on fault passage indicators (FPIs) and insulators.

The consumption level applications of smart grid are not only required for TSO and DSOs but also for consumers since they enable users to monitor and control their microgrids, home appliances, and smart devices by using HEMS. The WSN assisted smart grid applications can be listed as AMR and wireless smart metering, HEMS and residential power control, microgrid and Nano-grid monitoring, data acquisition for DSM, and remote monitoring for outage detection and fault protection. The wireless remote metering and monitoring facilitates interaction with smart home appliances throughout residential energy management system. Moreover, distributed generation and residential energy management are allowed by integration of WSN to HEMSs and smart grid components used in consumer side. The dynamic load control and remote monitoring enables DSOs to detect outages, faults, energy theft, and excessive consumption in order to manage DR and load shedding requirements [42].

As discussed in this section, the WSN integration in smart grid provides further control and monitoring opportunities with low-cost and improved mobility. Moreover, increased reliability and decreased latency facilitate power network management capabilities of TSO and DSOs. The recent studies are based on WSN security and communication protocols to improve interoperability of heterogenous devices and systems operating in the context of smart grid environments.

1.4 Evolution of internet of energy concept

The concept of "Internet of Energy" has been firstly suggested by Jeremy Rifkin in Third Industrial Revolution to define sharing of energy and increased use of RESs. The energy sharing is achieved by using a peer-to-peer (P2P) approach as in computer internet in Internet of Energy (IoE) where two-way power flow is provided by power distribution networks (PDNs) [43]. There is an analogy between network routers of computer internet and energy routers of IoE has been put forward. Thus, PDNs are associated over energy routers in this scenario. The interaction of WSNs, sensors, transducers, AMI technologies and related ICT components have promoted improvement of IoE [44].

The smart grid is defined as system of systems due to its complex and heterogenous infrastructure that is comprised by diverse power and communication technologies. However, smart grid infrastructure is operated by contribution of energy distribution and control systems at any nodes. The intelligent control and monitoring nodes of smart grid are quite similar to Internet of Things (IoT) approach connecting any device at any platform. The M2M and human-to-machine (H2M) interactions enable extending the internet concept for monitoring and control of objects, namely things, used in daily life. The integration of smart grid and IoT presents another description of IoE or Energy Internet (EI) provides to achieve broader connectivity by including cloud computing and big data management technologies. The internet communication improves smart grid applications providing interaction of various ICT and communication interfaces to be used in association. Moreover, M2M communication opportunities enable decentralized control and monitoring features for smart grid control systems [45]. EI presents remarkable differences according to current infrastructures in terms of transmission systems, decentralized energy exchange, energy router interaction at distribution nodes, large-scale integration of DERs, and diverse generator types directly connecting to utility grid without converter requirement. The EI provides widespread energy exchange in different types since it uses electrical energy as an intermediate for transmission. The energy routers interact with several DG connections, relays and reclosers to control energy transformation between islanded microgrids and utility grid. On the other hand, energy routers ensure fault detection and protection schemes due to its fast and accurate measurement capabilities. The increased interest on DG and Res integration to utility grid has fostered development of EI concept. China targets to cover around 30% of energy demand from RESs and discussed to increase this rate over 50% in near future for distribution system. In the EU, Germany embraces an aggressive policy to increase its RES share to 30% up to 2020 and to 50% by 2030 while United Kingdom sets a rate around 20% for RESs. The share of RESs in generation cycle is increased up to 30% in United States where California aims to reach 33% in a few years [46].

The recent trends in IoT communications have promoted EI developments by low-cost and efficient devices with lower power consumption as a next step. The communication technologies of smart grid is enriched with the use of web-based internet services that brought reliable connectivity between various devices. In addition to devices, internet protocols are also emerged regarding to requirements of residential and industrial smart grid applications. The novel IP and TCP based protocols such as IPv6 and IPv6 over low power WPAN (6LoWPAN) have been replaced with legacy ZigBee technologies using IEEE 802.15.4. on the other hand, the use of building automation and control networks (BACnet) have been widespread with the development of HEMS and residential monitoring and control systems. The 6LoWPAN is improved to increase system security of IEEE 802.15.4 based networks. It uses mesh topology to provide extensive scalability that is depended to routing protocols. The mesh network topology provides self-healing ability to the network if network traffic quickly increases and link is broken [45].

An illustration of EI infrastructure is presented in Fig. 1.17 including DER and RESs in addition to conventional bulk generation plants, residential and industrial loads producing energy demand, control center interacting with other sections of infrastructure, smart MDMS center managing metering and

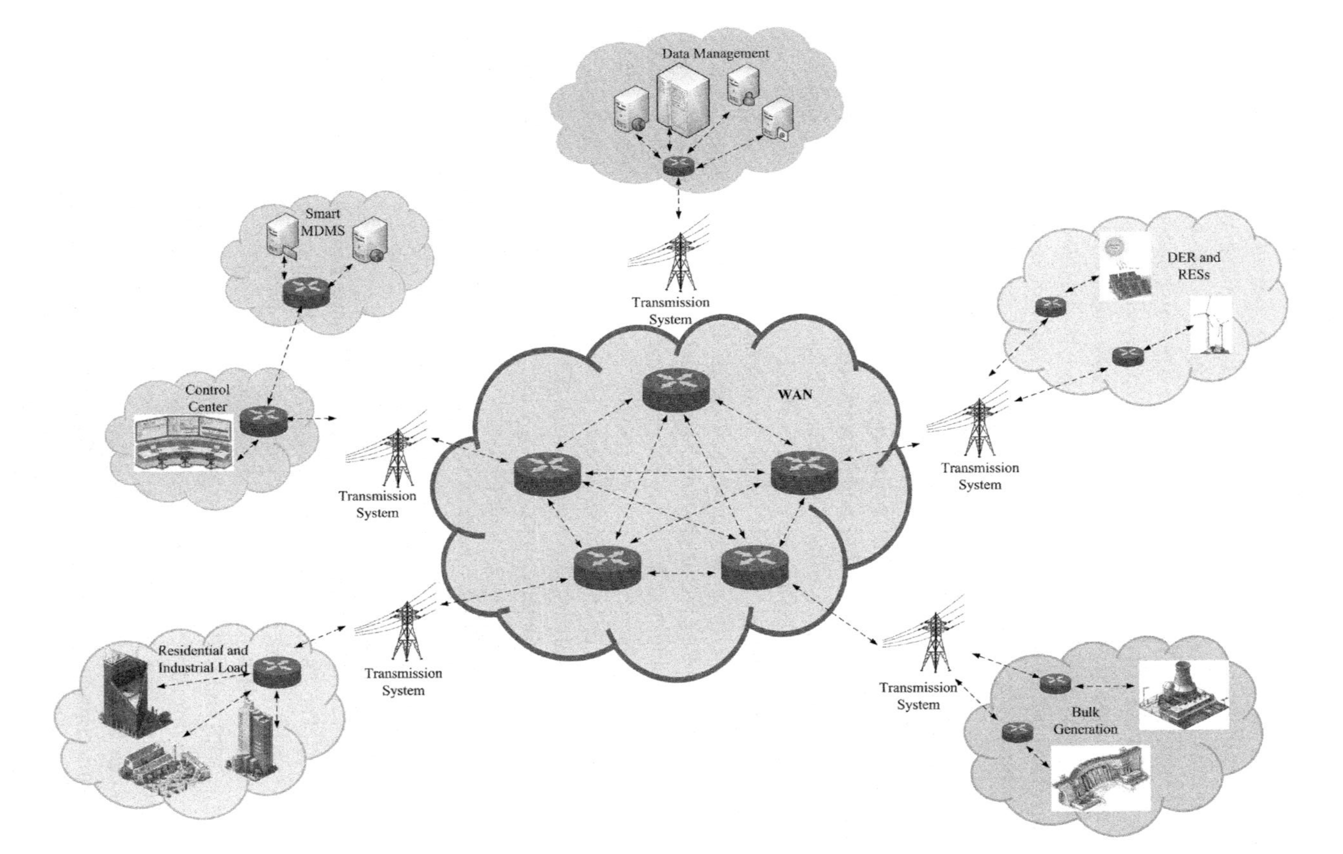

FIG. 1.17 EI infrastructure and communication interface.

measurement system, and data management center providing data storage, backup and data process. The energy routers illustrated in each section provide communication and energy transmission throughout the entire system. The energy exchange along different sections are also provided by interaction of energy routers located at each subsection and centralized WAN section. The control and data management sections facilitate operation of energy router while smart MDMS provides required data to determine demand rates and operate DSM and DR programs. Since massive data stacks are generated by such a system, data storage and management processes require artificial intelligence based analysis methods and cloud computing type database management systems. The power network concept of an EI infrastructure is much more complex and heterogenous comparing to regular smart grid applications. The efficient and reliable energy management is ensured by energy routers that instantly monitor and control generators, energy exchange rates, and consumption data [47].

The key components of IoE, or namely EI, evolution are energy router, data acquisition and data processing, and network technologies. The energy router is core of the presented infrastructure in Fig. 1.17 where the generation, distribution, consumption and storage are combined in a complex system. Regardless of generating plant type, energy router is responsible from two-way power exchange along the diverse sources and loads. Therefore, the importance of plug-and-play (PnP) systems are more significant to ensure high-quality power transmission. In addition to these, efficient and secure energy management tasks should be provided by energy routers. The energy router concept was firstly proposed by North Carolina State University with Future Renewable Electric Energy Delivery and Management (FREEDM) system. Later, several other names such as energy hub, e-energy and digital grid router (DGR) have been put forward, but energy router found a wide acceptance by major of community. The energy router is basically comprised by solid state transformer (SST), distributed grid intelligence (DGI) software and communication interface to interact with CPS.

Another featured energy router has been proposed by Swiss Federal Science with the name of energy hub that is designed for energy conversion and storage operations. The DGR term has been suggested by Japan researchers to define energy router interacting with utility network and ICT interfaces. The DGR is operated on IP based communication network connecting generators, converters, RES and DERs with utility grid [47]. The contributions of EI to current smart grid infrastructure have leveraged numerous achievements in terms of power sources and communication technologies. Moreover, a wide variety of DERs can be easily connected and adapted to CPS due to PnP and advanced communication technologies used in EI environment. We describe EI concept as Smart Grid 2.0 since that brings many improvements to existing smart grid perspective. The detailed description of EI services and application in the context of Smart Grid 2.0 are presented in the following section.

1.5 Energy internet as smart grid 2.0

The proposed EI concepts such as FREEDM, Digital Grid, Global Energy Internet (GEI) of China and *E*-Energy are related to energy transmission control and power conversion issues in the context of ICT applications. The preliminary projects have proposed several IP based communication interfaces and MV level energy generation and grid-interconnection approaches. The GEI differs among other with its ultrahigh voltage (UHV) support and global interconnection of other smart grid infrastructures to implement an energy backbone [48]. However, EI concept is now more than just energy generation and conversion but also ensuring the ubiquitous, available, shared and interoperable smart grid for producers and consumers. Similarly, energy is not only an asset in EI infrastructure but also a generalized resource that is used by any individual and business environments. Therefore, improvement of EI concept has been associated with industrial revolutions and even with smart grid evolution versions as we have proposed in this section.

It is noted that the key components of energy generation and consumption cycle are centralized and decentralized generation, DERs integration to generation and distribution networks, smart and integrated energy management. In addition to these major components, recent advances promoted researches on microgrid concepts that are shift consumers to prosumers by active participation to generation cycle, virtual power plant (VPP) technologies improving the reliability of power network along smart grid infrastructure, and smart energy management and control techniques. In addition to power and communication networks, the data management and processing requirements such as big data, privacy and security are extensively associated with EI ecosystem. All these contributions and improvements provided by EI are introduced in the following chapters in detail.

References

[1] Y. Kabalci, A survey on smart metering and smart grid communication. Renew. Sustain. Energy Rev. 57 (2016) 302–318, https://doi.org/10.1016/j.rser.2015.12.114.

[2] S.K. Salman, Introduction to the Smart Grid: Concept, Technologies and Evolution, The Institution of Engineering and Technology, London, 2017.

[3] Y. Kabalci, E. Kabalci, Modeling and analysis of a smart grid monitoring system for renewable energy sources. Sol. Energy 153 (2017) 262–275, https://doi.org/10.1016/j.solener.2017.05.063.

[4] S. Borlase, Smart Grids Infrastructure, Technology, and Solutions, CRC Press Taylor & Francis Group, Boca Raton, FL, 2013.

[5] H. Farhangi, The path of the smart grid. IEEE Power Energ. Mag. 8 (2010) 18–28, https://doi.org/10.1109/MPE.2009.934876.

[6] US Public Law 110-140, Energy Independence and Security Act of 2007, (2007).

[7] S.M. Muyeen, S. Rahman (Eds.), Communication, Control and Security Challenges for the Smart Grid, The Institution of Engineering and Technology, London, 2017.

[8] CEN-CENELEC-ETSI, Smart Grid Coordination Group Smart Grid Reference Architecture, CENELEC, 2012.

[9] NIST Framework, NIST Framework and Roadmap for Smart Grid Interoperability Standards, Release 1.0, 2010.

[10] The GridWise Architecture Council, Smart Grid Interoperability Maturity Model Beta Version, US Department of Energy, 2011. https://www.gridwiseac.org/pdfs/imm/sg_imm_beta_final_12_01_2011.pdf.

[11] C. Dänekas, C. Neureiter, S. Rohjans, M. Uslar, D. Engel, Towards a model-driven-architecture process for smart grid projects. in: P. Benghozi, D. Krob, A. Lonjon, H. Panetto (Eds.), Digital Enterprise & Design Management, Springer International Publishing, Cham, 2014, pp. 47–58, https://doi.org/10.1007/978-3-319-04313-5_5.

[12] IEEE Standards Committee, IEEE Guide for Smart Grid Interoperability of Energy Technology and Information Technology Operation with the Electric Power System (EPS), End-Use Applications and Loads, Institute of Electrical and Electronics Engineers, New York, NY, 2011. http://ieeexplore.ieee.org/servlet/opac?punumber=6018237. Accessed 5 September 2017.

[13] R.H. Khan, J.Y. Khan, A comprehensive review of the application characteristics and traffic requirements of a smart grid communications network. Comput. Netw. 57 (2013) 825–845, https://doi.org/10.1016/j.comnet.2012.11.002.

[14] B.M. Buchholz, Z. Styczynski, Smart Grids–Fundamentals and Technologies in Electricity Networks. Springer, Berlin, Heidelberg, 2014. https://doi.org/10.1007/978-3-642-45120-1.

[15] S. Mirsaeidi, X. Dong, S. Shi, B. Wang, AC and DC microgrids: a review on protection issues and approaches, J. Electr. Eng. Technol. 12 (2017) 2089–2098.

[16] F. Blaabjerg, Y. Yang, D. Yang, X. Wang, Distributed power-generation systems and protection. Proc. IEEE 105 (2017) 1311–1331, https://doi.org/10.1109/JPROC.2017.2696878.

[17] T. Adefarati, R.C. Bansal, Integration of renewable distributed generators into the distribution system: a review. IET Renew. Power Gener. 10 (2016) 873–884, https://doi.org/10.1049/iet-rpg.2015.0378.

[18] F. Blaabjerg, R. Teodorescu, M. Liserre, A.V. Timbus, Overview of control and grid synchronization for distributed power generation systems. IEEE Trans. Ind. Electron. 53 (2006) 1398–1409, https://doi.org/10.1109/TIE.2006.881997.

[19] A. Hooshyar, R. Iravani, Microgrid protection. Proc. IEEE 105 (2017) 1332–1353, https://doi.org/10.1109/JPROC.2017.2669342.

[20] I. Colak, E. Kabalci, G. Fulli, S. Lazarou, A survey on the contributions of power electronics to smart grid systems. Renew. Sustain. Energy Rev. 47 (2015) 562–579, https://doi.org/10.1016/j.rser.2015.03.031.

[21] A. Kaur, J. Kaushal, P. Basak, A review on microgrid central controller. Renew. Sustain. Energy Rev. 55 (2016) 338–345, https://doi.org/10.1016/j.rser.2015.10.141.

[22] J. Stephens, E.J. Wilson, T.R. Peterson, Smart Grid (R)Evolution: Electric Power Struggles. Cambridge University Press, New York, 2015. https://doi.org/10.1017/CBO9781107239029.

[23] E. Kabalci, Emerging smart metering trends and integration at MV-LV level, in: Smart Grid Workshop Certif. Program ISGWCP Int., IEEE, 2016, pp. 1–9.

[24] J.A. Momoh, Smart Grid: Fundamentals of Design and Analysis, Wiley, Hoboken, NJ, 2012.

[25] K.M. Muttaqi, A. Esmaeel Nezhad, J. Aghaei, V. Ganapathy, Control issues of distribution system automation in smart grids. Renew. Sustain. Energy Rev. 37 (2014) 386–396, https://doi.org/10.1016/j.rser.2014.05.020.

[26] N.S. Nafi, K. Ahmed, M.A. Gregory, M. Datta, A survey of smart grid architectures, applications, benefits and standardization. J. Netw. Comput. Appl. 76 (2016) 23–36, https://doi.org/10.1016/j.jnca.2016.10.003.

[27] F. Li, W. Qiao, H. Sun, H. Wan, J. Wang, Y. Xia, Z. Xu, P. Zhang, Smart transmission grid: vision and framework. IEEE Trans. Smart Grid 1 (2010) 168–177, https://doi.org/10.1109/TSG.2010.2053726.

[28] N. Kayastha, D. Niyato, E. Hossain, Z. Han, Smart grid sensor data collection, communication, and networking: a tutorial: smart grid sensor data collection, communication, and networking. Wirel. Commun. Mob. Comput. 14 (2014) 1055–1087, https://doi.org/10.1002/wcm.2258.

[29] L.M. Camarinha-Matos, Collaborative smart grids—a survey on trends. Renew. Sustain. Energy Rev. 65 (2016) 283–294, https://doi.org/10.1016/j.rser.2016.06.093.

[30] Y.Y. Chen, J.L. Liao, G.W. Chang, L.Y. Hsu, Y. Li, H.J. Lu, Applying IEEE Standard 1159.3 for Power Quality Analysis Platform Implementation. IEEE, 2016, pp. 1–5, https://doi.org/10.1109/PESGM.2016.7741942.

[31] IEEE Std 1379-2000, IEEE Recommended Practice for Data Communications Between Remote Terminal Units and Intelligent Electronic Devices in a Substation, (2001) 60.

[32] A. Kumar, V. Srivastava, M.K. Singh, G.P. Hancke, Current status of the IEEE 1451 standard-based sensor applications. IEEE Sensors J. 15 (2015) 2505–2513, https://doi.org/10.1109/JSEN.2014.2359794.

[33] T.S. Basso, R. DeBlasio, IEEE 1547 series of standards: interconnection issues. IEEE Trans. Power Electron. 19 (2004) 1159–1162, https://doi.org/10.1109/TPEL.2004.834000.

[34] X. Zong, P.A. Gray, P.W. Lehn, New metric recommended for IEEE standard 1547 to limit harmonics injected into distorted grids. IEEE Trans. Power Delivery 31 (2016) 963–972, https://doi.org/10.1109/TPWRD.2015.2403278.

[35] IEEE, Std C37.238-2011 IEEE Standard Profile for Use of IEEE 1588TM Precision Time Protocol in Power System Applications, IEEE, 2011, p. 66.

[36] IEEE, IEEE Std C37.111 Measuring relays and protection equipment—Part 24: Common Format for Transient Data Exchange (COMTRADE) for Power Systems, IEEE, 2013.

[37] IEEE, Std C37.118.2-2011 IEEE Standard for Synchrophasor Data Transfer for Power Systems, IEEE, 2011, p. 53.

[38] IEEE, IEEE Standard for Electrical Power System Device Function Numbers, Acronyms, And Contact Designations IEEE Std C37.2TM-2008, IEEE, 2008.

[39] D. Mah, P. Hills, V.O.K. Li, R. Balme, Smart Grid Applications and Developments, Springer, New York, 2014.

[40] Q. Sun, H. Li, Z. Ma, C. Wang, J. Campillo, Q. Zhang, F. Wallin, J. Guo, A comprehensive review of smart energy meters in intelligent energy networks. IEEE Internet Things J. 3 (2016) 464–479, https://doi.org/10.1109/JIOT.2015.2512325.

[41] S. Yinbiao, K. Lee, et al., Internet of Things: Wireless Sensor Networks, International Electrotechnical Commission, Geneva, Switzerland, 2014.

[42] E. Fadel, V.C. Gungor, L. Nassef, N. Akkari, M.G.A. Malik, S. Almasri, I.F. Akyildiz, A survey on wireless sensor networks for smart grid. Comput. Commun. 71 (2015) 22–33, https://doi.org/10.1016/j.comcom.2015.09.006.

[43] W. Hou, G. Tian, L. Guo, X. Wang, X. Zhang, Z. Ning, Cooperative mechanism for energy transportation and storage in internet of energy. IEEE Access 5 (2017) 1363–1375, https://doi.org/10.1109/ACCESS.2017.2664981.

[44] M. Jaradat, M. Jarrah, A. Bousselham, Y. Jararweh, M. Al-Ayyoub, The internet of energy: smart sensor networks and big data management for smart grid. Procedia Comput. Sci. 56 (2015) 592–597, https://doi.org/10.1016/j.procs.2015.07.250.

[45] N. Bui, A.P. Castellani, P. Casari, M. Zorzi, The internet of energy: a web-enabled smart grid system, IEEE Netw. 26 (2012) 39–45.

[46] Q. Sun, Y. Zhang, H. He, D. Ma, H. Zhang, A novel energy function-based stability evaluation and nonlinear control approach for energy internet. IEEE Trans. Smart Grid 8 (2017) 1195–1210, https://doi.org/10.1109/TSG.2015.2497691.

[47] K. Wang, J. Yu, Y. Yu, Y. Qian, D. Zeng, S. Guo, Y. Xiang, J. Wu, A survey on energy internet: architecture, approach, and emerging technologies. IEEE Syst. J. (2017) 1–14, https://doi.org/10.1109/JSYST.2016.2639820.

[48] K. Zhou, S. Yang, Z. Shao, Energy internet: the business perspective. Appl. Energy 178 (2016) 212–222, https://doi.org/10.1016/j.apenergy.2016.06.052.

Chapter 2

Smart metering and smart monitoring systems

Chapter outline

2.1 Introduction

The degraded structure of conventional power grid and increased energy demand has caused serious overloading, curtailments and blackouts. These kinds of faults induce power quality and capital losses due to underperforming generators, electric machines, pumping systems, industrial and critical loads. Many governments and local authorities have been forced to prevent such losses by modernizing nationwide grid infrastructure by including smart transmission and distribution systems with smart metering and smart monitoring capabilities. Thus, it would be possible to monitor and react to instant demand changes due to improved demand side management (DSM) and demand response (DR) programs. The DSM approach aims to ensure the balance between generation and consumption sections with applying DR programs to change habits of consumer, and to rehabilitate load capacity of transmission and distribution networks. The smart grid transformation of conventional power grid targets to increase efficiency, reliability and flexibility in addition to distributed generation (DG), decentralized control and integration of renewable energy sources (RESs) and distributed energy resources (DERs).

The DSM applications highly rely on smart metering and smart monitoring infrastructures in transmission and distribution networks in the context of smart grid. The increased capabilities and features of smart metering systems ensure

From Smart Grid to Internet of Energy. https://doi.org/10.1016/B978-0-12-819710-3.00002-8

instant monitoring of generation and consumption rates, load amount of substations, demand profile and other similar parameters of active grid [1]. An optimized and efficient demand side integration is based on two-way communication system, advanced metering infrastructure (AMI), automatic meter management (AMM) and comprehensive monitoring technologies. The smart metering is needed to accomplish these requirements that are not possible to be performed by mechanical and regular metering devices. Smart meters are assumed as milestone of smart grid evolution as well as smart power networks and improved power electronic devices [1, 2]. The smart metering requirements have been presented in the previous chapter where sensors, transducers, sensor networks and phasor measurement unit (PMU) have been introduced. This chapter deals with smart metering, AMI, PMU applications and smart monitoring systems.

The security and sustainability of smart grid infrastructure are highly depended to precisely organized metering and monitoring technologies that provide two-way communication and rapid response to operational conditions. It is noted that outages cost around $25– $180 billion to US economy [2]. It is obvious that huge amount of blackout and curtailment costs can be prevented by a certainly planned monitoring and metering infrastructure. In addition to DSM, energy management systems (EMSs) and distribution management systems (DMSs) are required to ensure power quality measurements for DERs and microgrid networks penetrating to utility grid. The EMS may be comprised by centralized or decentralized approaches that centralized EMS is based on intelligent algorithms and services and decentralized EMS is performed by using logical applications operating on entire network. The EMS is responsible for detecting active and reactive power balance against the varying conditions of DERs, demand, load profile, transmission and distribution losses, voltage disturbances, and power quality of utility grid [2].

A smart metering system provides measurement and recording of electricity, gas and water consumption in the infrastructure that has been located. The measured consumption data may be locally saved and transmitted to monitoring center in specified intervals of a few minutes to daily periods. Therefore, a smart meter should be equipped with several capabilities such as internal data storage, two-way communication, report generating, remote control for connection and disconnection, and tamper detection to prevent malicious interventions. The featured capabilities of smart metering system allow transmission system operators (TSOs) and distribution system operators (DSOs) to acquire instant demand data, to manage outage and curtailment faults, to operate service restoration and self-healing procedures, and to perform distribution network analysis, demand planning and billing operations. The recent advances in smart metering technologies enabled evolution of conventional automated meter reading devices to AMIs that allows managing mentioned processes and services by TSOs and DSOs. The key components of an AMI system are based on communication, database management and server technologies besides measurement interfaces [2, 3].

The communication infrastructure should be resistant against interference, latency, noises and fading sources. The component of a communication system along smart grid may be compliant with wireline or wireless transceivers regarding to operation area and requirements. The prominent wireline communication system used in smart metering applications is known as power line communication (PLC) while wireless communications are based on radio frequency (RF), microwave or mesh transceivers at various frequency and data rates. The smart metering and monitoring applications perform required control services due to AMI technology [2, 4, 5]. In this chapter, we firstly introduce smart metering concepts and systems and then present a deeper description of smart meters, AMI technologies, and smart monitoring systems. The evolution of smart metering, protocols and standards of smart meters and AMI technologies, and monitoring architectures have been presented in detail in the context of this chapter.

2.2 Smart metering concept and systems

The very early electricity meters were electromechanical and known as accumulation meter since they were recording the consumption by using a mechanical rotation triggering counters. The billing was performed by on-site detection of customer consumption and by monthly periods in this system. It was a primitive and unreliable management scheme against malicious interventions. The billing periods were not applicable for many industrial plants in conventional metering and billing system due to conventional meters. The first improvements on electromechanical or accelerometer metering infrastructure have been performed by using AMR systems that were capable to communicate in one-way from meter to reader. The next development in metering has been achieved around early 2000s with smart metering systems measuring, collecting, analyzing, and managing energy by use of information and communication technology (ICT). The residential and industrial applications have been integrated to two-way communication networks by the improvement of AMI systems that is fundamental component of smart metering. The development and evolution of smart metering systems have been illustrated in Fig. 2.1 where it is expected to achieve advanced smart metering as a next step of AMI technology. The smart metering technology have leveraged DSM and demand side integration to real-time data transmission and advanced control applications such as monitoring, metering, communication, and management.

Therefore, smart metering is rather than a single interfacing technology with its capabilities of integrating intelligent control and management between customer and system operators. It also provides estimation, decision-making and data management technologies. Smart metering creates a fundamental connection between grid, consumer loads, generation and transmission network, and asset management through integration of several type of area networks such as home area network (HAN), building area network (BAN), industrial area

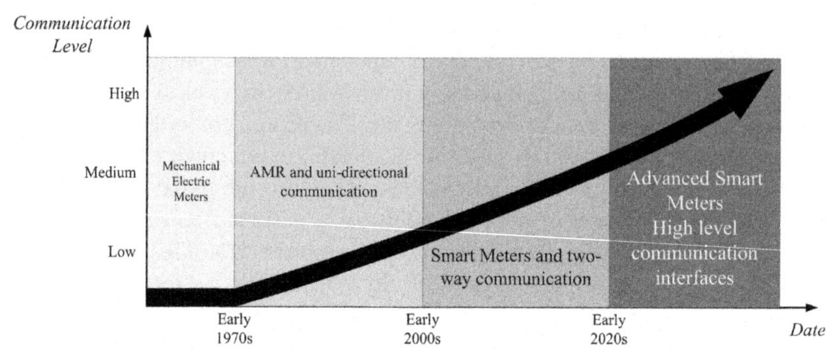

FIG. 2.1 Overview of traditional power grid.

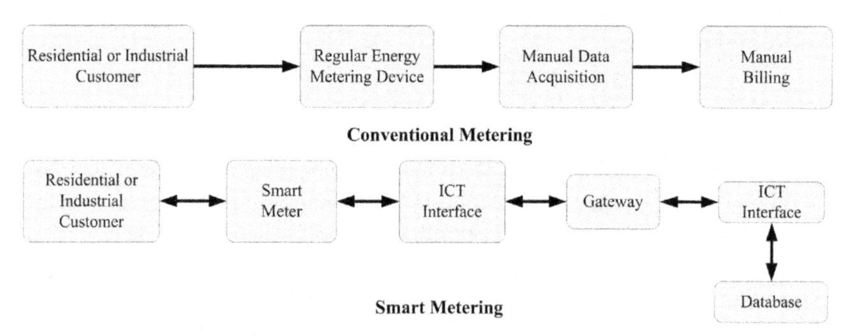

FIG. 2.2 Conventional and smart metering architectures.

network (IAN), field area network (FAN), and consequently wide area networks (WANs) [1, 4, 6]. The architectures of conventional metering and smart metering systems are illustrated in Fig. 2.2 where conventional metering is unidirectional and almost all the processes are done by manually. On the other hand, the smart metering operations are based on two-way data transmission between customer and monitoring center. The smart meter is interfaced with two-way communication devices interacting with gateways and database that are used to establish a communication channel.

In addition to ICT interfaces, smart meter utilizes several sensors and control devices to ensure security and reliability of transmission system. Moreover, smart meters can be programmed to perform a set of functions while conventional meters do not provide such an option. The functionalities of smart meters can be listed as to present instant and accurate consumption data, to communicate with remote monitoring center, to provide data base for analyzing and assessment of power quality, to receive control commands from monitoring and control station and to perform required processes such as turn off or scheduled operations, to operate schedule based tariffs, to interact with interfacing devices and home energy management system (HEMS), to detect and prevent power losses, tamper detection and energy theft interventions [1, 4].

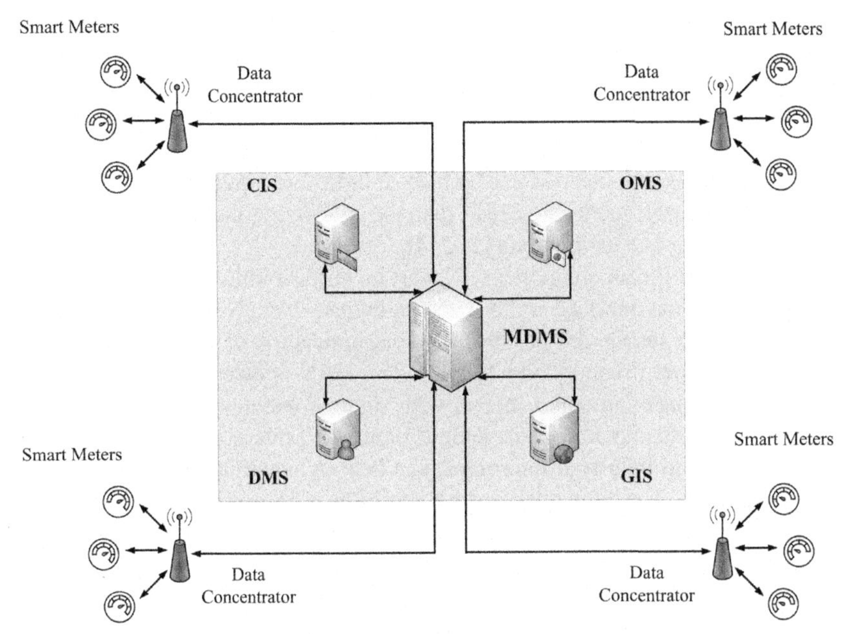

FIG. 2.3 A general centralized smart metering infrastructure.

Intelligent collector devices comprising a group of smart meters for measurement and data transmission comprise the ICT interface of smart metering architecture presented in Fig. 2.2. The intelligent collectors are widely known as data concentrators that inherit measurement data from smart meters and provide data exchange mostly over wireless gateway channels as shown in Fig. 2.3. The metering data management system (MDMS) is the core of smart metering architecture with its servers operating as customer information system (CIS), outage management system (OMS), geographical information system (GIS), and DMS. The data concentrators are responsible for connecting customer premises and smart meters with MDMS by performing data exchange between head ends.

The MDMS is located at distribution network level and used for data acquisition from smart meters. The inherited data are processed, evaluated, and stored to analyze actual status of distribution network. The auxiliary services are used for specific purposes that OMS monitor power quality parameters of grid and detects any fault or disturbance situation for reporting to MDMS. To this end, regular measurement data transmitted at high frequencies are neglected and low frequency detection signals are taken into consideration by OMS. The GIS is used to collect geographical data of smart meters and customer premises and generates valuable data for decision-making purposes. GIS service is a common platform to store, to analyze and to display geographical information on asset management and under or over loaded substations.

Moreover, GIS provides data to CIS to facilitate its operation during consumption rate detection and billing operation by including geographical location. The CIS is crucial to ensure customer services and providing reliable billing services. The system manages user accounts, consumption rates with time stamp indicators, and provides database for user interactions. The DMS can be considered as the supervisor of whole architecture where it is responsible for tracking power quality and load demand to provide the required data for decision making and estimations [1, 2, 4].

Another significant component of smart metering architecture is comprised by communication interface and ICT technologies. HAN, NAN, and WAN networks as shown in Fig. 2.4 perform the communication of entire architecture between customer premises and MDMS. The HAN is essential network integrating and connecting smart meters with smart home appliances, EMS systems, in-home display and management interfaces, micro generation sources in residential and industrial microgrid, and heating, ventilation and air conditioning devices. The fundamental task of HAN is to ensure interoperability of appliances in residential smart grid by using wireline and wireless communication infrastructures, protocols and services. In addition to comprise a communication platform, it also provides security and reliability of consumer data and entire metering system. Thus, a secure and widespread energy management system is achieved for the households.

HAN infrastructure provides to operate several services such as remote control, monitoring, multimedia systems, surveillance cameras and energy management applications. Another significant contribution of HAN to smart metering system is related with microgrid and plug-in electric vehicles (PEVs) that is required to be monitored by DNOs in the context of DSM and DR management issues. The internal residential networks are associated with the left part of utility grid by using NAN and WAN infrastructures. The fundamental objective of NAN is to perform consumption data of customers inherited from smart meters to WAN devices. It is assumed that data size of a residential smart meter is lower than 100 kB/day and multiple smart meter connection are considered while planning transmission line data rates. The massive data

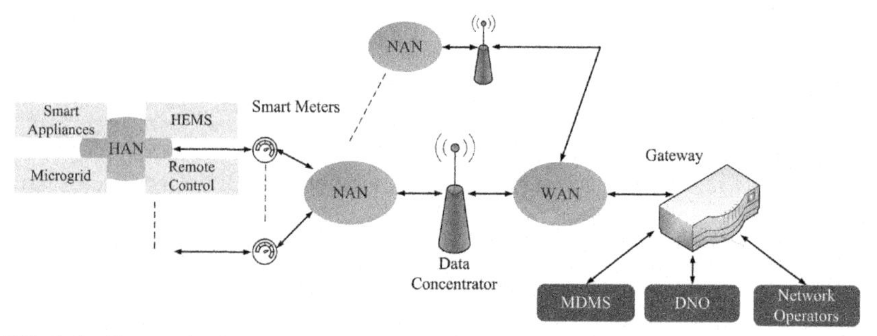

FIG. 2.4 Communication architecture in smart metering infrastructure.

transmission requires data concentrators between NAN and WAN connection nodes and data management is managed through concentrator devices. A WAN is comprised by several NANs exchanging information between smart meters and monitoring center. The communication medium of a WAN is mostly based on licensed frequency bands and technologies such as general packet radio service (GPRS), Worldwide interoperability for microwave access (WiMAX), Satellite and cognitive radio networks (CRNs) [2, 4, 6, 7].

The electricity meter is exploited to detect consumed energy rate of customer as its name implies. In addition to consumed electricity, the transmission and distribution charges are also calculated by DSOs regarding to consume quantity. The most widely and long-lasting used electricity meter is known as electromechanical accumulation meter that records consumption according to time intervals. The accumulation meters cannot be remotely reached or controlled and thus, billing was performed by specific period by manual readings. Since the billing periods are quite long as months, short intervals are required for billing and meter readings in recent years. To this point, some advanced electricity meters called interval meters have been improved with the capability of recording consumption data in short intervals such as hourly or once in a half an hour.

The interval meters have increased capabilities of DSOs by allowing them to schedule specific tariffs and DR programs to manage customer requirements. Moreover, DSM and DR programs have come into question to improve control and management abilities in the context of generation and consumption balance. The evolution of metering devices illustrated in Fig. 2.1 describes these improvements. The AMR applications have brought one-way communication option to monitoring and control centers that facilitated near real-time metering. The smart meters are more sophisticated comparing to their predecessors with their two-way communication capability. Besides, smart meters provide real-time measurement data at any interval on demand. Thus, DSM and DR program developments are completed in faster and reliable approaches comparing to previous applications. The comparison of AMR and AMI features has been listed in Table 2.1. The smart meters provide dynamic pricing regarding to real-time measurement and instant connection between customer premises and network operator. On the other hand, AMI technology provides utility management, DR programs, and emergency response options against monthly data reading option of AMR systems. In addition to billing and CIS features of AMR, AMI provides several key services such as customer data management and display, outage management, and emergency DR. The remote connection and disconnection feature of smart meters facilitate to improve DSM and DR programs on network operator side by enabling load shedding operations. Moreover, theft and losses are prevented due to remote control capability of smart meters.

Another contribution of smart meters is increased interaction of customers with network by exploiting DR programs, HEMS, and self-cost or billing control capabilities while none of these are available in conventional metering

TABLE 2.1 Comparison of AMR and AMI architectures

Feature	AMR—Conventional meters	AMI—Smart meters
Meter type	Electromechanical, electronic	Electronic with remote control and communication features
Data acquisition interval	Monthly	Real-time
Data recording method	Cumulative kWh	Dynamic pricing
Operational options	Monthly data reading	Pricing, utility management, DR, emergency response
Key contributions	Billing, CIS	Billing, CIS, customer data display, outage management, emergency DR
Customer interaction	None	DR programs, HEMS, cost management
Additional services	None	Smart appliances, microgrid and PEV integration
Outage and faults	Phone call	Self-detection, verification, self-healing

technologies. Additionally, smart meters promote customers to build their own microgrid plants due to grid-tie operation allowed by two-way power and communication capabilities.

Another key contribution of smart meters is penetration of PEV and energy storage system (ESS) to utility grid since smart meters enable grid-to-vehicle (G2V) and vehicle-to-grid (V2G) operations. The outage and fault detection are another crucial contribution of smart meters to utility grid and network operators. The sensor networks interacting with smart meters are convenient to detect any outage and fault occurred in customer network. The two-way control allows detection, localization, verification and fixing the outage ad faults at individual customer level [4, 6].

The improvement of AMI systems provides many advantages to customers and network operators at any level including distribution automation (DA), substation automation, fault detection and protection, energy efficiency, asset management, DR and DSM. The advantages can be classified as short-term and long-term due to their effects and contributions that are listed in Table 2.2.

The improved ICT infrastructure and two-way communication opportunities have leveraged evolution of smart metering systems that provide many

TABLE 2.2 Advantages of AMI

Phase	Customer side	Supplier and DSOs	All levels
Short-term	Energy saving	Decreased metering and commercial cost	Increased service quality, flexible pricing programs
	Frequent and accurate billing	Frequency and accurate meter reading	Increased penetration of DERs, flexible load profile
Long-term	Receiving pay back for DG integration	Reduced peak demand and reduced peak price due to DR programs	Increased reliability of energy supply chain
	Decreased bill costs with improved system	Better planning of generation, transmission and maintenance	ICT support for remote control, DG, and DSM
	Minimized energy costs	Real time system operation at distribution level	Increased integration of PEVs, ESSs, and DERs
	Increased comfort with HEMS integration	Additional service support as cable TV, multimedia and communication	Decreasing the peak demand increase rates

advantages for customers, energy suppliers, TSOs and DSOs. The prominent benefits are decreased energy consumption due to integration of remote monitoring and control features, more accurate and short-term billing for customers in short-term intervals. On the other hand, the improvements of smart metering processes allow customer to comprise their own DER based microgrid plants and to receive pay back against their consumption. Thus, the energy costs and bills are gradually decreased with the aid of HEMS integration to residential network. The benefits of energy suppliers and DSOs are also related to costs and billing issues where metering costs and efforts are seriously decreased due to instant and two-way connection.

The peak demand and peak price may also be decreased regarding to outputs of decision making and estimation algorithms. The evaluated consumption and demand data facilitates improvement of DSM and DR programs and provides better planning of generation, transmission and maintenance schedules. It also provides prevention on fraud and thefts [6]. The architectures and specific features of area networks will be presented in detail in the following chapter. The communication is crucial component of smart metering and has played an important role in the evolution of smart electricity metering that has been discussed in the following section.

2.3 Smart meters

The Smart Grid requires two fundamental devices that are sensor networks and smart meters to monitor grid situation. Both of devices are operated with wired and/or wireless communication infrastructures. The smart meters are robust devices for measuring, storing, processing and transmitting power consumption rates where they are used. Although the conventional meters are just used for measurement of consumed energy, smart meters provide additional information such as total harmonic distortion (THD) rates, frequency rates, power quality, peak load times and specific interval data. Moreover, smart meters are used for rapid detection of outages, assisting to troubleshooting and self-healing processes, and reporting reasons of outages. The acquired and processed data are transmitted to data management unit (DMU) over secure transmission channels by smart meters. The smart meters can inherit many measurement data form sensor networks and smart home appliances throughout a smart environment, and they can be remotely controlled for disconnection or reconnection of customer premises due to its two-way communication capabilities. The measurement data acquired from sensors are transmitted to smart meters and then the processed measurements are delivered to DMUs by using communication interface of smart meter [8, 9].

The remainder of this subsection deals with hardware and device structures, communication technologies, remote control features, DSM applications, theft and fraud control, and security and privacy issues of smart meters.

2.3.1 Hardware and accurate metering

Although the conventional electricity meters are based on electromechanical structures that measures consumption by revolutions of rotating magnetically conductive disc at a proportional speed of magnetic flux, the electronics meters do not include rotating equipments and perform measurement process by resistors or sensors. Despite the benefit of rotating component removal, electronic meters require many advanced control and calibration functions, and digital microprocessor technologies. Even though, electronic meters eliminate many drawbacks of conventional electromechanical meters due to increased and sophisticated device configurations. First, they provide real time pricing (RTP) that plays crucial role on improving DR programs and schedules even for each individual customer. On the other hand, smart meters allow modular configuration to enable improving tailored metering device with additional functions such as communication options, metering requirements, and so on except base configuration.

The fundamental function of a smart meter is to ensure regular and accurate metering that serves basic purpose of smart grid as reducing waste consumption and decreasing the demand at peak hours. The smart grid requires frequency measurements acquired from entire grid in short intervals as hourly or more shortly. There are many commercial smart meters can be given as examples to measurement periods based on hourly, half an hour, and even 15-min intervals. In addition to central smart meters, some featured smart meters provide multiple measurement channels for dedicated uses of main asset, DERs such as solar or wind turbines, and individual load types. This is called multi-channel metering and it prevents requirement for spare smart meter utilization for DG and loads. In addition to measurement frequency, data recording and warning is another important feature of smart meters. It provides information for both of utility operators and customers to know about consumption rates and usage limits. Some countries prefer prepaid metering systems to provide this function where customers charge a limited credit for their smart meter and they become able to consume electricity until the credit discharged. Another solution in contrast to prepaid usage is alarm-based warning function of smart meters those can be programmed to an exact usage limit and smart meter warns consumer when the limit is reached [1, 10].

In addition to RTP information, smart meters can store measurement data for usage history and cumulation of periodic usage including consumption, tariff programs, overall costs and so on. The stored data can be monitored over HEMS by reaching user interface of smart meter or provided interface software of smart meter vendor. Most of the service suppliers also provide web-based monitoring interface for their customers. It is obvious that smart meter deployments are leading the way in smart grid applications all around the world. The United States is also leading among countries with around 50% replacement of millions of electricity meters with smart meters [10, 11].

The advanced smart meters should meet smart grid requirements according to regional arrangements. In the United States, the most appropriate communication system for smart meters is low power radio frequency (LPRF) mesh network operating at Sub-1 GHz frequency band. On the other hand, it will be better solution to use wireline communications instead of wireless in some European countries such as Spain or France. The narrowband PLC technologies are widely preferred in these countries due to technical infrastructures. Since there is not any solution fitting to each situation, a smart meter requires combination of many alternatives in metering and communication interfaces. In addition to analog front end (AFE) and communication requirements, smart meters should comply with residential ICT and gateways. It is noted that ZigBee technology based on IEEE 802.15.4 standard is widely used in smart meter and HEMS integration at 2.4 GHz frequency in the United States while Sub-1 GHz RF and PLC combination is mostly preferred in smart metering applications in United Kingdom and Japan [11].

The block diagram of a three-phase four-wire smart meter has been presented in Fig. 2.5 as a very detailed example of commercial smart metering devices. Although current transformers (CTs) and voltage transformers (VTs) comprise the AFE, low cost implementations are also possible to comprise AFE by using current and voltage sensing resistors. An internal rectifier mostly

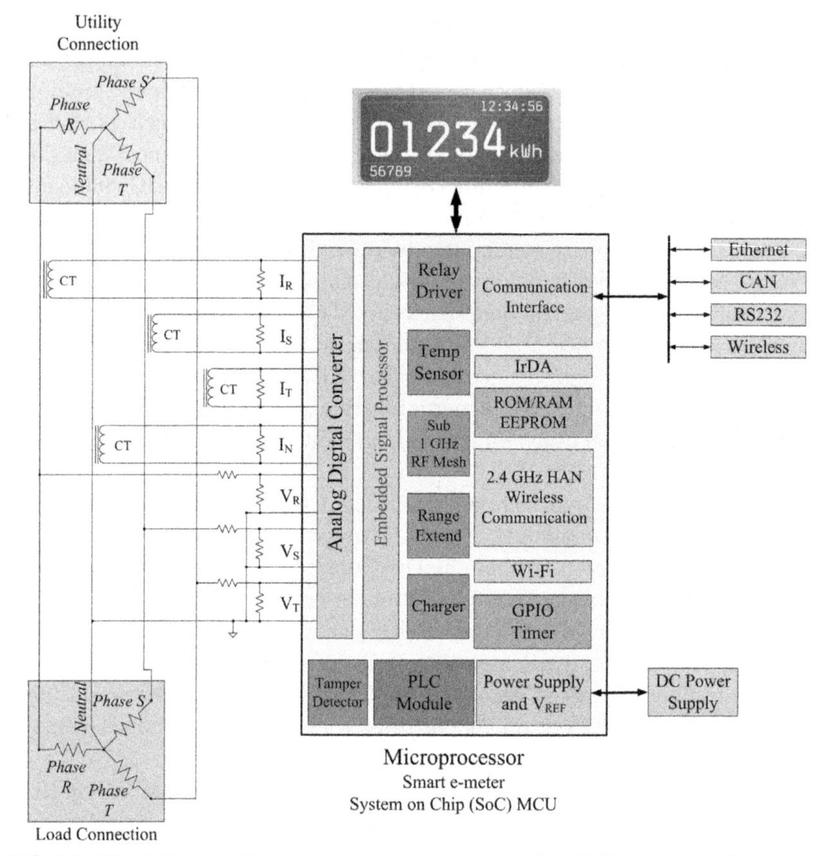

FIG. 2.5 Block diagram of a three-phase four-wire smart meter with its internal sections.

powers the smart meters and dc-dc converter device that is fed by utility line and a back-up section is placed for battery-supply. The auxiliary circuits such as power conditioning and filtering are integrated to main power supply section of smart meter to ensure supply reliability. The continuous monitoring of utility or power grid is performed by using direct or indirect measurement of current and voltage values. The indirect measurements are based on Ampere's and Faraday's principles where a voltage is induced on coils of sensor or Rogowski coil and the induced voltage is calculated referring to proportional current rate. However, direct measurements are based on resistor networks that are placed resistor divider or shunt resistor to measure voltage and current values.

Shunt resistors are placed series on the measurement line and are used for current metering without galvanic isolation. In case of isolation required, shunt resistor is interfaced with a pulse transformer before connection to microprocessor or microcontroller unit (MCU). The current sensing resistors should be reliable and accurate as much as possible to ensure secure measurement.

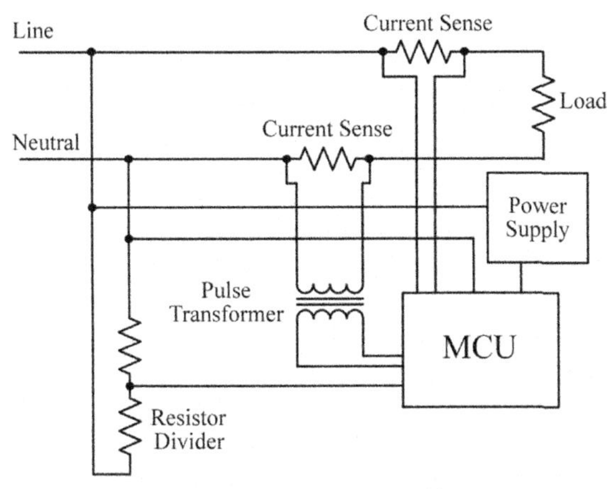

FIG. 2.6 Schematic diagram of a single-phase smart meter with current and voltage sense resistors.

In residential smart meters, very low value shunt resistors are placed in series with high current grid line at the smart meter inlets. The schematic diagram of a resistor based AFE is shown in Fig. 2.6. The current flowed through shunt resistor is calculated referring to Ohm's principle where grid voltage and resistor value are known. The multiplying measured voltage through divider resistor and current yields instantaneous power consumption at any time [11, 12]. In addition to fundamental measurement capabilities, most of the smart meters can calculate phase difference and frequency of current and voltage values regarding to their mathematical processing features. Substantially, embedded signal processor following analog digital converter (ADC) of smart meter as seen in Fig. 2.5 handles this process. The protection and connection/disconnection controls of smart meter are accomplished by tamper detector, relay driver, temperature sensors, and power supply block those all are operated in accordance due to embedded System on Chip (SoC) software.

The SoC is also responsible for operating several communication protocols and services including wired and wireless technologies such as Ethernet, CAN bus, RS232, RS485, PLC, infrared data association (IrDA), and 2.4 GHz wireless communication for integrating to HAN or BAN networks. Some featured smart meters include near field communication (NFC) functions for prepaid utilization in addition to widespread communication technologies. Moreover, microprocessor or MCU of the smart meter requires to support advanced functions like dynamic RTP, DR, remote connection and disconnection control, network security, and wireless update features for its firmware upgrades [1, 11].

2.3.2 Communication interface

The communication networks of a smart meter should comply with two-way data transmission needs of smart grid infrastructure. The two-way communication concept defines mutual connection procedure between all sides of customer and network operators. Therefore, the installed communication medium should ensure bidirectional data transmission between customers and network operators in addition to customer-network operator channel. The communication module of smart meters is equipped with several types of communication technologies and operates independent from AFE and recording sections of smart meter. Thus, an error occurred on connection module does not cause any disruption on smart meter functions. The communication interface of smart meter includes wireline or wireless transmission technologies as discussed earlier. Moreover, the communication technologies should cooperate with network architectures such as HAN, NAN, and WAN as depicted in Fig. 2.4 where area networks are based on many protocols and services on layer structure for ensuring secure and reliable communication. The layer structure of any network is comprised regarding to some standards that most widely accepted on is open systems interconnection (OSI) model comprised by seven layers. The initial layer of network architecture is physical layer (PHY) that is responsible for electrical, physical and optical transmission of data as its name implies. The second layer known as data link layer where medium access control (MAC) provides direct connection between two nodes and corrects erroneous data transmitted by PHY layer. The MAC layer enables Ethernet, IEEE 802.11 Wi-Fi, IEEE 802.15.4 and ZigBee type communications to operate at the data link layer. The third layer of OSI conceptual model is network layer that is responsible for ensuring connection between different network types of nodes. Many types of communication devices and technologies causing a heterogeneous architecture comprise the smart grid. Therefore, network layer enables each node with an address and communication technology to operate on the same area network [2, 10].

The utility grid requires a smart metering network including high and low bandwidth sections if we recall Fig. 2.4. The WAN meets high bandwidth requirements with its huge number of communication channels and connections with data concentrators and gateways. The communication networks that are used in smart metering are classified into three groups as shown in Fig. 2.7 and communication technologies are listed under each area networks with their network layer interaction.

2.3.2.1 Home area network

HAN is the fundamental network architecture deployed at customer premises to monitor and control integrated devices and to perform smart metering processes. It integrates many devices such as HEMS associated smart appliances, smart meters, smart sockets and plugs, and PEVs. HAN facilitates interaction of energy management systems and smart appliances to improve control and comfort of users and network operators. One of the most important benefits of HAN

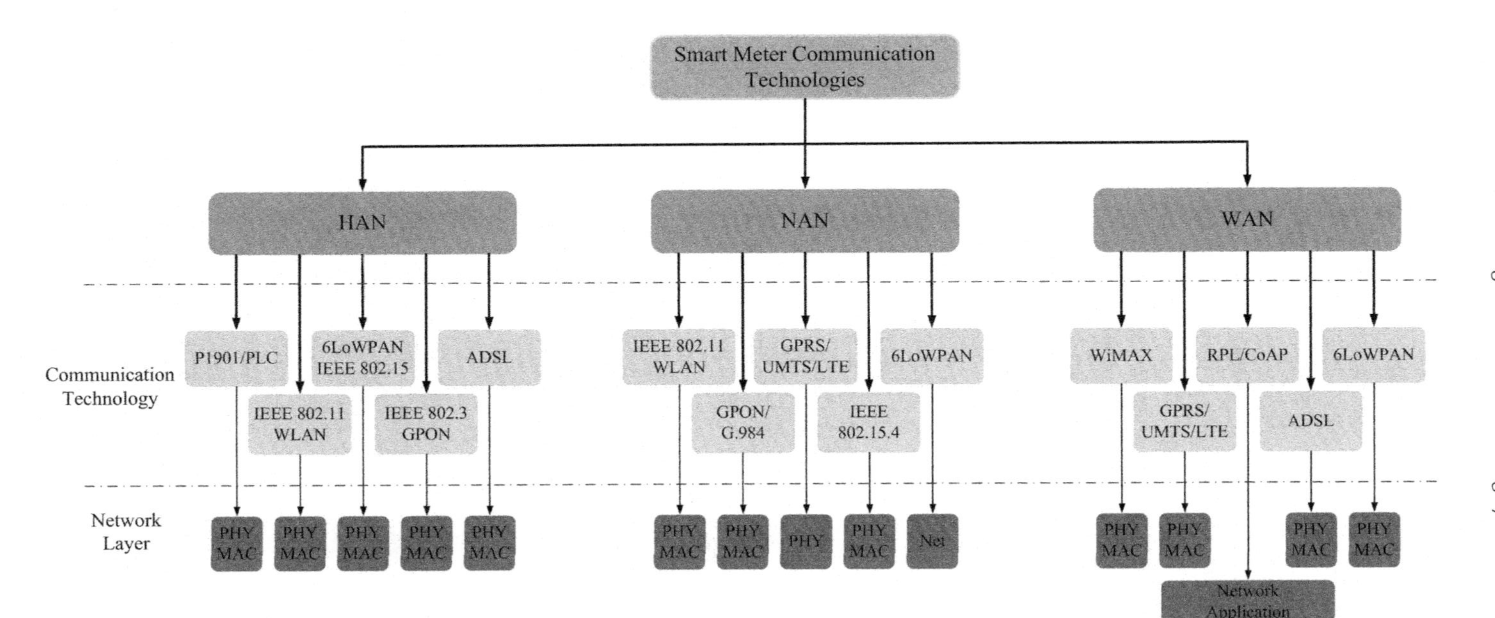

FIG. 2.7 Communication technologies and network layers used in smart meters.

for TSO and DSOs is to enable DSM by establishing a two-way communication infrastructure for smart meters. Another important contribution of HAN is its capability to get connected to outer world by NAN interactions.

A wide variety of technologies including PLC, IEEE 1901 broadband over power lines (BPL), ITU-T G.984 gigabit passive optical network (GPON), IEEE 802.3ah ethernet passive optical network (EPON), IEEE 802.11 wireless local area network (WLAN), IEEE 802.15.1 wireless personal area network (WPAN), IEEE 802.15.4 low rate WPAN (LoWPAN) and asymmetric digital subscriber line (ADSL) comprise the heterogenous environment of HAN architecture as common communication technologies.

The PLC is one of the most widely used wireline communication technology that is based on data transmission over power lines. Internet based infrastructures can be used in the context of BPL connections for wireline communication of first mile and last mile devices. The fundamental benefit of wireline communication is its reliability and robustness against interference. The ADSL and fiber optic technologies including GPON and EPON are also widely used besides PLC in wireline communication systems. The data transmission rates of DSL line vary between 10 Mbps and 10 Gbps while fiber optics extends these ranges up to 160 Gbps. Two major PLC technologies that are narrow band PLC (NB-PLC) and broadband PLC (BB-PLC) are classified according to their transmission frequencies.

The NB-PLC can be used in low and high voltage line up to 150 km transmission length and operates up to 500 kHz transmission frequency. On the other hand, BB-PLC allows high data transmission rates up to 200 Mbps and high transmission frequencies up to 30 MHz [2, 13]. Regardless of wireline or wireless, the communication technologies interact with HAN over PHY and MAC layers.

2.3.2.2 Neighborhood area network

The NAN is located between HAN and NAN to serve as a central bridge of smart metering infrastructure devices. The NB-PLC can be used in low and high voltage line up to 150 km transmission length and operates data transmission through HAN and WAN systems. In addition to communication technologies that are used in HAN, there are several additional technologies are used in NAN networks for communication. NAN allows use of open standard to increase reliability, security, interoperability and flexibility of system. Moreover, NAN enables internet-based communication technologies to be used in architecture. The transmission rate of wireless communication can be extended up to 1Gbps by the contribution of IEEE 802.11 WLAN. The MAC based LoWPAN enables to communicate in 868 MHz, 902–908 MHz, and 2.4 GHz license free bands. The IEEE 802.15.4 standard is a basis for ZigBee that is a low-cost and easily integrated device for smart metering applications. Moreover, several cellular technologies such as universal mobile telecommunications system (UMTS) and long-term evolution (LTE) are used in communication architecture of NAN and they are capable to interact with PHY and MAC layers.

2.3.2.3 *Wide area network*

WAN operates as a backbone between DA and power network operators by supporting two-way communication. The bulk generation, transmission and distribution systems including power plants, substations and distribution transformers communicate with utility grid over WAN architecture. The communication technologies of WAN network can be more than one of wireline and wireless technologies including IEEE 802.16 WiMAX, GPRS, UMTS, LTE, ADSL, LoWPAN, fiber optics and routing over low power and lossy networks (RPL). GPRS provides data transmission bandwidth up to 172 kbps while the latest cellular technology, LTE, increase this rate up to 300 Mbps in coverage area and 75 Mbps for mobile users. The WiMAX, which is an execution of IEEE 802.16 standard for metropolitan area networks (MANs) is one of the fundamental technology providing connection between data management point (DMP) and smart meters in WAN architectures.

WiMAX is based on orthogonal frequency division multiple access (OFDMA) that is the multi-user adaptation of regular OFDM digital modulation scheme. The multi-user structure is obtained in OFDMA by arranging the subsets of several subcarriers to unique consumers that allow simultaneous data transmission from a huge group of consumers in low data rates. The recent internet protocol (IP) that is called IPv6 brings a huge addressing space facilitating integration of millions of devices to WAN infrastructure. The 6LoWPAN denotes IPv6 enabled LoWPAN and operates between IPv6 and MAC layer of IEEE 802.15.4 to provide IPv6 support in data management. The constrained application protocol (CoAP) is a downscaled version of Hyper-Text Transfer Protocol (HTTP) on the User Datagram Protocol (UDP) for constrained resource applications to tackle challenges met in IPv6 transmission over LoWPAN. On the other hand, one of the fundamental contribution of CoAP is its support for machine-to-machine (M2M) communication and mutual HTTP conversion due to constrained internet protocol [2, 13].

2.3.3 Remote control features

Smart meters are not only used to transmit measured consumption data but also for special control requirements of smart appliances due to two-way communication capability. The smart meter enables users or service operators to turn on or turn off any specific residential device through communication link. Besides, smart appliances can be associated with HEMS and consumers can manage residential environment in terms of security, comfort or control by controlling surveillance cameras, heating, ventilating and air conditioning (HVAC) systems, washing machines or other devices on demand. The control operations are performed by wireless or internet-based user interfaces that also provides instant and previous consumption data of residential grid [10].

2.3.4 Demand side management

Besides the discussed functions, another important feature of smart meters is DSM that is accomplished thanks to detailed control and monitoring capabilities through two-way communication. The conventional grid structure was based on supply side management (SSM) since the generation and control processes were centralized and load management has been performed considering generation capacity. However, decentralized control and DG integration that are brought by smart grid to conventional grid enabled DSM considering consumer side requirements and DR programs. The most important target of DSM is to move energy demand of customers to out of peak periods and to decrease energy consumption at peak periods. Thus, load shifting processes are performed by some technical methods called peak clipping, valley filling and peak shifting. The peak clipping defines decreasing the energy consumption during peak time while valley filling refers to increasing the energy consumption during valley times since generation is sufficient to meet demand of load. The peak shifting that is combination of peak clipping and valley filling methods facilitate to ensure generation and demand balance through the utility grid. It is apparent that such a control operation requires rapid and accurate measurement system to ensure instant monitoring of generation and load sides. Moreover, other fundamental functions of smart meters such as data storage, alarm and warning, and data transmission are needed to enable DSM programs. Decision-making algorithms to generate an appropriate DR program for changing customer tendencies process the measured consumption data. The most widely used dynamic pricing methods are noted as time of use (ToU), incremental tariffs, critical peak pricing (CPP), critical peak rebate (CPR), and RTP [10]. These pricing programs are used to sustain balance between generation and load demand by the support of smart metering.

2.3.5 Theft and fraud control

The improved intelligent electronic devices (IEDs) such as digital relays, circuit breakers, fault passage indicators, and protective relays can be integrated to smart meters. Besides, fundamental sensors of smart meters can be used to detect unauthorized consumption, theft and fraud use of electricity. The advanced communication infrastructures facilitate detection and addressing of theft and fraud with the support of smart meters. The time-stamp feature of smart meters that is used to reporting consumption with exact logging can provide increased control on detection of losses and leakages through utility grid. Network operators can immediately detect any abnormal consumption or unauthorized use of electricity and can prevent utilization by remotely disconnecting the detected smart meter. To this end, network operators benefit from GIS and CIS services to identify corresponding customer and smart meter [10, 14].

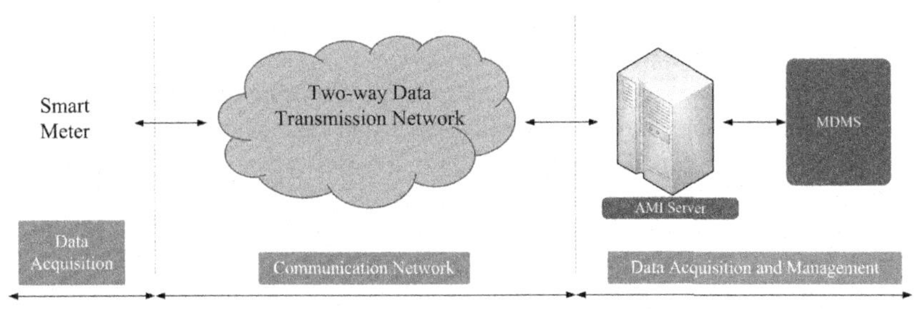

FIG. 2.8 General architecture of advanced metering infrastructure.

2.4 Advanced metering infrastructure

The AMI architecture is comprised by three main components that are smart metering section, communication network, and data acquisition and management section of utility network as shown in Fig. 2.8. The components of AMI system are associated to install a unique platform for DA, outage management, and customer services. The DA provides several contributions such as fault detection and localization, protection, isolation, energy efficiency, asset management and remote control. These benefits of DA rely on smart metering capabilities of entire network. AMI has been improved regarding to AMR to eliminate its disadvantages caused by one-way data transmission. The two-way communication capability of AMI also provides DMS that is an instant and online decision-making tool relying on data acquisition from utility grid and generating control commands for distribution system automation, circuit breakers, reclosers, voltage regulators and substation control. DMS requires an efficient and reliable communication infrastructure to accomplish these functions and widespread sensor networks for accurate data acquisition. The deployment of smart sensors, smart meters and AMI systems meet the requirements of a sophisticated DMS system [2, 4, 5, 14]. We have introduced three main components of an AMI system in previous section. The protocols, standards and security issues of AMI technologies are introduced in the following subsections.

2.4.1 AMI protocols and standards

The protocols and standards are defined to establish a regulation base for any system to enable interoperability of heterogenous hardware and software. The Institute of Electrical and Electronics Engineers (IEEE) defines interoperability term as "the ability of two or more components or systems to exchange information and use the information exchanged." Since AMI technology is based on an ICT architecture of many devices provided by diverse of vendors, there have been numerous standards defined by standard development organizations such as IEEE, International Electrotechnical Commission (IEC), American National

Standards Institute (ANSI), National Institute of Standards and Technology (NIST) and so on. We will deal with standards of two major standard development organizations since both have widespread acceptance by most of industry.

Even though each device integrated to any AMI system may be produced by different vendors, they should be capable not only transmitting the measurement or control data to receiving node but also should comply with data exchanging devices to receive and to accomplish the required process. In addition to diversity of standard development organization, there are many types of interoperability standards have been improved for specific applications such as follows;

- Multimedia related standards for microwave, fiber optics, Wi-Fi, wireline communication and cellular communications,
- Data transmission standards such as internet standards, IP, TCP, UDP and HTTP standards,
- Application related standards as IEC 61850, IEC 61968, common information model (CIM), ANSI C.12 and so on
- Security standards such as advanced encryption standards (AES) AES 128, AES 256, public key infrastructure (PKI), and certificates

IEC is one of the leading standard development organization and organized by several technical councils (TCs) that TC 57 develops standards of communication and interoperability for smart grid. The featured communication standards that are developed by IEC TC57 for electric power networks are illustrated in Fig. 2.9. The presented standards scheme includes standards for interoperability layers, domains and zones. The interoperability layers are listed starting from components layer, communication layer, information layer, function layer and business layer while domains refer for generation, transmission, distribution, DER and customer premises. The zones are process, field, station, operation, enterprise, and market. The standards developed by IED TC57 are classified according to zones and communication layers as follows;

- IEC 61850 is improved for mapping, service models, and object models in the field zone. It is associated with substation automation, DER and DG including photovoltaics, wind turbines and similar RESs, DA applications, supervisory control and data acquisition (SCADA) communications, and PEVs
- IEC 61970 for generic interface definition (GID) and CIM at control centers for transmission and distribution abstract modeling, application and data-based integration,
- IEC 61968 for CIM, distribution management and AMI back office interface
- IEC 62351 that is improved for support and security services that focuses on IEC protocols, management operations of network and system, and authorized access control.

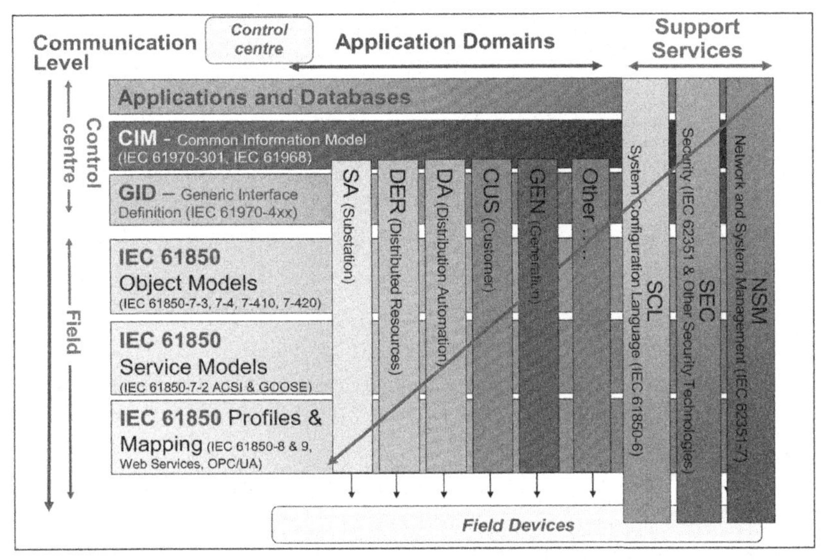

FIG. 2.9 Common Information Model (CIM) and IEC 61850 models of IEC [15]. *(With the permission of International Electrotechnical Commission (IEC).)*

The smart meters are compliant with the specifications of some IEC standards such as 61000-4 standard group, IEC 62052-11 and IEC 62053-11 standards [14–16].

In addition to aforementioned standards, IEC 62056 series are specific standards for electricity metering and data exchange for metering, tariffs, and load controls. Starting from 2008, the AMI communication standards have been shifted from PHY layer to applications layer since application level protocols are more robust on isolate configurations and implementations. The PHY layer AMI communication standards have been defined by ANSI C12.18 standard while device level standard is ANSI C12.21 and application level standard is ANSI C12.22. ANSI has also working groups for specific standards as other standard development organizations. One of the most widely used ANSI standard for AMI is ANSI C12.18-2016 *Protocol Specification for ANSI Type 2 Optical Port* defines a standard interface for communication of C12.18 devices and C12.18 Client systems over optical port. The C12.18 Client refers to any electronic device including computer, mobile reader, master station system or other communication devices. The standard relies on seven-layer OSI network infrastructure and is designed for data transmission formats of tables for delivering data to transport layer through network. Another ANSI standard that is also listed in Table 2.3 is C12.19 *Utility Industry End Device Data Tables*. ANSI C12.19 standard provides a common data structure definition for end devices and particularly for smart meters in data transferring applications. The standard has been put through after cooperation with meter producers and AMR services.

TABLE 2.3 Widely used ANSI standards for communication and AMI applications

Standard	Network layer	Function
ANSI C12.18 *Protocol specification for ANSI type 2 optical port*	Communication protocol	Data transfer by using point-to-point (P2P) protocol
ANSI C12.19 *Utility industry end device data tables*	Data model	Metering data model in tables
ANSI C12.21 *Protocol s pecification for telephone modem communication*	Communication protocol	Modem based P2P data transfer protocol
ANSI C12.22 *Protocol specification for interfacing to data communication networks*	Communication protocol	C12.22 network-based data transfer

The standard data structure of transmitted information is defined as sets of tables in C12.19 standard that each table is grouped into sections named as decades. ANSI C12.21 *Protocol Specification for Telephone Modem Communication* is designed similar to C12.18 where the transmission medium is telephone networks instead of optical ports. C12.21 defines communication points as C12.21 Device and C12.21 Client where client may be a computer, a master station or electronic communication device integrated to switched telephone network to install a connection with C12.21 Device. Later, C12.22 standard has been developed to accommodate data communication for AMI applications as its name *Protocol Specification for Interfacing to Data Communication Networks* implies. C12.22 standard defined protocol specifications considering four communication modes in operation. The communication modes deal with End Device or node definition, point-to-point (P2P) interface, one-way transmission, and End Device communication over optical ports. The first operation mode requires device or node implementation that complies with C12.22 descriptions. In the second operation mode, interface requires a C12.22 device such as smart meters and a C12.22 Communication Module as network device. The third operation mode is based on data acquisition, translation and transmission in one-way. The fourth operation mode is essential for communicating with C12.18 Type 2 optical local ports [5, 14].

2.4.2 AMI security

The deployment of an AMI infrastructure relies on deciding to possible technology, installation, operation and maintenance. The metering operations should ensure accurate, precise and reliable data acquisition, and should be able to communicate with a wide variety of devices. The development and increased deployment of AMI systems have raised concerns on data protection and privacy. One of the critical challenges of smart metering and AMI system deployment is related to cyber security of entire infrastructure according to Electric Power Research Institute (EPRI). The increased utilization of recent metering technologies and integration with communication technologies make the smart metering systems much more vulnerable to cyber-attacks and incidents. The use of wireless communication networks and widespread data acquisition nodes have caused exposures to potential intrusions and attacks on metering infrastructure [17, 18].

The attackers target to intrude communication network, to achieve unauthorized control ability, to modify metering and billing data, and to manipulate load and operation conditions. The precautions for preventing intrusions include authentication, authorization, and privacy controls. Although the privacy control methods are well known and widely used for several years, recent AES and PKI technologies provide increased security and high performance in privacy control. Another significant privacy control method has been introduced as Triple Data Encryption Algorithm (3DES) as a robust solution, but NIST reports that 3DES will be an insecure method by 2030. The wireless communication technologies are protected by widespread standards such as IEEE 802.11i and IEEE 802.16e with diverse security levels. On the other hand, wireline communication systems are secured by firewalls, virtual private networks (VPN), and IP Security (IPSec) methods. The higher-level protections require improved protection mechanisms like Secure Shell (SSH), Security Socket Layer (SSL) and Transport Layer Security (TLS) that are more convenient than regular protections [18]. Theses security technologies are based on key management requirements and entire smart grid infrastructure. However, it should be noted that millions of devices, organizations, and deployments comprise whole system.

The communication networks of AMI and smart grid infrastructure differs in specific applications since each application or devices rely on various functionality, architecture and structure. AMI is one of six major applications of smart grid where the left is related with DR, wide area monitoring systems (WAMSs), DER and ESSs, EVs, and DMS. The security requirements and applications of AMI are quite like CPS requirements in terms of confidentiality, integrity, availability, and accountability in terms of component security. On the other hand, communication networks such as HAN, BAN, and NAN may be posed to threats [5, 18, 19]. Therefore, we deal with AMI security in two main titles as follows.

2.4.2.1 AMI security related to components

AMI security is one of the crucial topics in smart grid infrastructure. The AMI security specifications are widely researched and several developments have been proposed by standard development organizations as discussed earlier. Moreover, device manufacturers and vendors pay serious attention to security to ensure reliability of devices. AMI components are one of the key systems for increasing entire system security by ensuring availability, reliability, and conformity to requirements. The AMI infrastructure is comprised by a wide variety of devices with network and communication capabilities. Besides the smart meters, customer gateways, communication networks, and AMI headend are key components in an AMI ecosystem. The component related security issues are confidentiality, integrity, availability, accountability, and authorization.

Confidentiality refers to customer privacy on personal informations such as consumption behavior and patterns. The metered data and consumption rates, periods, and durations may become an indicator about daily life of consumer. In this context, the measurement and metering results should be securely stored and should remain confidential by preventing unauthorized accesses. The AMI communication networks should also prevent data transmission between consumers to ensure confidentiality. The lack of confidentially may expose AMI system to cyber-attacks and intrusions. An AMI system interfaced with HAN over customer gateway should be seriously handled and privacy of system should be ensured.

Integrity is related to security of metered and transmitted AMI data to utility center. Integrity implies protecting data and commands of AMI system from unauthorized accesses. The AMI systems are located at customer premises and physically protected. However, cyber-attacks target to reach measured metering data on transmission line between smart meters and monitoring center. The smart meters may be posed to cyber-attacks of intruders to manipulate or hack the measurement data. Therefore, smart meters transmission system should detect and prevent cyber-attacks.

Availability ensures connectivity and ubiquitous data transmission through the communication infrastructure between smart meter and utility control center. The communication failures or physical problems prevent availability of AMI system. In order to operate AMI system in a reliable way, availability should be guaranteed by predefined controls or signaling procedures. Attacks targeting the availability of system prevent accessibility and may cause critical results since the system is not reachable on demand. Cyber physical system (CPS) intrusion detection and automatic diagnostic algorithms perform the detection methods of such attacks.

Accountability refers to reliability of remote control and metering compatibility. It is also known as nonrepudiation and is related to financial applications. The metering results should be converted meaningful data with time stamps and audit logs should be generated to ensure accountability. The continuous time synchronization is crucial to acquire customer information, time of use data, load and generation data. All the smart meter parameters such as control commands, tariffs, and stored data should be accountable for accurate billing processes.

Authorization enables users, customers and devices to get right to access specific resources and applications. The authorization control ensures prevention of intentional attacks targeting system security and unauthorized intrusions to the system. The authorization process is accomplished by assigning specified roles to user types, devices, and several different privileges are set. The AMI infrastructure will be exposed to violating attacks in case of authorization control lacks. The digital certificates, PKI, and AES algorithms can be used for authorizing users and devices through AMI system [5, 18, 19].

The major vulnerabilities of AMI system are listed as unencrypted area network traffic, bus sniffing, tampering, improper cryptography, denial-of-service (DoS) attacks, and authentication weakness.

2.4.2.2 Security threats of AMI networks

The security threats targeting AMI networks are investigated regarding to each network layer in seven-layer OSI model as a communication network. Therefore, the layer attacks are listed as physical layer attacks, link layer attacks, network layer attacks, transport layer attacks, application layer attacks besides IP based and other specific attacks. PHY attacks include tampering, damaging or breaking down of metering device and AMI infrastructure. These threats can be coped with tamper resistant and damage-free devices transmitting alerts and alarms during intrusion detection.

The link layer enables multicasting by allowing nodes to connect and disconnect to network, and security is enhanced by this way. MAC protocols are in charge for channel allocating and deploying available resources to nodes. The MAC layer attacks and intrusions are prevented by specified protocols that are operating identification filtering. The network layer attacks mostly target traffic flow by changing routing tables that are required to deliver transmitted message to destinations. The attacker changes routing table to change original traffic flow to a specific route and then generates modified message or information to cause traffic jam. The major network layer attacks include DoS attacks, routing black holes, and wormholes. The DoS attacks aim to cause disturbances on network traffic and eavesdropping on transmitted data. The routing black holes are based on hacking a single node and then directing all network traffic to hacked node. The wormholes target to create tunneling to change network direction to enable attacker to monitor network traffic. The IP is responsible to sustain confidentiality and authorization in smart grid communication networks. Therefore, cyber-attacks against confidentiality target IP stacks for spoofing, convergence and spying. Transport layer attacks are like IP attacks since it is managed by TCP and UDP protocols. The security of AMI networks can be ensured by use of data encryption methods as discussed earlier. The data encryption is performed by using symmetric or asymmetric key cryptography that transmitter and receiver uses the same key in symmetric cryptography while transmitter and receiver uses public or private keys in asymmetric key cryptography [5].

Attackers may perform cyber-attacks to generation, transmission, distribution and consumption levels of smart grid infrastructure. The most vulnerable section of entire utility network is reported as distribution and control levels due to numerous node deployments. The cyber-attacks specifically target consumption level to manipulate load level and DSM operations. The internet-based control and monitoring operations provide intrusion chance to attackers due to previously discussed layer structure. The data manipulation of measurement may include load rate increments, overflow alerts and malfunctions in monitoring operation. These attack types are generally named as IP based load-altering attacks that attempt to control and manage the load types. It is noted that IP access to can be performed for three types of loads that are data centers and server loads, direct load control (DLC), and indirect loads. The server type loads are quite flexible and load rate depends on the operation load of data center. Therefore, attackers target to reach server centers and to increase load rate by producing overloaded traffic for increasing consumption. The DLC attacks aim to access industrial and residential loads to cause malfunctions and changing operation conditions. The DLC is directly related with DSM operations such as decreasing peak demand, load shedding, and valley filling and so on. Therefore, DLC attacks seriously affect reliability and security of utility network. The indirect load control attacks target IP based HEMS applications since consumers control their residential loads to manage consumption data. The control of smart appliances is essential to decrease energy costs but intrusions can disturb control operations and may inject false command or price data to HEMS. The attacks are prevented by using protection and authentication procedures that are presented earlier [20].

2.5 PMU applications in smart grids

The PMUs are essential devices used in WAMS that provides real time monitoring of generation and transmission lines through wide areas. The WAMS would prevent effects of major blackouts met in several countries if they had used in utility grid. Smart grid infrastructure requires DA applications for remote control and monitoring applications to take rapid action against grid disturbances and to perform flexible power arrangements. The utility automation facilitates DR applications during outages by routing generated energy from various power plants. Thus, the critical load management can be accomplished in a safer way. This management and control operation is known as self-healing functionality of utility. Another key contribution of DA operation is effective management of feeders and feeder networks to enable multiple stations and distribution lines to be used on demand. The global positioning system (GPS) assisted PMUs have been introduced in Chapter 1 that first applications of these devices date back to 1980s. Nowadays, PMUs are critical devices used in synchronized measurement technology (SMT) based systems for utility applications. The use of SMT enables to achieve highly accurate time stamps for any measurement signal and provides improvement of conventional measurement systems that are based on SCADA. This transformation allows achieving more intelligent and improved metering and measurement systems to be used in geographically spanned areas [21, 22].

WAMS improves capabilities of conventional SCADA system in terms of complex system integration, monitoring of long utility networks and transmission lines, and real time measurement. It extensively improves monitoring and control operations to improve reliability, resiliency, efficiency and security of utility network. The PMU is improved to detect phase differences and measurements are performed by using fast Fourier transform (FFT) or discrete Fourier transforms (DFT). The Fourier transforms are used to detect phase patterns and differences at the desired frequencies. The phasors are time stamped at the measurement and receiver sections by the support of GPS synchronization and thus, the differences can be easily calculated. The phasor calculations are based on sampling rates, filtering features, window length, frequency estimations, and phase calculation algorithms.

PMUs are mostly installed at high voltage transformers and power generation plants. In a three-phase generation or transmission network, each phase is equipped with a voltage and a current transformer as shown in schematic diagram of a PMU connection in Fig. 2.10. The voltage and current transformers are required to attenuate high power input of transmission network to transmit attenuated signals to ADC of PMU. The processed measurement data are transmitted by using satellite and GPS transceiver that operates as receiver and transmitter.

The hardware architecture and components of a basic PMU system has been illustrated in Fig. 2.11. The initial section is comprised by analog inputs where a low-pass filter (LPF) is used for anti-aliasing of measured signals. The

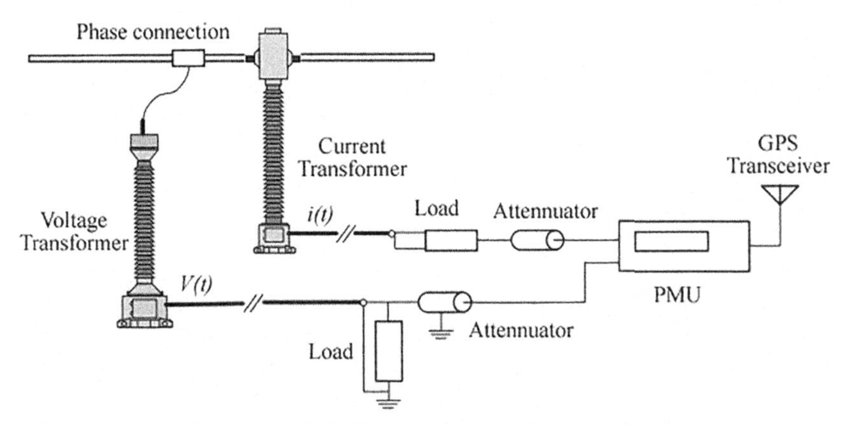

FIG. 2.10 Schematic diagram of a PMU connection through transmission line.

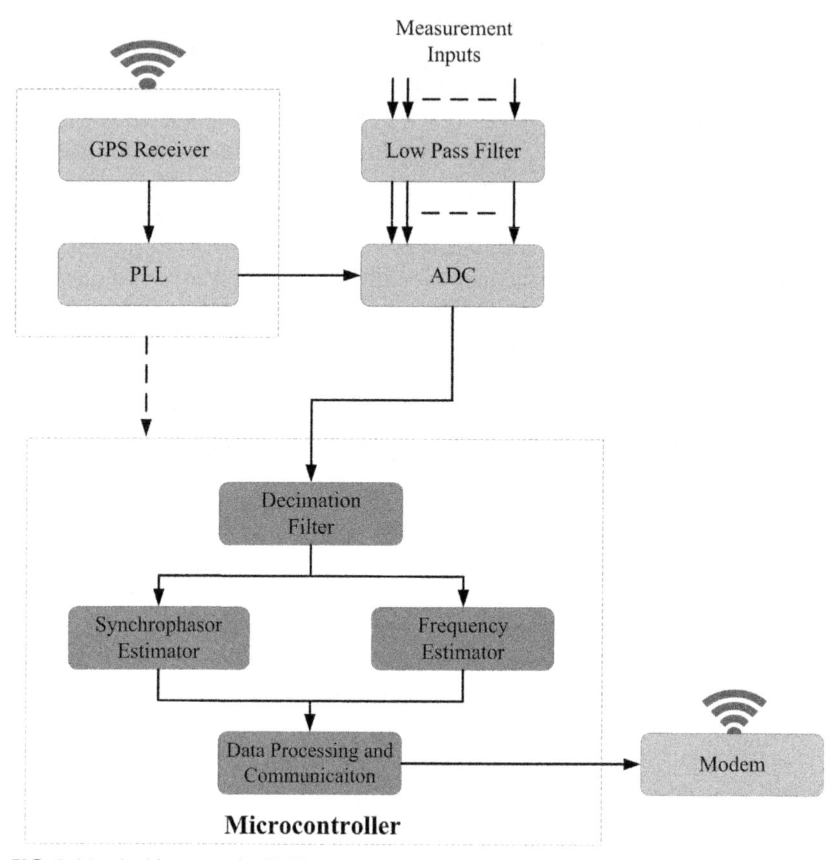

FIG. 2.11 Architecture of a PMU system.

anti-aliasing operation may be achieved by using one of two LPFs that one is regular analog filter with higher cut-off frequency while the other one is with higher sampling rate in addition to regular analog filter. The combination of analog and digital filters provides increased filtering performances comparing to analog LPFs. Moreover, the higher sampling rate allows high precision phase calculations. The MCU section of PMU includes calculation, detection and processing blocks for measured voltage and current phasor signals. The measured and processed data are arranged with time stamp generated by GPS signals and transmitted to receivers by modem section [22].

One of the most important measurements of PMUs is synchrophasor measurement since it provides detailed information about status of utility network. Power network operators will be able to acquire complex measurement from utility network by using PMUs and phasor data concentrators (PDCs) to prevent grid disturbances. PDCs are essential devices to combine measurement data of many PMUs to unify a transmission data. WAMS use synchronized phasor measurement that is assisted with reliable communication and monitoring systems to perform real time control and management operations. There are several devices such as digital fault recorder (DFR) are developed to realize protection tests and control of protection relays [21]. There are two specified standards exist for PMU data communication that are IEEE C37.118 and IEC 61850. Although IEC 61850 standard is developed for substation monitoring applications, it is accepted as a key standard for real time data acquisition requirements. The C37 standard includes a set of definitions for synchronized phasor measurement systems including synchrophasor, frequency, and rate of change of frequency (ROCOF) measurements. Moreover, it describes time tagging and synchronization requirements for measurement of mentioned quantities. The precision time protocol of GPS communication is defined by IEEE C37.238 while IEEE C37.118 describes phasor measurements. On the other hand, IEEE C37.242 provides a set of definition on PMU calibration, test and installation procedures. PMU and synchrophasor data transfers are also defined by IEEE C37.118 standards, IEEE C37.244 defines PDC requirements, and IEEE C37.111 describes common format data exchange requirements in PDC data storage applications.

A detailed list of WAMS applications and their PMU and communication resource requirements are shown in Table 2.4. The PMU requirements are classified according to low, medium, and high in terms of numbers while communication requirements are evaluated into three groups considering their resource requirements [22]. The list presents specified and widely used PMU applications in smart grid infrastructure.

TABLE 2.4 WAMS applications and requirements

PMU requirements	Communication requirements	WAMS application type
Low	Low	Steady state estimation
Low	Low	Thermal line monitoring
Low	Medium	Phase and frequency monitoring
Low	Medium	Event monitoring and detection
Low	Medium	Oscillation monitoring and detection
Medium	Low	Dynamic parameters estimation
Medium	Low	Post event analysis
Medium	Medium	Voltage stability monitoring

Continued

TABLE 2.4 WAMS applications and requirements—cont'd

PMU requirements	Communication requirements	WAMS application type
Medium	Medium	RES integration and monitoring
Medium	Medium	Congestion management
Medium	High	Adaptive protection
Medium	High	Dynamic state estimation
High	Medium	Power system restoration
High	High	Automated wide area monitoring
High	High	Reliability monitoring and control
High	High	Transient stability monitoring
High	High	Wide area stability monitoring

2.6 Smart monitoring systems

The recent smart grid applications require a comprehensive monitoring and remote-control technologies. These infrastructures are widely implemented regarding to AMR, AMI and ICT technologies that have been introduced in this chapter. The diagnostic and troubleshooting technologies are based on accurate monitoring to detect critical conditions and changes in smart grid. Moreover, monitoring infrastructure should also be considered in planning stages to achieve a predictive smart grid architecture. A detailed monitoring and diagnostics system is based on continuous measurement devices located at field assets, ICT interface to transmit measured data to MDMS, a robust database system to store mass of data, and analyzing software to perform data mining, estimations and decision-making algorithms to generate valuable data by using repositories. The measurements are provided by sensor networks and IEDS at the measurement nodes in the field. The efficient and proactive grid management rely on advanced grid monitoring, diagnostic software, and predictions [23, 24].

As previously discussed in PMU application section, wide area monitoring is a key component for grid monitoring that is based on time-synchronized measurements. The improved features of such as system is defined as wide area monitoring, protection and control (WAMPAC) system. WAMPAC is based on synchronized measurements acquired over utility network and transmission of the signals as phasors. The monitoring of utility network relies on fundamental bus voltage and current measurements in phasor type data that are used to determine phase differences, frequency, ROCOF, power factor and similar parameters. The time synchronization is ensured by GPS and satellite-based time stamping that use coordinated universal time (UTC) to provide highly accurate measurements. The smart grid infrastructure is comprised by large penetration of RESs in addition to conventional sources and power plants. Moreover, advances in power electronics and grid regulations have facilitated integration of PEVs to utility network. Thus, a widespread utility network can be illustrated as seen in Fig. 2.12 where RESs, transmission and distribution lines, substations, and generation plants have been shown [23, 24]. Smart grids are expected to be self-healing against power system faults and failures by sustaining its reliable operation capability to supply and distribute the generated power. Although the utility networks are planned and installed in a resistant structure to faults, the RES penetration and DERs affect the generation capacity due to their intermittent structure. The RES integration may be performed at transmission, distribution and even consumption levels. The utility network requires to disconnect these integrations in any fault, failure or isolation conditions to ensure healthy operation of grid. The dynamic system response is realized when generators or loads rapidly react during grid disturbances. Therefore, WAMPAC system should monitor the grid conditions and perform the required operating procedures on demand. This protection scheme required a widespread communication interface to transmit instant measurements and process the data that are acquired from different distant locations. Since the smart grid infrastructure requires large integration of RES and storage systems, the limits of EMS and DMS systems will be forced to tolerate fluctuations of generation

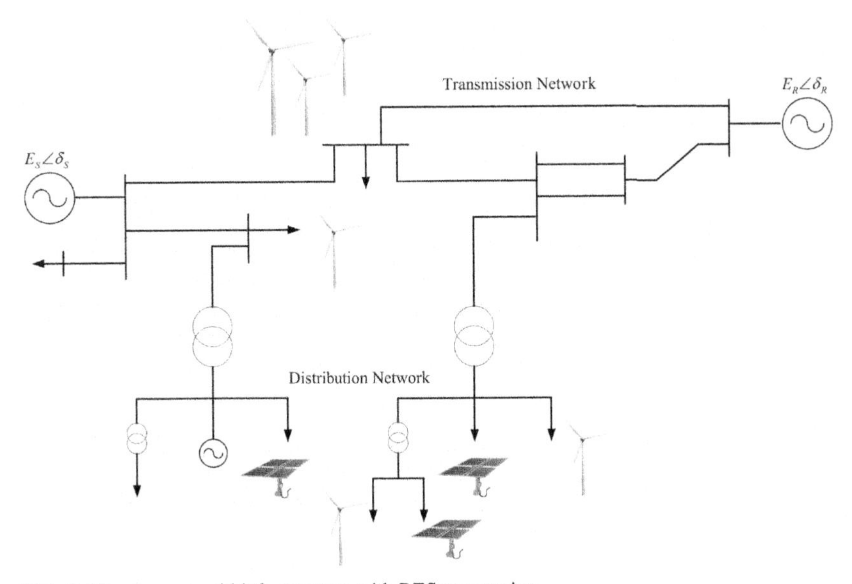

FIG. 2.12 A smart grid infrastructure with RES penetration.

and loads. The transient power fluctuations caused by intermittent structure of generators should be handled by an advanced monitoring system at central and decentralized control levels at substations.

The general requirements of monitoring and troubleshooting processes of a self-healing system are based on collecting, analyzing, diagnosing, predicting, and prioritizing the data. The data collection enables availability of data in any database for detailed analysis while the data analysis provides information on asset conditions and related failure or fault risks comparing to known situations. The identified risks are evaluated in diagnosis step to develop a self-healing program, and verifications are performed in prediction step.

The monitoring architectures rely on data acquisition systems such as SCADA and remote terminal units (RTUs) for condition estimations. The acquired informations on system level of smart grid are provided by PMU devices that ensures highly reliable and synchronized input are obtained. A drawback of current monitoring system is related to RTUs since they do not provide synchronized measurement and causes less accurate data acquisition. The drawback of RTU is tackled by using GIS system to provide more accurate localization [14]. The monitoring system is also equipped with smart sensors and sensor nodes in field assets. In addition to data acquisition, monitoring infrastructure generates alert signals to indicate fault or failures. The monitoring system should provide real time and historical data for specific purposes. Some of these sensor-based measurement and monitoring systems are shown in Fig. 2.13. The most widely used sensors include CTs, voltage transformers (VTs); temperature, optical, pressure and vibration sensors. They can be deployed at any level of transmission and distribution line at medium voltage (MV) or high voltage (HV) applications to monitor surge arresters, circuit breakers,

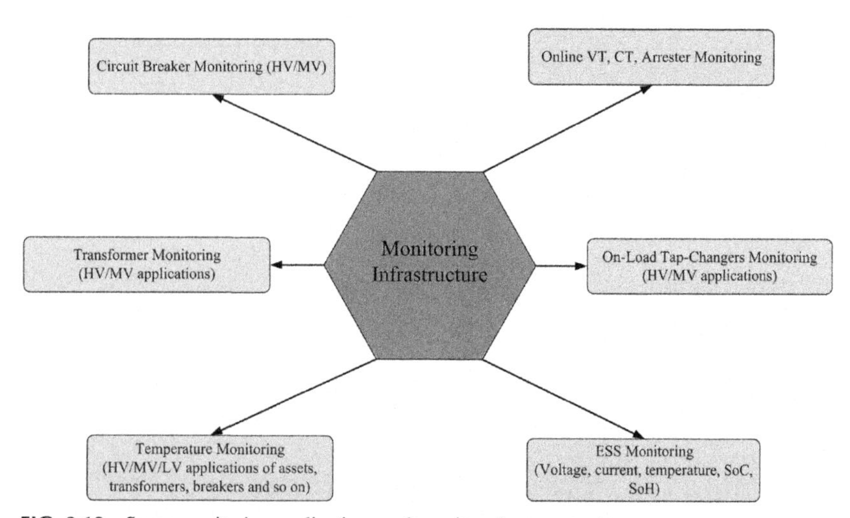

FIG. 2.13 Some monitoring applications and monitored components.

substations and transformers, on-line tap-changers (OLTCs), temperature changes and ESSs. In addition to direct measurement, several calculations for active and reactive power, power factor, total harmonic distortion (THD) rates, and frequency can be performed regarding to measurement data.

The monitoring architectures can be classified as local, substation, and centralized according to sensing and analysis operations. The local monitoring and data processing requirements are met by smart sensors that is a specified standalone device with embedded measurement and communication capability. It is widely used for transformer monitoring, feeder and asset current measurements, and fault passage indicators (FPIs). These types of applications are mostly seen in SCADA systems for acquiring fault detection and warning purposes. The substation monitoring requires distributed smart sensors to acquire data from outside of substation. The sensors are hierarchically located and interacts with gateways. A substation controlling server is responsible for managing gateways in this monitoring structure. The centralized or central monitoring is used for system-wide monitoring as its name implies. The monitoring data are acquired from field sensors and stored in a central database for real time and historical monitoring requirements. The central database provides required data to asset management, diagnostic applications, and real time management applications [25].

References

[1] E. Kabalci, Emerging Smart Metering Trends and Integration at MV-LV level, IEEE, 2016, pp. 1–9, https://doi.org/10.1109/ISGWCP.2016.7548264.

[2] Y. Kabalci, A survey on smart metering and smart grid communication, Renew. Sustain. Energy Rev. 57 (2016) 302–318, https://doi.org/10.1016/j.rser.2015.12.114.

[3] J. Lloret, J. Tomas, A. Canovas, L. Parra, An integrated IoT architecture for smart metering. IEEE Commun. Mag. 54 (2016) 50–57, https://doi.org/10.1109/MCOM.2016.1600647CM.

[4] M.S. Thomas, J.D. McDonald, Power System SCADA and Smart Grids, CRC Press.pdf, (2015).

[5] Y. Xiao, Security and Privacy in Smart Grids, CRC Press Taylor & Francis Group, Boca Raton, FL, 2014.

[6] J.B. Ekanayake, N. Jenkins, K. Liyanage, W. Jianzhong, A. Yokoyama (Eds.), Smart Grid: Technology and Applications, first ed., John Wiley & Sons, Ltd, Chichester, 2012.

[7] H. Mouftah, M. Erol-Kantarci (Eds.), Smart Grid: Networking, Data Management, and Business Models, CRC Press, 2016https://doi.org/10.1201/b19664.

[8] Y. Kabalci, E. Kabalci, A low cost smart metering system design for smart grid applications, in: 2016 8th International Conference on Electronics Computers and Artificial Intelligence ECAI, 2016, pp. 1–6, https://doi.org/10.1109/ECAI.2016.7861078.

[9] J. Zheng, D.W. Gao, L. Lin, Smart Meters in Smart Grid: An Overview. IEEE, 2013, pp. 57–64, https://doi.org/10.1109/GreenTech.2013.17.

[10] Q. Sun, H. Li, Z. Ma, C. Wang, J. Campillo, Q. Zhang, F. Wallin, J. Guo, A comprehensive review of smart energy meters in intelligent energy networks, IEEE Internet Things J. 3 (2016) 464–479, https://doi.org/10.1109/JIOT.2015.2512325.

[11] O. Monnier, A Smarter Grid With the Internet of Things, Texas Instruments, 2013. http://www.ti.com/lit/ml/slyb214/slyb214.pdf. Accessed 4 August 2017.

[12] Bourns Inc., Smart Meter Power, Measurement and Communications Port Protection, United States, 2013.

[13] M.F. Khan, A. Jain, V. Arunachalam, A. Paventhan, Communication technologies for smart metering infrastructure, in: IEEE Students Conference on Electrical Electronics and Computer Science SCEECS 2014, IEEE, 2014, pp. 1–5.

[14] S.K.S. Salman, Introduction to the Smart Grid: Concepts, Technologies and Evolution. Institution of Engineering and Technology, 2017, https://doi.org/10.1049/PBPO094E.

[15] SMB Smart Grid Strategic Group (SG3), IEC Smart Grid Standardization Roadmap, IEC, Zurich, 2010.

[16] R. Morello, C. De Capua, G. Fulco, S.C. Mukhopadhyay, A smart power meter to monitor energy flow in smart grids: the role of advanced sensing and IoT in the electric grid of the future, IEEE Sensors J. 17 (2017) 7828–7837, https://doi.org/10.1109/JSEN.2017.2760014.

[17] E.E. Queen, EIE, Smart Meters and Smart Meter Systems: A Metering Industry Perspective an EEI-AEIC-UTC Whitepaper, Edison Electric Institute, Washington, DC, 2011.

[18] Y. Yan, Y. Qian, H. Sharif, D. Tipper, A survey on smart grid communication infrastructures: motivations, requirements and challenges. IEEE Commun. Surv. Tutorials 15 (2013) 5–20, https://doi.org/10.1109/SURV.2012.021312.00034.

[19] R. Rashed Mohassel, A. Fung, F. Mohammadi, K. Raahemifar, A survey on advanced metering infrastructure. Int. J. Electr. Power Energy Syst. 63 (2014) 473–484, https://doi.org/10.1016/j.ijepes.2014.06.025.

[20] S. Goel, Y. Hong, V. Papakonstantinou, D. Kloza (Eds.), Smart Grid Security, Springer, London, 2015.

[21] Y. Xiao, Communication and Networking in Smart Grids, CRC Press Taylor & Francis Group, Boca Raton, FL, 2012.

[22] S.M. Muyeen, S. Rahman (Eds.), Communication, Control and Security Challenges for the Smart Grid, The Institution of Engineering and Technology, London, 2017.

[23] E. Kabalci, Y. Kabalci, A wireless metering and monitoring system for solar string inverters, Int. J. Electr. Power Energy Syst. 96 (2018) 282–295, https://doi.org/10.1016/j.ijepes.2017.10.013.

[24] S. Borlase, Smart Grids Infrastructure, Technology, and Solutions, CRC Press Taylor & Francis Group, Boca Raton, FL, 2013.

[25] S. Borlase, Smart Grids Advanced Technologies and Solutions, second ed., CRC Press Taylor & Francis Group, Boca Raton, FL, 2018.

Chapter 3

Smart grid network architectures

Chapter outline

3.1 Introduction

The comprehending of smart grid (SG) architecture helps to understand essential requirements for SG communication networks. For the development of conceptual SG architectures, diverse standardization bodies and organizations (e.g., the U.S. Department of Energy (DOE) [1], the State of West Virginia [2], the National Institute of Standards and Technology (NIST) [3] and the Institute of Electrical and Electronics Engineers (IEEE) [4]) have played a vital role. Nevertheless, the IEEE 2030-2011 standard has been extensively acknowledged as the most popular standard in order to meet SG architecture, configuration and other requirements [4]. This standard has recommended a reference model called Smart Grid Interoperability Reference Model (SGIRM) that is developed to address challenges related to interoperability between various parts of energy systems, communication technologies and information systems. This model undertakes the task of guidance between communication infrastructures and different levels of SG systems [5]. A typical communication network model of SG systems is illustrated in Fig. 3.1 according to the IEEE 2030-2011 standard. One of the most crucial parameter for the SG systems is end-to-end (E2E) communication capability that needs to be handled exactly while designing communication architectures. In order to ensure E2E communication channels, the IEEE 2030-2011 standard suggested a layered structure. The first layer is about security that protects the data while transmitting through the network. In next layer, to faithfully achieve the quality of service (QoS) requirements, a network management is responsible for controlling of the communication

From Smart Grid to Internet of Energy. https://doi.org/10.1016/B978-0-12-819710-3.00003-X

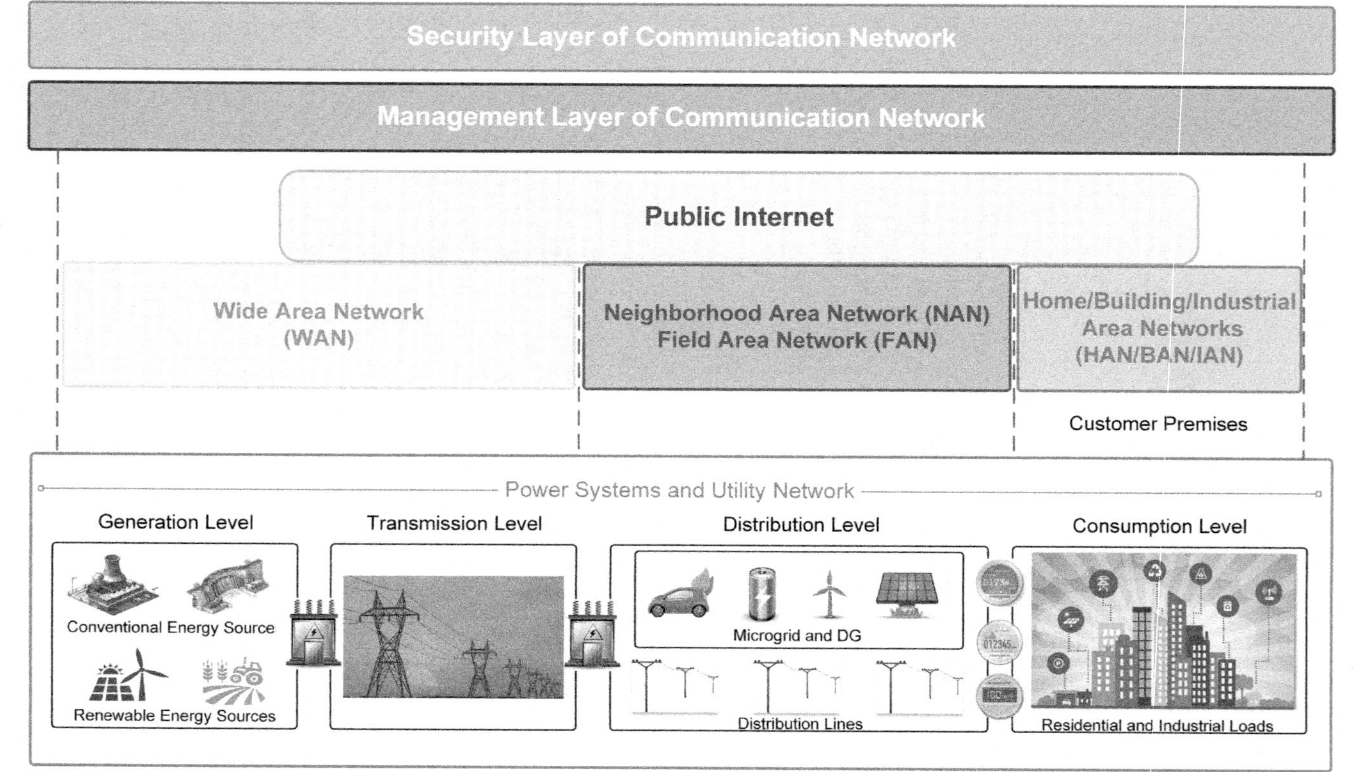

FIG. 3.1 Block diagram of E2E SG communications model.

connections. Lastly, the standard expresses a chain of area networks that are superposed over the energy/power domains. Connection of devices across the whole SG network is basically provided by this chain of area network. The model of E2E communication proposed by the standard is also depicted in Fig. 3.1, which includes the public internet as an interface on top of the area networks.

Although communication architecture of the SG is defined by IEEE 2030-2011 standard in a hierarchical arrangement, actual challenge is to develop harmony between various technologies and to adapt them into limited the information and communications technology (ICT) infrastructures. Therefore, three subnetworks have been defined for the SG communication systems. First subnetwork consists of building area networks (BAN), home area network (HAN), and industrial area networks (IAN) that are well-defined to be used at consumption site. Second subnetwork named the local area network (LAN) that is used at distribution site of the power network comprises of neighborhood area network (NAN) and field area network (FAN). Third and last subnetwork is the widest network among the three and termed as wide area network (WAN) which allows communication ranging up to several km. The WAN consists of a number of LANs, virtual private networks (VPNs) and data management systems (DMSs) for transmission level operations [6, 7].

The connection along applications and conceptual reference model of SG architecture is provided by the communication networks as shown in Fig. 3.1. The key attributes of SG are the generation, transmission, distribution and consumption, which may perform even better if the overall communication infrastructure use in the SG is split into private networks, core networks and WANs. For the control and management of power plants and substations, digital subscriber line (DSL) and fiber optic wireline connections behave as the ICT backbone network. The WAN, LAN, NAN, FAN, BAN, HAN and IAN type networks are used for the particular control and management of applications from the generation level to the consumption level [6, 8]. Distributed energy resources (DER) integration is also used at the generation level of SGs along with traditional power generation, and the generation level is associated with the transmission level by using step-up transformer as shown in Fig. 3.1. The WAN networks provide the ICT interface between generation and transmission levels in order to guarantee the connection of domains and supervisory management system. The LAN, NAN, and FAN networks are employed for connecting the distribution and micro grid management systems to distribution level and substations. The distribution level contains micro grid and substation distribution that assure two-way power flow. The consumption level is classified into residential, industrial, and substation loads. The HAN is used as communication interface for management system in case of residential loads whereas BAN and IAN are employed for industrial and substation loads. The WAN plays a vital role to establish connection among substations, energy storage systems (ESSs), DERs, feeders, transformers and other bulk equipment. The WAN provides long distance data communication and control signal transmission with very low latency. However, bandwidth requirement of the WAN is the highest one among network types utilized in SG systems. The NAN is responsible to take care of consumer measured data like demand rate, consumption level, and power quality transmitted by AMI networks, whereas FAN is employed to build a communication medium between backhaul and distribution network services. Therefore, FAN networks guarantee data transmission between management system and distribution substations, feeder points, and services. An alternative AMI network throughout the SG concept is placed at consumption level where data transmissions such as smart meter (SM), home energy management system (HEMS), electric vehicles (EVs) consumption, and micro grid generation rates of customers are obtained by utilizing HAN networks. The details of network architecture schemes utilized in SG systems will be explained by the following sections.

3.2 Premises network schemes

The premises network aims to collect sensor data within the customer premises from different types of smart devices and appliances, and sends management commands to them for better management of energy consumption. Houses, apartments, residential and commercial buildings, and plants are the targeted coverage areas of premises network schemes. HAN, BAN and IAN are also known as private (premises) networks because those are the first network scheme related to customer characteristics. Many applications like remote control, light control, smart energy management, security and safety can be realized by deploying home automation networks including numerous type of sensors and actuators [9, 10]. For example, in light control applications, a light can be turned on/off by a switch, or actuated by remote control, or can be controlled automatically by using sensors, and it is automated on the basis of demand request by the utility company in case of SG systems. The BAN and IAN terms are employed to mention similar network structures when a network is realized in buildings and industrial areas, respectively.

3.2.1 Home area networks (HANs)

In customer premises, the use of HANs is required to accomplish monitoring and control processes of home appliances and smart devices. New services such as automatic metering infrastructure (AMI) and demand response (DR) are also implemented over these networks. The HAN is considered as the most significant networking technology for SG systems on the basis of bidirectional data communication for administrating DR by utilities [11]. In addition, a HAN is responsible to collect sensor data from different appliances within a house or a residential dwelling, and it intends to send control commands to the appliances to ensure efficient energy management. A residential control center can observe and manage smart home appliances (e.g., refrigerators, air conditioners, kitchen stoves, heaters, dishwashers, washing machines, dryers, chargers for electric car and so on) for making perfect the power supplies and energy consumptions. The best part of a HAN is that it can share energy consumption rates with in-home monitors, and it is able to offer functionality to turn off cycling heaters or air conditioners during peak load conditions. Besides that, the HAN also provides a card-activated prepayment scheme and managing the charging and discharging processes of plug-in electric vehicles (PEVs).

HEMS is a very important component of a HAN which not only authorizes users to follow how much power is being consumed by their home appliances at any specific point of time but also in a wide time frame. The HAN comprises of the SM, actuators, smart sensors, smart appliances, intelligent electronic devices (IEDs) and home gateway (HGW). The HEMS interconnects with a SM mounted in user site to expedite SG applications regarding DR, time of use (TOU)-based energy management and so on, and undertakes a communication gateway (GW) role to transmit information including price of real-time energy, energy consumption, and control command among the HAN and the utility. The HEMS also authorizes the users for managing their home appliances in order to reduce their energy consumption and electricity bill. Generally, a star topology is used in HAN scheme for both wired technologies (e.g., Ethernet and power line communication (PLC)) and wireless technologies (e.g., ZigBee and Wi-Fi) [12]. Normally, HANs allow approximately 10–100 kilobits per second (Kbps) data rates and cover areas of up to 200 m^2. On the other hand, all home appliances can be connected together using wireless connections (generally using the unlicensed 2.4 GHz frequency) by deploying smart metering concept. Whole home appliances can be controlled by SMs and they provide a complete data about the power usage of every appliance [13]. The comparison of HAN communication based on wired and wireless technologies is shown in Table 3.1.

There are certain applications which are very important for the HANs like home automation, building automation, and industrial automation. Authorizing and control are two essential and inevitable functionalities offered by a HAN scheme. Authorizing is responsible to detect and manage different piece of equipment that are forming a self-organizing network whereas control function of a HAN tries to provide promising interoperability among the different communication sections of the SG.

The SG (or utility center) and the end user relationship are of utmost importance. Therefore, an interface between the utility and the consumer called energy service interface (ESI) is used which behaves as secure bidirectional communication interface between them. Various kinds of interfaces may be supported by the ESI for ensuring secure bidirectional communications. The main purpose of utility secure interactive interface is to present secure two-way communication whereas utility public broadcast interface is being utilized in order to provide price signaling and one-way reception of event at user devices. For exchange of metering information, the ESI may be interconnected

TABLE 3.1 Comparison of popular communication technologies for HAN applications

Technology	Standard	Operating frequency	Data rate	Distance (m)	Security	Cost
Ethernet	IEEE 802.3x	125 MHz	Up to 10 Gbps	100	High	High
PLC	IEEE 1901	2–100 MHz	Up to 200 Mbps	200	High	Medium
Bluetooth	IEEE 802.15.1	2.4 GHz	Up to 721 Kbps	100	High	Low
Wi-Fi	IEEE 802.11x	2.4 GHz	Up to 600 Mbps	100	High	Low
ZigBee	IEEE 802.15.4	2.4 GHz	Up to 250 Kbps	100–1600	High	Low

to a SM by using wired connections or over the HANs, which share this information to the utility. Real Time Pricing (RTP) from the utility is also provided to the ESI through the AMI infrastructure and the RTP information is shared with the customers. The customers have the option to employ a monitoring panel connected to the ESI or any web-based consumer energy management system (EMS) (be located in the SM, an individual GW, or a third party), and answer to pricing signals from the utility. There are some control-enabled devices at the customer site. Through the ESI and smart devices, the utilities will be able to implement their load control programs by accessing these devices.

Superiorities of the use of HAN scheme can be summarized as follows. The HAN allows involvement of end user in SG infrastructure in order to facilitate the utilities to manage peak loads. It also provides information to utilities about the energy consumption of each and every end user and ensures centralized access for utility centers to control all the devices at the end user premises. The key objective of HAN is to ensure that SG is meeting the reliable and seamless energy requirements and protect grid from any unwanted blackout by controlling or shifting the loads. Consumer has option to minimize its electricity bill by shifting loads from peak timing to normal load timings. On the other hand, main challenges of the HAN scheme can be sorted out as follows. The biggest challenging task is to integrate different technologies into the HAN.

Interoperability of technologies for a widespread purpose like HEMs is a significant apprehension due to the utilization of different technologies in SG, and also their solutions have to be acknowledged by the market. Security and privacy of the end user are critical problems that have to be handled.

3.2.2 Building area networks (BANs) and industrial area networks (IANs)

Monitoring and controlling user devices and exchange of information with the utilities are the main tasks of a BAN similar to the HAN. However, coverage area of BANs is larger than the HANs since they have to cover whole building which further comprise of several apartments and offices. The BAN can be considered as a collection of HANs connected through a building SM that is usually placed at point from where power is fed to the building. Electronic household appliances, several SMs and BAN GW are the components of a BAN. The BAN GW collects, analyzes, and stores data from SMs and then transmits to a NAN GW. Monitoring of the power demand and usage of residents of the corresponding building can be done by BAN GW. The BAN can also transmit information received from NAN GW to the SMs. Since cost is very important, in order to accomplish data transmission with cheaper methods, data are combined such a way that wireless communication done in BAN-HANs become economical. Through this combined data, more broad coverage can be achieved by passing through cellular carriers in the form of groups. Generally, HAN SM is used where computational and storage requirements are usually low, therefore often considered as resource-constrained device, whereas BAN SM is used as central node server because of its capabilities to provide more high spectacles. This network scheme may also serve to micro grids where electricity can be generated by exploiting renewable energy sources (RESs) such as biomass, tidal power, solar energy and wind energy. Because of the abundant number of network components, data rate requirement of a BAN is higher when compared to a HAN.

An IAN is a communication network design for industrial environments that comprises of controllers, connected sensors, and specialized software designs for building management. Building automation and energy management applications for industrial buildings or multi-buildings are tackled by the IANs in order to optimize energy, economic and environmental performance of whole active devices. A micro-grid is an essential element of IAN likewise to the BAN. Nonetheless, micro grids used in this network scheme possess higher capacity and complexity when compared to the BANs. Moreover, SMs employed in IANs should have the ability to read supplementary information such as quality of power, voltage fluctuations, and phase measurements due to refined application requirements of the industry.

Several joint characteristics and design principles are shared by the HAN, BAN, and IAN schemes although there are a number of dissimilarities between

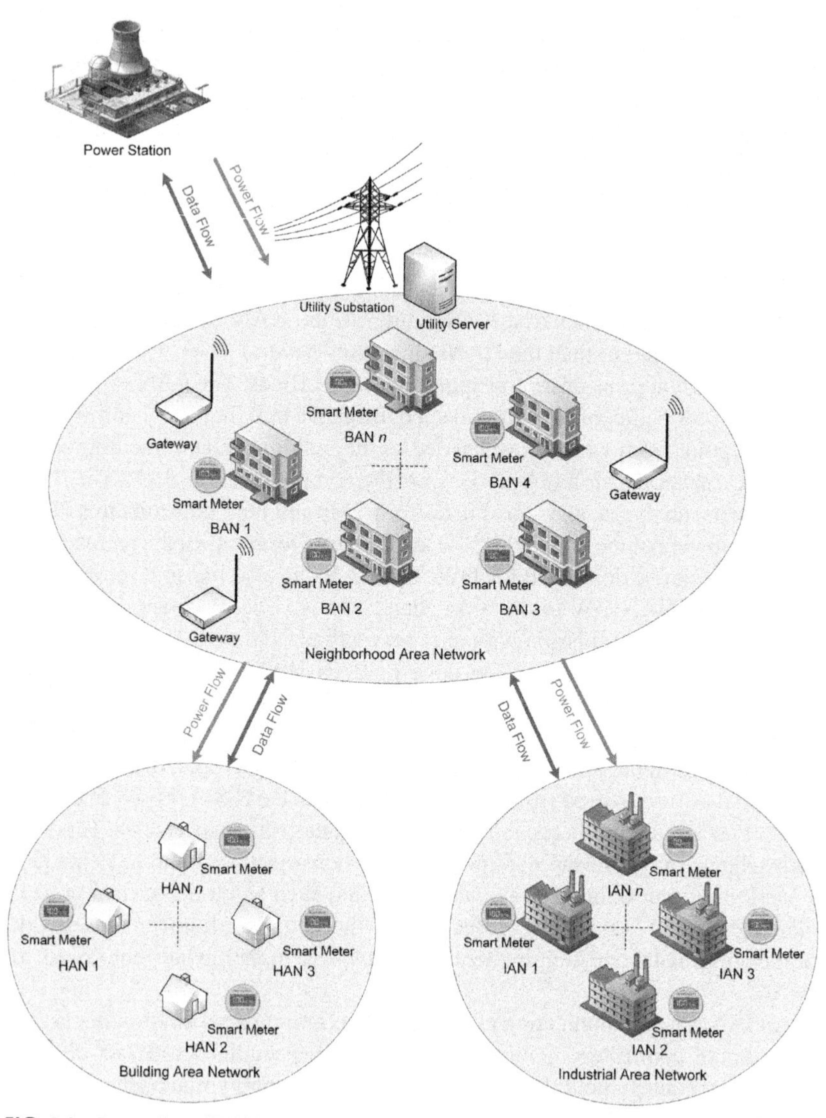

FIG. 3.2 Integration of BAN and IAN schemes into the SG communication network.

them. In order to achieve short-range communications among different network elements for applications like monitoring and control, they are typically utilized in indoor mediums. The common thing in all these networks is the SM that acts as a GW to link these network schemes with other network stages and the utilities. Integration of premises network schemes into the SG communication network is illustrated in Fig. 3.2.

The network deployed at customer side will be more effective and efficient only if it has features of simplicity, cost efficiency, less power consumption and security. The most important advantage of wireless communications utilized in HAN/BAN/IAN is that it can be established anytime, anywhere and without spending too much money for installation. Besides that, wireless communication has also the provision to add on and remove the devices. For these reasons, wireless communications are better options than wired ones for these network schemes. Moreover, it is unrealistic to choose wired communication methods in the scenarios where the nodes are highly dense like in sheer volume of home automation networks. However, wireless communication has to undergo through multipath environment since many wireless devices such as mobile phones, mobile routers, remotes, different type of sensors and microwaves are used at home which may face interferences due to reflections through walls.

3.3 Neighbor area networks (NANs)

The NAN is responsible of managing flow of communication signals between WANs and consumer premises area networks, which can be done either through wired or wireless communications. In a neighborhood, energy information received from customers are transmitted to a utility company through the NAN. The SMs are the endpoints of NANs that are assumed to be at the core of SG systems. For AMI applications, the SMs are connected to local access points through NAN. It gathers information from numerous household appliances in its vicinity and conveys these data to the WAN. The SMs are endpoints of NAN installed on the outdoor of single family houses or apartment buildings. After a specified regular interval, a SM records electrical, water, and/or gas usage of consumer, and then provides electronic means to read this data. The SMs also facilitate in real-time or specified-time sensing, notifications on power outage situation, and monitoring of power quality. Commonly, coverage area of a NAN is about several square kilometers (e.g., 1–10 km). On the basis of topology of power grid and the deployment of communication technologies and protocols, the number of SMs in every NAN scheme changes from several hundred to several thousands. The data rate requirements of every SM may be different based on the application types. For instance, low data rates (as much as bps per SM) is required for interval and on-demand meter reading applications whereas higher data rates (typically tens of Kbps per SM) may be necessary in order to maintain advanced applications such as fault detection, advanced distribution automation and so on. Since NAN is responsible for carrying a massive amount of diverse data types and deploying control signals among utility companies and an enormous number of devices mounted at customer premises, the NAN is considered as a crucial part of the SG communication networks.

The NAN plays a vital role for the connection of distribution side appliances. A mesh can be created by the deployment of SMs entirely or may be formed by combination of SMs and some GWs to transmit data. Normally, a wireless network is formed in a region among the SMs of a home cluster. A NAN can potentially utilize a multi hop mesh method since the transmission range is typically more than 500 m. Communication technology, protocol and standards with stability, security, power efficiency, low-cost and low latency characteristics are required for the NAN which also have the ability to empower asynchronous upstream data traffic. The NANs use both wireless technologies such a Wi-Fi, radio frequency (RF) technologies, Worldwide Interoperability for Microwave Access (WiMAX), cellular (3G and 4G), and Long Term Evaluation (LTE) and wired technologies such as, PLC, Ethernet, Data over Cable Service Interface Specification (DOCSIS) [9]. The most common applications of the NANs are meter reading, distribution automation (DA), DR, outage resource management, TOU pricing, service-based switching operation, prepaid payment, electric transmission and distribution monitoring, end user information and message alerts, buildings network admin, utility updates, program and configuration updates, etc. [14].

Energy consumption information from all appliances are gathered by the meter reading process and collected information are shared with the utility centers by exploiting two-way communications. By observing the power usage of each individual end user from utility center, SM readings support better management of energy utilization. An application of on-demand meter readings permits the end user to know about their energy consumption queries whenever required. Software-based utility application is being adopted by SM interval that gathers information from SMs after well-defined intervals regularly in a day by utility centers.

One of the most important features of TOU is broadcasting of price information to user that authorizes the end user to manage their loads to minimize their energy bills. Due to sudden variations in load, RTP provides short-term fluctuating price information (e.g., variation of price in 10 or 30 min). Pricing information at very high peak demand is given by critical peak pricing (CPP). One of the most important task of the SG is DR that permits utility centers to govern load in relative to peak timings at the end users premises such as controlling of thermostats, electrical vehicle charging, air conditioning, and so on. All the important services and operations such as connection or disconnections to improve the reliability of SG are done by utilities by following DR.

One of the other important SG operations is outage and restoration management (ORM) that immediately senses outages of power. Using SM readings, one can understand the issues of low voltages and high voltages immediately. The issues arise in electrical vehicle charging can be fixed. The SG become more intelligent due to the ORM and alleviates the performance by improved utility management. The background software that utilize to operate SG and fixes the errors for improving performance is regularly updated by Firmware. End users may be informed about their energy consumption through customer information and messaging information.

In order to implement NANs for SG communications, network standards such as Wi-Fi (IEEE 802.11), ZigBee Alliance (IEEE 802.15.4), WiMAX (IEEE 802.16), and WSNs may be deployed depending upon the requirement. IEEE 802.15.4g and IEEE 802.11s are two latest IEEE standards that may be used for SG-based NANs [15]. The IEEE 802.15.4g maintains Physical Layer (PHY) and Medium Access Control (MAC) layer architecture of SG communication networks, whereas IEEE 802.11s takes care issues related to network operation in the SG. The low data rate wireless communication in outdoor mediums and wireless Smart Utility Network (SUN) requirements are addressed by the IEEE 802.15.4g. The SUN is used for the very large scattered network having low power requirements for operation. The SUN (IEEE 802.15.4g) consists of numerous wireless devices that are distributed over a large area and use efficient routing algorithms for data communication [16]. Since the SUN exploits the unlicensed frequency bands (2.4 GHz), it has to resist interference caused by any of wireless communication system (IEEE 802.11) that is working in the same frequency band. The IEEE 802.11s is an improved version of IEEE 802.11, and it enhances the RF parameters at MAC layer since better packet delivery ratio and route selection for multi-hop networks are possible in this standard. It also provides the features of on-demand routing protocol and tree-based proactive routing protocol. Highly secure and high-speed data transmission with better routing for wireless NAN applications can be achieved by IEEE 802.11s [17].

Topology of the network should be designed carefully so that it can support the routing algorithms which are very efficient in nature, therefore routing protocols for SG NANs needs to be developed in a well-designed way. Obviously, the requirements of SG communications may be addressed by well-designed routing algorithms. Since the NANs utilize mesh topology schemes as communication network in the SG, the characteristics of these networks schemes should also be taken into account while designing the routing algorithms. A mesh network which is compatible with IEEE 802.15.4g is known as a Low Power and Lossy Network (LLN) [18, 19]. Network devices work with constrained processing power and memory whereas low data rates are required for operation of communication connections in LLNs. Internet Engineering Task Force (IETF) Routing over LLNs (ROLL) working group has designed Routing Protocol for LLNs (RPL) that can be employed as a routing protocol in SG NANs. Nonetheless, certain revisions may be needed to address the needs of the SG operation. Since revised version of hybrid wireless mesh protocol (HWMP) of IEEE 802.11s provides stable, high-speed and secure data transmission, many researchers have proposed to employ it as a candidate routing algorithm for SG NANs [17].

A special type of tree-based routing protocols [18] is the RPL that uses destination-oriented directed acyclic graphs (DAGs) as the abstractions of one network topology, where every node is allotted with a private index (may be also referred as rank value) to preserve its status in the DAG. The relation of a node with the other nodes in the DAG is given by the index value of a node. A DAG has a directed graph and tree-like structure, where whole edges are directional and having no cycles (i.e., closed path). In addition, every DAG has a root which is basically a GW node in wireless mesh networks (WMNs). In order to prevent any routing loop, the index value of nodes at any path to a DAG root should be varied regularly. The DAG is constructed on the basis of control information announcement whose details are available in [18, 19]. It is important to note that many issues need to be investigated in SG applications. However, one of the most crucial issues is how to customize the RPL to address particular needs of SG NANs.

The RPL is implemented on AMI system in [19] where authors of the paper deployed a stationary multi-hop wireless AMI network that contains many SM nodes and one GW node. For the proper implementation of network, first of all it is required to provide network information to every node present in the network which is compulsory to be store and uphold by each node. Later, incoming and outgoing unicast forwarding rules are defined for all the nodes. These rules are also considered as data traffic rules. An expected transmission time (ETX) depending on the DAG scheme is also presented in this study where ETX value of each connection may vary on the basis of data traffic of communication links. Novel DAG establishment and maintenance methods were also developed in this proposal, where ETX variation of a node will broadcast over the whole network by using control messages, and activates the maintenance technique of whole nodes for adaptation to ETX. In this way, incoming unicast traffic problems can be resolved by ensuring that E2E transmission has high reliability. However, to resolve the challenges of routing for outgoing unicast traffic, a method known as reverse path recording is recommended. Every node can save the source and last-hop node of inward data transmission via this new method, without any additional protocol overhead.

A WMN having single GW node may not be appropriate for SG NANs. An improved RPL solution for the SG NANs is suggested in [18]. Since the network topology is in the form of multiple trees, every tree belongs individual GW and functions in its private channel in order to prevent interference with each other. In these scenarios, SM nodes have to be capable to discover automatic GW and detect connectivity loss, however, primary RPL does not support this feature. Two components into the routing tables of every node were included to overcome this issue. First component is a data record structure where an array is employed to accumulate channel number and related optimal index value information. In the routing process, on the basis of routing messages received, a node will save and update the record array, and search for the best available channel for data transmission. The second one is a counter that is responsible to store number of failures occurred in transmissions. The present connectivity of channel is missing, if number of failures goes beyond the predefined threshold value. Hence, a channel scan process should be initiated by the node.

The use of a hybrid routing rule and airtime cost metric for static WMNs in the SG NANs are proposed by HWMP of IEEE 802.11s that is also a tree-based routing protocol [17]. Routing trees of HWMP are constituted on the basis of systematically broadcasting root announcements. Currently, there are two types of research going on about the HWMP. First one is to develop a practical routing link cost metrics, because SG features are not reflected by the default metric. A very crucial problem in such a network is to understand how traffic can be balanced among multiple GWs [20].

For route selection, air cost metric is used in the HWMP. Any node can improve route selection process by relating air cost information of obtained packets with calculated air cost information. The air cost problems shown in SG applications are taken into account in [17]. Number of MAC retransmissions of every packet was preferred to find out failure rate of every node (i.e., air time cost). Since different sizes of data packets may require different MAC retransmission levels, this situation will be not practical. For different packet sizes, use of the same retransmission level may decrease throughput of the network. This should be tackled in SG applications since size of data packets may be different for different services. A different method for calculation of error ratio at every node was also suggested in which retransmission of a large packet is considered less significant than retransmission of a small packet since smaller packets probably have lower bit errors in most of the scenarios. Designing of air cost metric using this technique is more desirable for SG NANs.

Multi-GW routing for SG NANs is explored in [20] where every node (every SM) has a tree-table to store tree information. It examines the tree-table to observe if there exists identical data aggregation point (DAP) information, when it gets a DAP GW notice. If it is available, then the node updates its tree-table on the basis of the pre-specified methods. Otherwise, a novel tree is generated on the basis of the obtained DAP information and included to the tree-table. In addition, a reserve path multi-GW diversity routing method is also developed to enhance reliability of routing scheme. Firstly, the optimal route for packet transmission is selected by a source node. In the event of a link is broken, the second route is exploited. However, if the both links are failed then on-demand route of HWMP is initialized, but another backup HWMP buffer is included at every node in order to record self-created and conveyed packets. Hence, a timer is required to systematically cancel stored backup packets. Network reliability can be advanced by using this routing algorithm on the cost of additional delay. For its ability to stabilize traffic among multiple GWs, a packet scheduling algorithm based on back pressure approach was also proposed. In this method, a conveying node chooses its following hop node depending on a novel principle that examines order size and optimal path metric value of its adjacent nodes. This process of choosing next hope will come to an end when one of GWs is designated as the optimal neighbor. This scheduling scheme enhance throughput of the whole network by equalizing traffic among GWs. However, the scheduling scheme may increase computation complexity.

3.4 Field area networks (FANs)

In power distribution level of the SG, high voltage electricity is transformed into low voltage electricity by using step-down transformers in order to fulfill the needs of commercial, industrial and home users [8, 21]. The transformers can also insulate faults and are utilized to provide a voltage regulation point as well as voltage transformation process. A FAN is responsible to ensure connectivity of smart devices in substations, distribution and transmission grids. Smart devices that may be used in FAN can be power line monitors, voltage regulators, breaker controllers, capacitor bank controllers, smart transformers, data collectors, and so on. In order to advance quality and reliability of power systems, these devices are employed to rapidly detect irregularities and failures. In the SG distribution domain, sufficient number of Remote Terminal Units (RTUs) along with Phase Measurement Units (PMUs) and IEDs will be essential to execute several substation automation functions. In addition, the distribution feeders can be exploited as a point of common coupling (PCC) for the connected DERs and micro-grid components. In order to ensure power delivery to the customer premises, the distribution feeders utilize power lines, cable poles and towers. In addition, deployment of wireless sensors along with these components would be essential to improve distribution supervisory applications. The key task of a FAN is to ensure exchange of information among numerous monitoring, control and protection applications available in between the feeder level equipment, distribution substations and applications.

A connection between backhaul of a utility service and a particular service stage of the distribution grid can be realized through the FAN communication network. Typically, a FAN presents the communication connections among the substation stage and customer stage by exploiting several collector combinations, access points and data concentrators. FAN channels having extremely robust, stable and low bandwidth features connect sensors/data collectors with a centralized GW. International Electromechanical Commission (IEC) 61850 standard is commonly employed for DA and substation within the FAN, which offers interoperability among IEDs and machine-to-machine (M2M) communications. Latency demands of the FANs for crucial applications may change from 3 to 10 ms according to the IEC 61850 standard [21, 22].

The applications utilizing the FAN schemes can be grouped into two categories as field based and customer based applications [11]. The first group, which has connection with sensors, transmission lines, voltage regulators and so on, consists of outage management system (OMS), supervisory control and data acquisition (SCADA) applications, distributed energy resource (DER) monitoring and control. The second group, which has connection with end-users such as houses, buildings, industrial users and so on, cover AMI, DR, load management system (LMS), metering data management system (MDMS). Both types of these applications functioning in the distribution domain possess several important needs. For instance, customer based applications call for highly scalable communication network among the utility and the user since more user and application demand may be required in the future. It is considered that time sensitivity is not a big issue for such type of applications. However, field based applications are regarded to have time-sensitive characteristics. Therefore, the utilities may have an option to dedicate separate communication networks for every applications or they can use single shared communication network for both groups. In addition, development cost can be minimized through a shared FAN whereas a dedicated network provides real-time communication capability and further security.

Moreover, the FAN provides access opportunity to field devices through smart devices such as laptops, notebooks, tablets and smart phones so as to gather and investigate data for fault detection, troubleshooting and service maintenance. The FANs also accommodate a plenty number of devices and serve in wide areas likewise to the NANs. The coverage of FANs and NANs may be overlapped since several smart devices are connected to these networks for implementations of many novel applications. For instance, the SMs should be reachable by both of the FANs and NANs to provide that the distribution grid is able to attain critical information from customer premises in real-time to perform effective volt/VAR control. Thus, numerous design rules and communication technologies are shared by the NANs and FANs. On the other hand, it is adequate to merely address the NAN scheme since it can be taken into account the representation of them.

3.5 Wide area networks (WANs)

A WAN is able to connect numerous NANs for collecting data from them and then it forwards gathered data to utility private network that is a central controller. The WAN makes possible long-haul communications between various DAPs of power generation systems, distributed energy resources, transmission and distribution systems, management systems and so on. The network scheme also provides bidirectional communication infrastructures for enabling several utility applications such as AMI, DR, DA, monitoring of power quality and demand-side management. The coverage area of WAN is very wide that covers approximately a few thousands of square miles, and its data rate is about 10–100 Mbps. In addition, these networks are required to utilize high bandwidths to ensure operation and control of these networks. The WAN aimed to establish communication connections among the utility systems and SG applications. Therefore, it covers two different network schemes that are called core network and backhaul communication network. While a core network is associated with metropolitan network of the utility and substations, the backhaul network is associated with DAPs (NANs) of the network. These network schemes are appropriate for data acquisition, status monitoring, fault detection, control and management of power grid [23–26].

Since the size of transmitted data and transmission environments deeply affect performance of SG communication networks, each section of SG communication networks have to employ different communication technologies. Generally, utilized communication technologies in the WANs are based on public networks including wired broadband technologies and cellular networks [27]. However, there exist several potential communication technologies for WAN applications such as PLC, fiber optics, Ethernet, DSL, IEEE 802.15.4, IEEE 802.11/Wi-Fi, Cellular (3G, LTE, LTE-A 5G), IEEE 802.16/WiMAX and so on. Also, it is important to note that there are several criterions, which may be sorted out as installation and maintenance costs, required data rate, coverage, power consumption, deployment characteristic of network and scalability of network scheme, to decide which technology is appropriate for interested applications in any SG communication network. The authors in [28] are investigated performance of a ZigBee based communication network in a SG system. Another approaches based on Bluetooth, Wi-Fi and WiMAX for SG communications are researched in [29–32]. In addition, utilization of Internet of Things (IoT) concept in SG communications is reported in [33] while the improvements on the advanced metering is presented in [34]. PLC based remote monitoring applications for SG systems are reported in [35–37]. On the other hand, security is a critical concern for public networks employed in WAN architecture of the SG systems. The use of VPN that aims to combine public and private network architectures has been recently considered as a solution for dealing with the security issues. In addition, the use of VPNs may also provide advantages for minimizing infrastructure costs.

The utilization of PLC technologies in SG WANs presents the advantage of covering communication network with low-cost by exploiting existing power lines that are essentially designed for energy transmission. The PLC technology can also provide full control depending on the wide coverage that is merely managed by the utilities. The PLC systems aim to provide low latency by supplying direct routes among control systems and subsystems. Since several types of electrical devices are connected to the power lines, which behave as noise sources, there are various noise kinds adversely affecting performance of PLC systems. In addition, varying impedance of power lines and electromagnetic interference problem lead to important problems in terms of distortion and signal attenuation. These issues are considered important problems for PLC based SG systems. Another choice for the SG WANs is the use of IP based technologies. The IP based communication systems are able to provide wide coverage advantage to connect entire components of SG system. Differentiated services and multi protocol label switching method can be utilized in these systems to ensure high QoS and secure connections. In addition, the IP based systems employ advanced security systems such as Internet protocol security (IPSec). On the other hand, latency is an important problem for IP based communication systems since slave nodes cannot transmit data in master/slave configuration. This issue is considered as main disadvantage of IP based communication systems in SG applications. The main advantage provided by wireless networks in the SG WANs is the presenting very wide coverage with relatively low cost than that of the previous explained communication technologies. Furthermore, wireless communication technologies, especially cellular communication technologies, can provide very high data rates for enabling SG applications. Moreover, some of the wireless technologies are able to support mesh network scheme to improve reliability of the network. Before data are conveyed, a connection to the wireless network is needed that may generally be considered as a difficulty in the event of emergencies and outages. Long-term latency may cause remarkable problems for real-time services of SG systems. These issues are taken into account as the main problems of wireless technologies in SG WANs.

3.6 QoS requirements for SG networks

The SG communication networks should promote several developing applications to authorize monitoring, fault detection, protection and management transactions [38]. If a fault occurs in any subsystem, it will cause a number of failures since the SG is composed of many subsystems. The reliability of power systems is crucial due to the fact that numerous contemporary systems need power grids to maintain their functions. Therefore, power outages will frequently lead to a series of failures for the systems requiring a stable power [39]. As a result of this situation, there exists a growing need for monitoring and control of power grids. On the other hand, SG applications and their environments call for specific QoS performance requirements. The important QoS metrics for SG applications can be sorted out as bandwidth, data rate, latency, reliability and security.

One of the most important QoS metrics is data rate that is related to how speed the information can be transmitted among the SG equipment and communication networks. The reachable data rate can be estimated through Shannon capacity theorem that is determined by both bandwidth and signal-to-noise ratio (SNR). Measurement, control and management transactions, which operate based on the SG communication networks, should be realized to accomplish grid automation. If a data packet is conveyed over the communication network, it experiences a delay due to the followed routes [26]. Therefore, a latency is occurred for SG systems which is another significant issue for meeting the QoS requirements. Even though the effect of latency in communication networks may be tolerated by some applications such as metering applications, there also exist time-critical applications such as fault detection, protection and control applications. Although several systems are insulated from each other in contemporary power systems, a fault may lead to a series of failure. Therefore, reliability of SG systems is very important QoS metric. In addition, the reliability of SG systems should be ensured for both communication network sites and power system sites. It is important to note that the reliability of communication systems is not merely limited with connection outages since

TABLE 3.2 QoS requirement comparison of SG application types

Application type	Data rate	Latency	Reliability	Security
Smart metering	10 Kbps	Up to 10 s	Medium	High
Demand response	Low	500 ms–1 min	High	High
Connection and disconnection	Low	100 ms to a few minutes	High	High
Synchrophasor	100 Kbps	20–200 ms	High	High
SCADA	10–30 Kbps	Up to 200 ms	High	High
Substation communication	–	About 20 ms	High	High
Video surveillance	Several Mbps	A few seconds	High	High
FLIR	10–30 Kbps	100 ms	High	High
Power system optimization	2–5 Mbps	25–100 ms	High	High
Workforce access	250 Kbps	150 ms	High	High
Micro grid management	–	100 ms–1 min	High	High

long latency times may also cause to instability on system performances. On the other hand, security is another most critical issue for meeting QoS requirements of SG systems since the SG systems have to cope with physical attacks to preserve privacy of SG systems and communication networks. Some examples for data traffic and QoS requirements of SG communication networks are listed in Table 3.2.

One of the most critical data traffic types of SG systems is meter reading that is required to detect energy consumption of users. Therefore, SMs are employed to periodically collect energy consumption rates. In addition, the collected data can be used to predict data traffic requirement of the metering application. Since the SMs are located on consumer site, the latency of this application is long. For instance, energy consumption values gathered at quarter hour intervals are transmitted to utility service provider either each 4 h within the day or each 8 h at night. Frequency of data collection and the number of this reading process directly specify the size of required bandwidth for metering application. Typically, data rate requirement of every SM is nearly 10 Kbps while the latency varies between 2 and 10 s.

Home control devices (e.g., thermostats, load controllers, energy control panels etc.) can be controlled by the utilities through the DR technologies that provide users the opportunity to decrease or change their power utilization in periods of peak demand. The bandwidth requirement of the DR technologies is significant low since they only transmit commands to control devices such as switch on/off. On the other hand, latency requirement of these technologies may change in a broad interval from milliseconds to minutes.

Connection and disconnection circumstances, which have need of various response times, may occur in several situations. If the users are mobile, long latencies can be compensated. Nonetheless, if the connection or disconnection transaction will be employed as a response to grid circumstances, the latency has to be in the range of milliseconds.

One of the significant measurement technologies for wide-area situational awareness (WASA) is Synchronized Phasor Measurements (Synchrophasor) that is related to technologies developed to enhance the power system monitoring over wide geographic regions. Data traffic requirement of this application may change between 20 and 200 ms, and bandwidth requirement is in several hundreds of Kbps depending on the technical specifications. Data traffic of a SCADA system is depend on the number of IEDs located on substations that are systematically polled by the master. The data rate requirement of SCADA systems is predicted in the range of 10–30 Kbps while the latency need changes up to 200 ms. On the other hand, inter-substation communications are managed by Generic Object Oriented Substation Event (GOOSE) whose latency requirement is quite low. Communications among multiple substations for new SG applications may be realized by the GOOSE, and the latency requirement value may vary in the range of 12–20 ms.

Substation surveillance is the another popular application requiring a specified QoS. This type application needs relatively high bandwidth in the level of Mbps since the surveillance is generally realized by video systems. It is worth noting that the utilized equipment and targeted resolutions directly affect bandwidth requirement of this application. The latency characteristics of the surveillance systems is in several seconds.

Fault location, isolation, and restoration (FLIR) systems are utilized for protecting power systems in the distribution level. These systems should have low latency characteristics and the value of mentioned latency is in milliseconds. On the other hand, bandwidth requirement of this application type changes between 10 and 30 Kbps. In addition, protection processes are also ensured for microgrid systems. The latency values between 100 ms and 10 s should be considered for microgrids. Also, distribution grid optimization is a vital process which handles optimization transactions such as power quality optimization, volt/VAR optimization on distribution level of the SG. While the latency characteristic varies approximately 100 ms, the bandwidth requirement of this process is around 2–5 Mbps. In addition to latency and bandwidth needs, SG communication networks should present superiorities in terms of stability and security.

References

[1] U.S. Department of Energy, The Smart Grid: An Introduction, https://goo.gl/anJNgP, 2010. Accessed 22 January 2019.

[2] U.S. Department of Energy, West Virginia SG Implementation Plan (WV SGIP) Project: APERC Report on Assessment of As-Is Grid by Non-Utility Stakeholders, https://goo.gl/8t7L82, 2009. Accessed 22 January 2019.

[3] C. Greer, D.A. Wollman, D.E. Prochaska, P.A. Boynton, J.A. Mazer, C.T. Nguyen, G.J. FitzPatrick, T.L. Nelson, G.H. Koepke, A.R. Hefner Jr., V.Y. Pillitteri, T.L. Brewer, N.T. Golmie, D.H. Su, A.C. Eustis, D.G. Holmberg, S.T. Bushby, NIST Framework and Roadmap for Smart Grid Interoperability Standards, Release 3.0. National Institute of Standards and Technology(2014). https://doi.org/10.6028/NIST.SP.1108r3.

[4] IEEE, IEEE Standard Communication Delivery Time Performance Requirements for Electric Power Substation Automation, IEEE Std 1646-2004. (2005). https://doi.org/10.1109/IEEESTD.2005.957480_1-24.

[5] X. Fang, S. Misra, G. Xue, D. Yang, Smart grid—the new and improved power grid: a survey. IEEE Commun. Surv. Tutorials 14 (2012) 944–980, https://doi.org/10.1109/SURV.2011.101911.00087.

[6] Y. Kabalci, A survey on smart metering and smart grid communication. Renew. Sust. Energ. Rev. 57 (2016) 302–318, https://doi.org/10.1016/j.rser.2015.12.114.

[7] S.M. Muyeen, S. Rahman (Eds.), Communication, Control and Security Challenges for the Smart Grid, The Institution of Engineering and Technology, London, 2017.

[8] R.H. Khan, J.Y. Khan, A comprehensive review of the application characteristics and traffic requirements of a smart grid communications network. Comput. Netw. 57 (2013) 825–845, https://doi.org/10.1016/j.comnet.2012.11.002.

[9] N. Saputro, K. Akkaya, S. Uludag, A survey of routing protocols for smart grid communications. Comput. Netw. 56 (2012) 2742–2771, https://doi.org/10.1016/j.comnet.2012.03.027.

[10] C. Gomez, J. Paradells, Wireless home automation networks: a survey of architectures and technologies. IEEE Commun. Mag. 48 (2010) 92–101, https://doi.org/10.1109/MCOM.2010.5473869.

[11] W. Wang, Y. Xu, M. Khanna, A survey on the communication architectures in smart grid. Comput. Netw. 55 (2011) 3604–3629, https://doi.org/10.1016/j.comnet.2011.07.010.

[12] R. Yu, Y. Zhang, S. Gjessing, C. Yuen, S. Xie, M. Guizani, Cognitive radio based hierarchical communications infrastructure for smart grid. IEEE Netw. 25 (2011) 6–14, https://doi.org/10.1109/MNET.2011.6033030.

[13] E. Kabalci, Y. Kabalci, Introduction to smart grid architecture. in: E. Kabalci, Y. Kabalci (Eds.), Smart Grids and Their Communication Systems, Springer, Singapore, 2019, pp. 3–45, https://doi.org/10.1007/978-981-13-1768-2_1.

[14] M. Kuzlu, M. Pipattanasomporn, S. Rahman, Communication network requirements for major smart grid applications in HAN, NAN and WAN. Comput. Netw. 67 (2014) 74–88, https://doi.org/10.1016/j.comnet.2014.03.029.

[15] W. Meng, R. Ma, H.-H. Chen, Smart grid neighborhood area networks: a survey. IEEE Netw. 28 (2014) 24–32, https://doi.org/10.1109/MNET.2014.6724103.

[16] IEEE, IEEE P802.15.4g/D5, May, 2011: IEEE Draft Standard for Local and Metropolitan Area Networks Part 15.4: Low Rate Wireless Personal Area Networks (LR-WPANs) Amendment: Physical Layer (PHY) Specifications for Low Data Rate Wireless Smart Metering Utility Networks, IEEE, 2011. http://ieeexplore.ieee.org/servlet/opac?punumber=5976969. Accessed 22 January 2019.

[17] J.-S. Jung, K.-W. Lim, J.-B. Kim, Y.-B. Ko, Y. Kim, S.-Y. Lee, Improving IEEE 802.11s wireless mesh networks for reliable routing in the smart grid infrastructure, in: 2011 IEEE Global Conference on Communications Workshops (ICC), IEEE, Kyoto, Japan, 2011, pp. 1–5, https://doi.org/10.1109/iccw.2011.5963578.

[18] P. Kulkarni, S. Gormus, Z. Fan, B. Motz, A self-organising mesh networking solution based on enhanced RPL for smart metering communications, in: 2011 IEEE International Symposium on a World of Wireless, Mobile and Multimedia Networks, IEEE, Lucca, Italy, 2011, pp. 1–6, https://doi.org/10.1109/WoWMoM.2011.5986178.

[19] D. Wang, Z. Tao, J. Zhang, A.A. Abouzeid, RPL based routing for advanced metering infrastructure in smart grid. in: 2010 IEEE International Conference on Communications Workshops, IEEE, Cape Town, South Africa, 2010, pp. 1–6, https://doi.org/10.1109/ICCW.2010.5503924.

[20] H. Gharavi, B. Hu, Multigate communication network for smart grid. Proc. IEEE 99 (2011) 1028–1045, https://doi.org/10.1109/JPROC.2011.2123851.

[21] N. Myoung, Y. Kim, S. Lee, The design of communication infrastructures for smart DAS and AMI. in: 2010 International Conference on Information and Communication Technology Convergence (ICTC), IEEE, Jeju, Korea (South), 2010, pp. 461–462, https://doi.org/10.1109/ICTC.2010.5674796.

[22] N.S. Nafi, K. Ahmed, M.A. Gregory, M. Datta, A survey of smart grid architectures, applications, benefits and standardization. J. Netw. Comput. Appl. 76 (2016) 23–36, https://doi.org/10.1016/j.jnca.2016.10.003.

[23] L. Chhaya, P. Sharma, G. Bhagwatikar, A. Kumar, Wireless sensor network based smart grid communications: cyber attacks, intrusion detection system and topology control. Electronics 6 (2017) 5, https://doi.org/10.3390/electronics6010005.

[24] T. Hartmann, F. Fouquet, J. Klein, Y. Le Traon, A. Pelov, L. Toutain, T. Ropitault, Generating realistic smart grid communication topologies based on real-data. in: 2014 IEEE International Conference on Smart Grid Communications (SmartGridComm), IEEE, Venice, Italy, 2014, pp. 428–433, https://doi.org/10.1109/SmartGridComm.2014.7007684.

[25] P.P. Parikh, M.G. Kanabar, T.S. Sidhu, Opportunities and challenges of wireless communication technologies for smart grid applications. in: IEEE PES General Meeting, 2010, pp. 1–7, https://doi.org/10.1109/PES.2010.5589988.

[26] Y. Yan, Y. Qian, H. Sharif, D. Tipper, A survey on smart grid communication infrastructures: motivations, requirements and challenges. IEEE Commun. Surv. Tutorials 15 (2013) 5–20, https://doi.org/10.1109/SURV.2012.021312.00034.

[27] V. Terzija, G. Valverde, P. Deyu Cai, V. Regulski, J. Madani, S. Fitch, M.M. Skok, A.P. Begovic, Wide-area monitoring, protection, and control of future electric power networks. Proc. IEEE 99 (2011) 80–93, https://doi.org/10.1109/JPROC.2010.2060450.

[28] N.C. Batista, R. Melício, V.M.F. Mendes, Layered smart grid architecture approach and field tests by ZigBee technology. Energy Convers. Manag. 88 (2014) 49–59, https://doi.org/10.1016/j.enconman.2014.08.020.

[29] R. Amin, J. Martin, X. Zhou, Smart grid communication using next generation heterogeneous wireless networks, in: Smart Grid Communications (SmartGridComm), 2012 IEEE Third International Conference on, IEEE, 2012, pp. 229–234.

[30] S. Bera, S. Misra, M.S. Obaidat, Energy-efficient smart metering for green smart grid communication. 2014 IEEE Global Communications Conference, IEEE, Austin, TX, USA (2014) pp. 2466–2471, https://doi.org/10.1109/GLOCOM.2014.7037178.

[31] S. Kaebisch, A. Schmitt, M. Winter, J. Heuer, Interconnections and communications of electric vehicles and smart grids. in: 2010 First IEEE International Conference on Smart Grid

Communications, IEEE, Gaithersburg, MD, USA, 2010, pp. 161–166, https://doi.org/10.1109/SMARTGRID.2010.5622035.

[32] B. Wang, M. Sechilariu, F. Locment, Intelligent DC microgrid with smart grid communications: control strategy consideration and design. IEEE Trans. Smart Grid 3 (2012) 2148–2156, https://doi.org/10.1109/TSG.2012.2217764.

[33] E. Spano, L. Niccolini, S.D. Pascoli, G. Iannaccone, Last-meter smart grid embedded in an internet-of-things platform. IEEE Trans. Smart Grid 6 (2015) 468–476, https://doi.org/10.1109/TSG.2014.2342796.

[34] J. Garcia-Hernandez, Recent progress in the implementation of AMI projects: standards and communications technologies. in: 2015 International Conference on Mechatronics, Electronics and Automotive Engineering (ICMEAE), IEEE, Cuernavaca, Morelos, Mexico, 2015, pp. 251–256, https://doi.org/10.1109/ICMEAE.2015.43.

[35] Y. Kabalci, E. Kabalci, Modeling and analysis of a smart grid monitoring system for renewable energy sources. Sol. Energy 153 (2017) 262–275, https://doi.org/10.1016/j.solener.2017.05.063.

[36] E. Kabalci, Y. Kabalci, A measurement and power line communication system design for renewable smart grids. Meas. Sci. Rev. 13 (2013) 248–252, https://doi.org/10.2478/msr-2013-0037.

[37] E. Kabalci, Y. Kabalci, Multi-channel power line communication system design for hybrid renewables, in: Power Engineering, Energy and Electrical Drives (POWERENG), 2013 Fourth International Conference on, IEEE, 2013, pp. 563–568.

[38] G. Rajalingham, Y. Gao, Q.-D. Ho, T. Le-Ngoc, Quality of service differentiation for smart grid neighbor area networks through multiple RPL instances. in: Proceedings of the 10th ACM Symposium on QoS and Security for Wireless and Mobile Networks—Q2SWinet '14, ACM Press, Montreal, QC, 2014, pp. 17–24, https://doi.org/10.1145/2642687.2642695.

[39] S.V. Buldyrev, R. Parshani, G. Paul, H.E. Stanley, S. Havlin, Catastrophic cascade of failures in interdependent networks. Nature 464 (2010) 1025–1028, https://doi.org/10.1038/nature08932.

Chapter 4

Power line communication technologies in smart grids

Chapter outline

4.1 Introduction

Power line communication (PLC) systems utilize existing electrical power systems as a communication medium to enable data transmission over power lines, whereas the main task of the power lines is the delivery of AC (at 50 or 60 Hz frequency) or DC electric power from energy generation plants to the users/customers. The main superiority of the PLC systems is evident that these systems can save new channel establishment cost because of the fact that power lines are in use all over the world. Therefore, in last two decade, PLC systems have attracted much interest in the field of communication and smart grid (SG) systems, and in several application areas such as home automation, in-vehicle communication, automatic meter reading (AMR) and demand response and so forth [1–10].

Although the PLC systems have become more popular in the last two decades, the idea behind of the PLC dates back to the 1800s. The first PLC

From Smart Grid to Internet of Energy. https://doi.org/10.1016/B978-0-12-819710-3.00004-1

applications were related to remote meter reading and remote load management, and the first two patents of PLC regarding these applications were available in 1898 and 1901 [11, 12]. Various applications related to measurement, control and protection via power lines have been appeared after the patents, and researchers have been widely focused on the applications over medium voltage (MV) and high-voltage (HV) power lines. One of the most popular applications was ripple control that was a load management system for protecting and distribution of loads [13–15]. A ripple control system (RCS) was enabled one-way communication with low data rates. It operated between 125 Hz and 3 kHz frequencies for passing signals over distribution transformers. The most important disadvantage of the RCS was its high power requirement. For instance, a few megawatts may be needed for data transmission. A number of important developments have been made for RCSs until the 1950s [15, 16]. The employed modulation schemes in these advanced systems were amplitude shift keying (ASK) and frequency shift keying (FSK). According to utilizing digital modulation schemes, not only the high power requirements for data transmission was eliminated, but also higher data rates have been obtained. Afterwards, at the beginning of the 1980s, several studies were carried out to realize bidirectional PLC systems that were based on PLC systems developed for automation and AMR applications [15, 17]. In 1984, Enermet Melko was introduced as a bidirectional data transmission system that served between frequencies of 3.025 and 4.825 kHz with phase shift keying (PSK) modulation scheme. In 1990, Echelon introduced local operation networks (LonWorks) system that was a networking platform. In addition, LonWorks was built on a special protocol to meet control application requirements. The frequency bands defined by International Telecommunication Union (ITU) and the utilization of these bands in the PLC applications are illustrated in the Fig. 4.1. The frequency bands are defined as *super low (SLF)*, *ultra low (ULF)*, *very low (VLF)*, *low (LF)*, *medium (MF)*, *high (HF)*, *very high (VHF)*, *ultra high (UHF)*, *super high (SHF)*, *extremely high (EHF)* and *tremendously high frequency (THF)*, respectively.

The achievement of narrowband (NB) PLC systems supported the improvement of broadband (BB) PLC systems that are principally aimed to be utilized for internet service and home area network (HAN) applications. In 1997, the primary internet applications that are related to internet access and service providing via PLC systems were introduced in Europe. However, the obtained results were disappointing for the idea of internet access over the power lines. Therefore, the interest was changed to home applications and industrial communications in the early 2000s. Some industrial alliances such as the HomePlug Powerline Alliance (HomePlug), Universal Powerline Association (UPA), High Definition PLC (HD-PLC) Alliance, and The HomeGrid Forum have played important roles in the acceleration of the process. In the last decade, various standards have been defined to specify the implementations such as TIA-1113, ITU-T G.hn, IEEE 1901 FFT-OFDM, and IEEE 1901 Wavelet-OFDM

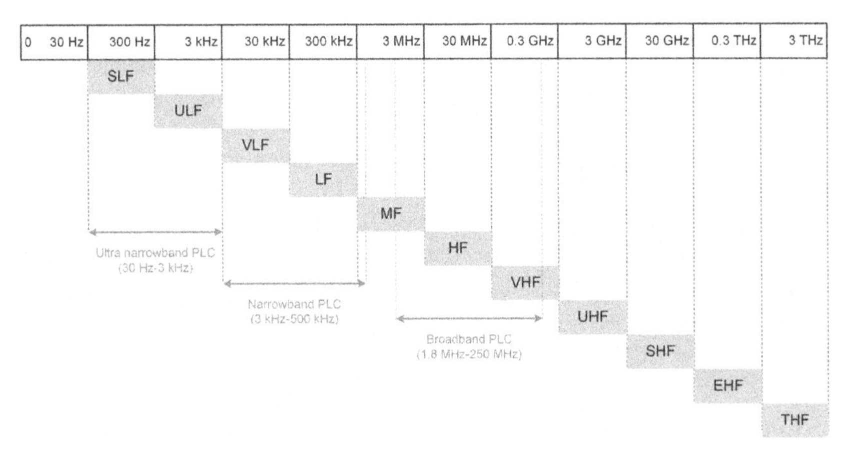

FIG. 4.1 Frequency bands utilized in PLC applications.

[17–19]. Even though various products that present physical layer (PHY) data rates of 14 Mbps (HomePlug 1.0), 85 Mbps (HomePlug Turbo), and 200 Mbps (HomePlug AV, HD-PLC, UPA) are developed, none of these technologies is capable of working together. On the other hand, the BB PLC systems that can be regarded as the complementary of Wi-Fi at home networking could not achieve a significant market share yet [1].

Although the PLC applications have been generally realized over AC power systems, there are several applications over DC power systems, i.e., vehicles. Vehicle control and further applications can be also managed via in-vehicle PLC systems where power cables of vehicles are employed as a communication channel medium. This idea has been investigated for many different types of vehicles [20–24]. Furthermore, several attempts and developments have been performed in order to enable communication among plug-in electric vehicles (PEVs) and charging stations [25–27]. Currently, one of the important application areas of the PLC systems is SG communication since remote metering, monitoring and control operations are the most important application areas of the SGs [1, 28].

4.2 Classification of PLC systems

PLC technologies are generally classified according to the exploited bandwidth in the technology. In a similar way, electromagnetic compatibility (EMC) regulations also consider used bandwidth amount for classifying these technologies. Generally, the PLC technologies are classified into three categories:

Ultra-narrow band (UNB) PLC technologies: These technologies utilize ULF band (0.3–3 kHz) or the upper part of the SLF band (30–300 Hz). In other words, the frequencies below 3 kHz are used by the UNB PLC technologies where the reachable data rate is around 100 bps. The UNB PLC technologies that have been exploited metering applications are based on proprietary solutions.

Narrowband (NB) PLC technologies: These technologies operate at frequencies between 3 and 500 kHz, and the NB PLC systems are composed of two subcategories called "low data rate (LDR)" and "high data rate (HDR)". The NB PLC standards differ from country to country, and the NB frequencies are specified by European Comité Européen de Normalization Électrotechnique (CENELEC) in Europe, Federal Communications Commission (FCC) in USA, Association of Radio Industries and Businesses (ARIB) in Japan, and Electric Power Research Institute (EPRI) in China.

- **Low data rate (LDR):** LDR NB PLC technologies that use single carrier modulation schemes are widely preferred for smart metering and indoor automation applications. Some popular examples of this category are IEC 61334-5 based metering systems, X10 and HomePlug C&C.
- **High data rate (HDR):** HDR NB PLC technologies can present up to hundreds of kbps data rates by using multicarrier modulation schemes such as orthogonal frequency division multiplexing (OFDM). Powerline Related Intelligent Metering Evolution (PRIME) Alliance and G3-PLC Alliance developed PRIME and G3-PLC, respectively. These industrial specifications have been recently involved into the ITU-T Recommendations G.9903 and G.9904.

Broadband (BB) PLC technologies: BB PLC systems operate in a very wide frequency range from 1.8 to 250 MHz. They employ multicarrier modulation schemes similar to the HDR NB PLC systems. In addition, this technology can reach data rates at the level of Mbps since it has been developed for high-speed indoor applications. The BB PLC technology based devices are compatible with TIA-1113, IEEE 1901 and ITU-T G.hn standards.

As shown in Fig. 4.2, the PLC technologies can be summarized in terms of operating frequencies, data rates, mapping schemes, regulations and standards. Further information for the specifications and standards of UNB PLC, NB PLC, and BB PLC systems are presented in Section 4.4.

4.2.1 Ultra-narrow band PLC

The first application for load control in the UNB frequencies was the RCS that was based on unidirectional communication feature with very low data rates. Other applications are AMR Turtle System and Two-Way Automatic Communications System (TWACS). While data rate of former system is extremely low as much as 0.001 bps, data rate of TWACS is maximum at most twice the mains frequency, i.e., 100 bps in Europe and 120 bps in North America [17]. Even

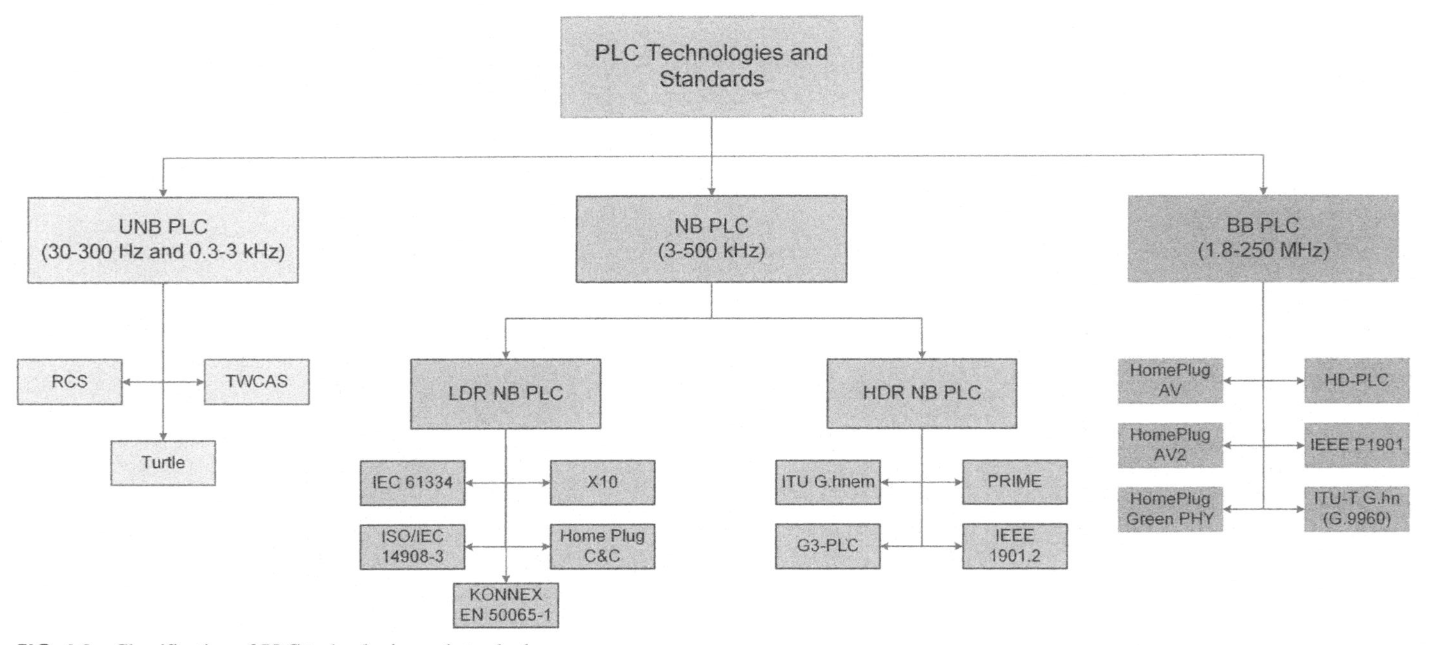

FIG. 4.2 Classification of PLC technologies and standards.

though the UNB PLC technologies have the advantage of very long operational range (typically >150 km), these technologies suffer from low data rates. Even though data rate of each link is very low in this group of PLC technologies, systems operating in this band can benefit several parallelization and addressing methods to improve scalability. On the other hand, these solutions have been employed by power utilities for more than 20 years in spite of being proprietary.

4.2.2 Narrowband PLC

NB PLC systems operate at frequencies between 3 and 500 kHz. These frequency bands are covered by CENELEC from 3 to 148.5 kHz, by FCC from 10 to 490 kHz, by ARIB from 10 to 450 kHz and by Chinese band from 3 to 500 kHz. The frequency allocation schemes of the NB PLC systems are illustrated in Fig. 4.3. The CENELEC band, which is divided into four categories, is characterized by EN 50065-1 in the EU [29]. The CENELEC-A band covers frequencies from 3 to 95 kHz and this band is only allocated for the use of utilities. While differential mode voltage level of the signals between 9 and 95 kHz frequency band is restricted to 134 dBμV, the power spectral density (PSD) is limited to 120 dBμV/200 Hz [30]. The frequencies within 95–148.5 kHz are reserved for end-user applications and are referred as CENELEC-B, CENELEC-C and CENELEC-D bands as can be seen from the Fig. 4.3. The CENELEC-B band covers 95–125 kHz frequency range, while CENELEC-C band contains 125–140 kHz frequency range and devices operating on this band should support carrier sense medium access/collision avoidance (CSMA/CA) protocol. The other one, CENELEC-D, covers frequencies between 140 and 148.5 kHz.

Since no special regulation has been made for frequencies above 150 kHz, EMC limits of the EN 55022 are principally adopted. Thanks to Directive 2004/108/EC, manufacturers can evaluate EMC of devices according to European

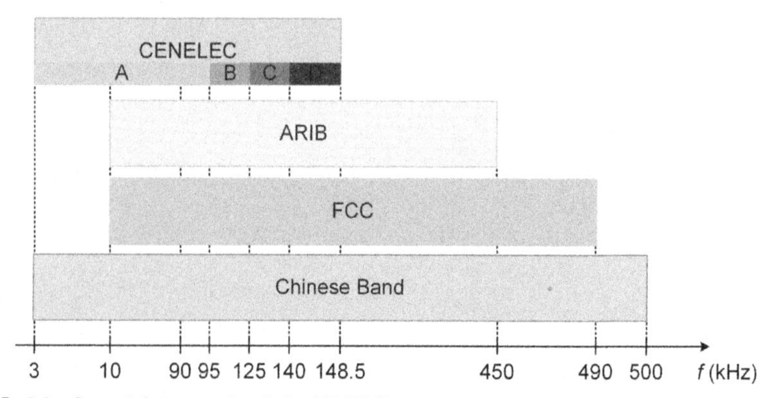

FIG. 4.3 Several frequency bands for NB PLC systems.

harmonized standard or a special procedure that should be approved by an independent notified body. The IEEE P1901.2 specification can be given as a special procedure example [30]. In the United States, the FCC regulates PLC emissions within 9–490 kHz frequency band through the Code of Federal Regulations, Title 47, Part 15 (47 CFR §15) where the PLC systems are divided into two subgroup as "power line carrier systems" and "carrier current systems". The second subgroup aims to create a connection among distribution substations and customers for enabling smart grid applications. Out-of-band conducted emissions limits and in-band radiated emission limits are described in §15.107(c) and §15.209(a) paragraphs, respectively [31]. In Japan, Standard T84 of the ARIB authorizes the use of 10–450 kHz frequency band for the NB PLC systems. In-band and out-of-band conducted emission limits are also defined with the measurement procedures. Even though the EMC regulations of China for the NB PLC systems are not accessible, it is reported that there are several PLC devices working below 95 kHz or up to 422 kHz [30].

The LDR NB PLC systems are compatible with several recommendations such as ISO/IEC 14908-3 (LonWorks), ISO/IEC 14543-3-5 (KNX), CEA-600.31 (CEBus), IEC 61334-3-1, IEC 61334-5 (FSK and Spread-FSK) and so forth. In addition, there are some non-Standards Developing Organizations (SDOs) based examples for LDR NB PLC systems such as Insteon, X10, HomePlug C&C, SITRED, Ariane Controls, BacNet. Pico Electronics developed X10 standard in 1975, which is an international protocol for communication between electronic devices employed in home automation. It uses low-voltage power lines for signaling and control operations. Digital data are transmitted by using 120 kHz carrier at the zero crossings of the AC signal and one-bit data is conveyed at each zero crossing. Following the X10, KNX and LonWorks are developed for several physical communication media such as twisted pair, power lines, infrared and Ethernet. LonWorks devices are able to work in one of two different frequencies based on the application type. While the CENELEC-A band is utilized for electric utility applications, the CENELEC-C band is exploited for commercial, in-home and industrial applications. Depending on the used frequency band, data rates are in the level of few Kbps. KNX is standardized through EN 50090 and ISO/IEC 14543-3-5 [32] while LonWorks is standardized by ISO/IEC 14908-3 [33, 34]. The most common PLC technologies used today that employ FSK or Spread-FSK are specified in the IEC 61334 (especially in IEC 61334-5-1 and IEC 61334-5-2).

On the other hand, HDR NB PLC systems operating in the CENELEC, FCC and ARIB bands are based on multicarrier technologies such as OFDM, and can present higher data rates typically up to 500 Kbps. When compared to the LDR systems, the HDR NB PLC systems are more appropriate for SG applications such as smart metering, remote control, and energy management. ITU-T G. hnem and IEEE 1901.2 are important standards of HDR NB PLC systems while PRIME and G3-PLC are non-SDO-based examples. It is important to note that several coexistence mechanisms have been developed to ensure interoperability between HDR and LDR NB PLC technologies. The standards, technical specifications and application areas of LDR and HDR NB PLC systems are listed in Table 4.1 [35].

4.2.3 Broadband PLC

The success of the NB PLC systems encouraged development of the BB PLC systems, and primary application studies have been performed for internet access and HAN applications. The first application that aimed to enable Internet access to residential customers over power lines were developed by Nortel and Norweb Communications in Europe. The developed technology has been tested in a network composed of homes and schools, and >1 Mbps data rates have been achieved thanks to performed experiments [17, 36]. Unfortunately, the project has been terminated in 1999 due to cost and EMC issues. Other projects conducted by Siemens and Ascom experienced similar situations. The Open PLC European Research Alliance (OPERA) project that was funded by the European Commission provided significant improvements in the field of the BB PLC.

After the first unsuccessful period of BB PLC, home applications and industrial communication applications of BB PLC systems have become more popular for industrial areas in the early 2000s. The HomePlug, UPA, HD-PLC, and The HomeGrid Forum have played important roles for BB PLC systems. In 2005, HomePlug AV was proposed for indoor environments in order to enable multi stream networking with high quality feature. Depending on its advanced PHY and MAC layer characteristic, this standard is capable to achieve 200 Mbps data rates over AC power lines in-home networks. Therefore, it provides high quality broadband services containing audio, video and data over existing power lines of home areas. Furthermore, HomePlug AV can support High-Definition TV (HDTV) and Voice over Internet Protocol (VoIP) applications owing to ensuring high data rates. The most popular commercial devices are compatible with HomePlug AV and HD-PLC standard. The MAC layer of these devices exploits time-division multiple access (TDMA) method and CSMA/CA protocol. In addition, it is important to note that the PHY layer of the IEEE P1901 standard has been created based on the HomePlug AV and HD-PLC standard. In 2010, HomePlug Green PHY that has been released as a special type of HomePlug AV standard presents low cost and low power consumption features for potential SG applications. Specifications of popular BB PLC standards and technologies are listed in Table 4.2 [35].

4.3 Characteristics of power lines

Power lines were exploited to carry electrical power in every place of the world until developments in signal processing and communication systems demonstrated that power lines can not only employed to deliver the electrical power,

TABLE 4.1 Standards, technical specifications and applications of the NB PLC systems [35]

Technology/standard	Spectrum	Modulation	Bit-rate	MAC	Carrier	Application
Insteon/proprietary	CENELEC C	BPSK	2.4 Kbps	–	Single	Home automation
Konnex/EN50090 EN13321-1 ISO/IEC14543	CENELEC B	Spread FSK	1.2 Kbps	CSMA	Single	Home automation
X10/proprietary	CENELEC B	Pulse-position modulation	50 or 60 bps	CSMA/CD	Single	Home automation
CEBus/EIA-600	CENELEC C, FCC, ARIB	Spread spectrum	8.5 Kbps	CSMA/CD	Single	Home automation
LonWorks/Echelon	CENELEC A, C, FCC,	BPSK	up to 5.4 Kbps	CSMA/CA, CSMA/CD	Single	Home automation
UPB/proprietary	CENELEC A	Pulse-position modulation	240 bps	–	Single	Home automation
HomePlug C&C/HomePlug consortium	CENELEC A, C, FCC, ARIB	Differential chaos shift keying	0.6–7.5 Kbps	CSMA/CA	Single	Command and control
Meters & More/proprietary	CENELEC	BPSK	Up to 4800 bps	–	Single	Advanced metering infrastructure
G3-PLC/ERDF	CENELEC A, FCC	OFDM DQPSK DBPSK	34–240 Kbps	CSMA/CA	Multiple	Advanced metering infrastructure
PRIME/prime alliance	CENELEC A	OFDM D8PSK DQPSK DBPSK	128 Kbps	CSMA/CA TDMA	Multiple	Advanced metering infrastructure
G.Hnem ITU-T 9955	CENELEC A, B, C,D, FCC	OFDM QPSK 16-QAM	up to 1 Mbps	CSMA/CA	Multiple	Advanced metering infrastructure
IEEE P1901.2	CENELEC A, B, C,D, FCC	–	–	–	Multiple	Advanced metering infrastructure

TABLE 4.2 Standards, technical specifications and applications of the BB PLC systems [35]

Technology/ standard	Spectrum	Modulation/coding	Bit-rate	MAC	Carrier
HomePlug AV	1.8–30 MHz	OFDM (with adaptive bit loading)/turbo codes	200 Mbit/s	TDMA-CSMA/CA	Multiple
HomePlug AV2	1.8–86 MHz	OFDM/turbo codes	500 Mbit/s	TDMA-CSMA/CA	Multiple
HomePlug Green PHY	1.8–30 MHz	OFDM/turbo codes	3.8–9.8 Mbit/s	CSMA/CA	Multiple
HD-PLC	2–28 MHz	Wavelet OFDM/Reed-Solomon and Convolutional or LDPC	Up to 240 Mbit/s	TDMA-CSMA/CA	Multiple
IEEE P1901	2–28 MHz 2–60 MHz	OFDM or Wavelet OFDM/Reed-Solomon, Convolutional, LDPC	540 Mbit/s	TDMA-CSMA/CA	Multiple
ITU-T G.hn ITU-T G.9960	1.8–80 MHz	OFDM (with bit loading)/LDPC	>200 Mbps Up to 1 Gbps	TDMA-CSMA/CA	Multiple

but also can be utilized to transmit communication signals. After this evolution of the power lines, the PLC systems have been attracted increasing interest and these systems are considered as innovative technologies for several application areas such as internet access in indoor and/or outdoor, home automation systems, in-vehicle communication, SGs, advanced meter reading, demand response and so on. The most important benefit provided by the PLC systems, when compared with other wired and wireless communication systems, is that there is no requirement to establish an external communication channel [2, 4, 6–8].

Even though these systems provide installation cost minimization, the power lines expose a destructive channel effect since they have not been originally created for voice, data and image transmission. Furthermore, disruptive effects of power lines (i.e., high-level frequency-dependent attenuation, variable impedance and different type noises) are highly dynamic different from other wired communication systems. Noise characteristics of power lines cannot be defined through only additive white Gaussian noise (AWGN). Even though noises in power lines are complex, they can be categorized into five classes as *narrowband noise*, *colored background noise*, *periodic impulsive noise synchronous to the main frequency*, *periodic impulsive noise asynchronous to the main frequency*, and *asynchronous impulsive noise*. The narrowband noise, periodic impulsive noise asynchronous to the main frequency and colored background noise usually exhibit stationary behavior for a few seconds and minutes. Therefore, these noises are generally taken into account as background noise. On the other hand, periodic impulsive noise synchronous to the main frequency and asynchronous impulsive noise change in the level of microsecond swiftly depending on the on/off state of devices connected to the grid. According to these mentioned disruptive effects, PLC channels are characterized as a combination of *frequency-selective* and *time-selective* channels. In addition, it is worth noting that the power line channels behave more destructive channel environment than wireless communication channels. Therefore, secure transmission of high-frequency communication signals over power lines is a hard process, and to handle this task the PLC channel should be firstly described as precisely as possible.

In spite of the fact that there exist various methods to define the PLC channel characteristics, the most popular models can be classified into two categories as *bottom-up* and *top-down* approaches [3, 7, 37–47]. The bottom-up method defines the PLC channel model through transmission line theory and it requires some prior information regarding power grid infrastructure such as power line topology, characteristics of utilized power cables and connected loads to network to figure out impedance variations [37–42]. The other approach, top-down method, tries to model PLC channel depending on the long-term measurements of power line characteristics. In top-down method, curve-fitting techniques are utilized to acquired channel measurements for deriving mathematical expression of PLC channel variations. Hence, this modeling approach do not require any prior information about impedance variations of loads, employed power cables and network topology [3, 7, 43–47]. Deriving a general and universal channel model for power lines is not possible since there are many parameters affecting characteristic behavior of the PLC channel when compared to other traditional wired channel types. Furthermore, signal attenuation and interference of PLC channels may be affected by structural characteristics of buildings such as single-family houses, high-rise buildings, multi-flat buildings with riser or common meter room and so forth [48, 49]. Moreover, infrastructures of electrical transmission and distribution systems may be different in countries [50–52]. As a result of these differences, the PLC channels have very dynamic characteristics and present dissimilar behaviors in different locations.

Hensen and Schulz proposed a basic model to define channel transfer function of power lines in [43]. In this model, channel attenuation was only modeled as an increasing parameter with higher frequencies. The drawback of this channel model was that it was not taken into account multipath propagation effect. After Hensen and Schulz's channel model, Philipps [44] and Zimmermann [45] identified different PLC channel model that takes into account multipath effects. Philipps model defined the channel impulse responses through Dirac pulses that symbolize the sum of the signals coming from N various branches. The mathematical definition of Philipps model can be expressed as follows:

$$H(f) = \sum_{i=1}^{N} \rho_i \cdot e^{-j2\pi f \tau_i} \tag{4.1}$$

where ρ_i stands for reflection parameter while τ_i shows delay time. Another channel model, which was introduced by Zimmermann and Dostert [45], included an additional parameter to Philipps model for describing channel attenuation. The analytical expression of this model can be given as follows:

$$H(f) = \sum_{i=1}^{N} g_i \cdot e^{-\left(a_0 + a_1 f^k\right) \cdot d_i} \cdot e^{-j2\pi f \left(d_i / v_p\right)} \tag{4.2}$$

where g_i, d_i and v_p denote weighting parameter, line length and propagation speed, respectively. In addition a_0, a_1, and k parameters show frequency dependent attenuation [53]. While first exponential function in Eq. (4.2) represents attenuation, second one defines delaying time. After these channel model proposals, novel PLC channel models with various approaches that were suggested by considering structural differences of power systems of countries also introduced by researchers. These new models were based on two-port transmission line presentation of electrical network [37–42], statistical modeling approaches unlike the deterministic models [46, 47] and computation of multipath propagations via matrix approach [54–56].

When the PLC channel models suggested both for special countries and environments are taken into account, they can be also grouped according to the definition domains. While first approach of channel modeling is realized in time domain [17, 54–56], the second method defines the channel in frequency domain

[3, 17, 37–47]. In time domain modeling method, channel characteristics are usually identified with respect to multipath effects since high-frequency signal propagation cannot be accomplished through only one direct path among transmitter and receiver unit. The multipath effects are experienced when impedance mismatch or disconnection problems occur in the PLC channels. Thus, multipath effects should be taken into account during time domain channel modeling process [17, 54–56]. In the frequency domain method, channel modeling operation can be carried out by means of deterministic methods where exploiting transmission line theory or realistic channel measurement campaigns are preferred [3, 17, 37–47]. The use of transmission line theory presents advantages in terms of computational complexity compared to time domain method, since computational complexity of this approach does not depend on the topological structure and mismatched node connections of power systems. Nonetheless, the use of transmission line theory requires some priori information such as characteristics of cables employed in connections, impedance values connected to each node [3, 17, 37–41]. In realistic channel measurement campaigns method, experimental measurements are accomplished in different places, and measured data are collected for establishing a measurement database [3, 17, 54–56]. Later, channel-modeling task is performed by means of this generated database. Since this method is only based on long-term channel measurements, there is no any priori information requirement regarding power lines and power cables. In order to increase the accuracy of this modeling approach, experimental measurements should be realized for a long time in different locations.

The multipath effect, signal propagation and noise characteristic of power lines are closely related to infrastructure of power grid, network topology, cable characteristics, and impedance variations of loads connected to the network. In addition, channel behavior can dynamically change within very short time intervals due to varying impedances of the network. In the following subsections, we will present an outlook for infrastructural factors affecting channel characteristics.

4.3.1 Electricity grids and regional differences

The power grids have been evolved from their first installation to today, and they are being widely exploited throughout the world. The main components of traditional power grids are placed at generation, transmission, distribution, and consumption sections. The generation part of the power grids combines various power generation systems such as hydro-generators, thermos-generators, combined heat and power (CHP) plants, nuclear power sources and so on. An entire block diagram of a typical modern power grid is shown in Fig. 4.4 where the sections of power grid are entitled as *generation*, *transmission and distribution*, *energy storage and distributed generation*, and *residential and industrial consumers*. Energy conversion among stages is achieved by means of transformers. The generation stage contains both traditional and renewable energy sources (RESs). While the modern electricity grid is illustrated with

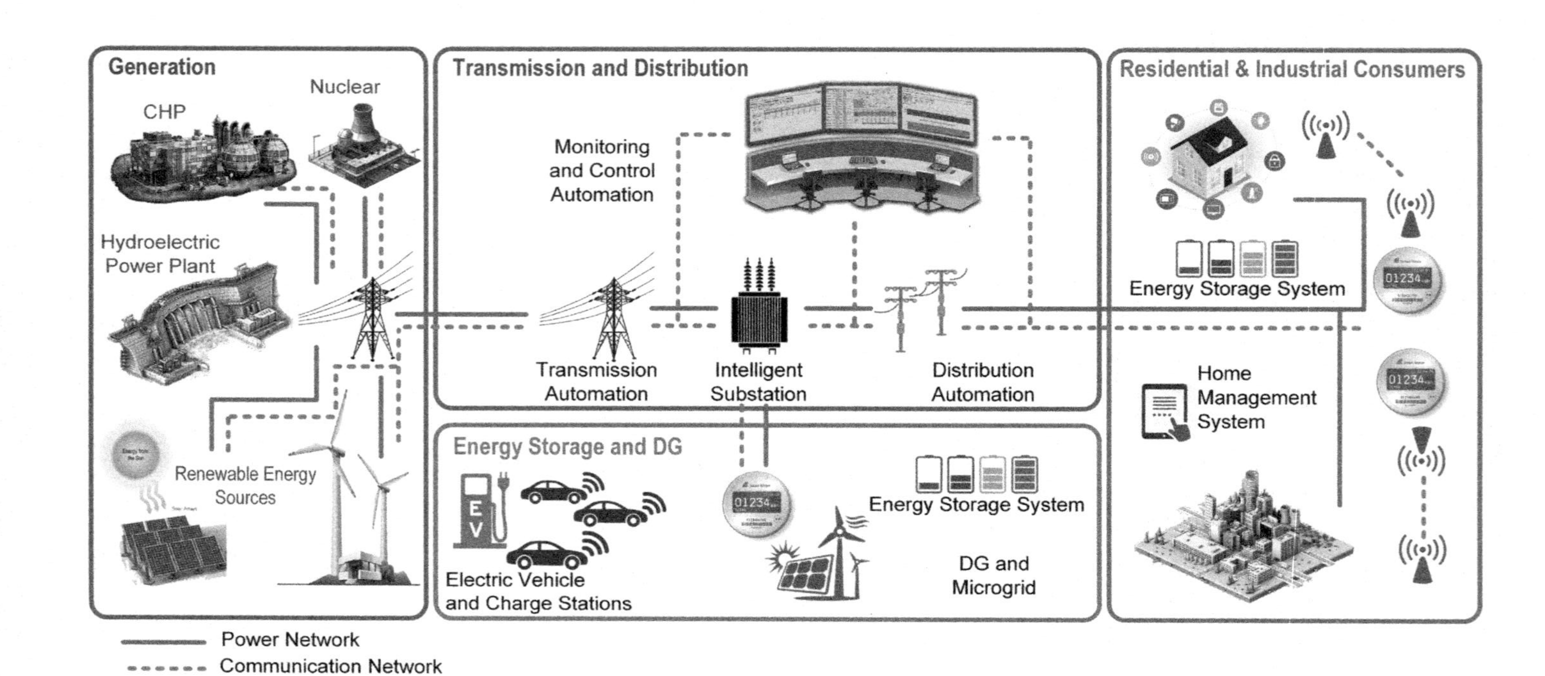

FIG. 4.4 The block diagram of a modern power grid with its main components.

red solid line, communication network that is included into the conventional power grid for evolving it to SG is illustrated with blue dotted line.

The HV lines are utilized to deliver electricity in the range of 110–380 kV at transmission stage of power grids. Since they are composed of long overhead lines without branches, the HV lines can be regarded as wave-guides due to less attenuation features when compared to MV and low-voltage (LV) lines. Arcing, corona noise and coupling costs are appeared the most important problems of the HV lines when they are considered for the PLC applications. Therefore, the BB PLC applications of HV lines have been limited until today.

The MV lines that transfer electricity in the range of 10–30 kV are connected to HV lines over primary transformer substations. This type of transmission lines that may be overhead or underground cables are employed to distribute power among industrial customers, towns and cities. In the event of mixed installation of overhead and underground cables at MV level, an increased attenuation may be occurred due to impedance mismatch. In addition, the MV lines can directly linked to intelligent electronic devices (IED) because of containing a small number of branches. Since monitoring and automation applications are generally required low data rates, NB PLC applications can be efficiently realized over MV lines.

The LV lines that deliver electricity in the range of 110–400 V are connected to MV lines over secondary transformer substations, and they are last stage of power network reaching customers. If a communication signal on MV lines passes to LV lines over secondary transformers, it suffers from an extreme attenuation, approximately in the level of 55–75 dB [57]. Channel characteristics of LV lines are extremely dynamic than that of the HV and MV lines since there exist many randomly connected/disconnected loads with different impedance values. The noises in the LV lines are unpredictable because of the dynamic impedance variations. Therefore, well-designed repeaters and coupling devices are widely needed to ensure a reliable PLC system with high data rate.

On the other hand, network connection topology, used electrical cables and other electrical devices and characteristics of power networks lead differences throughout the world. It is important to note that there are remarkable differences between power networks of countries that closely affect PLC channel behavior. The comparison of European and North American power networks is a good example to understand differences among countries. The European MV levels can change in the range of 10–24 kV by using overhead and underground distribution systems. Typically, the LV lines are connected over secondary transformer substations in the Europe. The secondary transformer stations are constituted in star topology with 400 V phase-to-phase and 230 V phase-to-neutral features. While a secondary transformer substation is composed of two or three transformers, these transformers may approximately serve for 150–300 customers. When the North American model case is considered, the MV levels are in the range of 4 and 34 kV. The secondary transformers in North America feed LV lines with 120 or 200 V according to load types. Most of LV lines in there are composed of overhead lines, and they only serve for 10–15 customers [58]. In addition, there is main frequency differences throughout the world. For instance, the main frequency is determined as 50 Hz in Europe and in many parts of Asia while American countries utilize 60 Hz main frequency. One another important factor affecting the PLC channels is that the characteristic behaviors of electrical devices and components change over the years due to varying impedances. Moreover, there are important differences among the buildings as mentioned before. Generally, the buildings can be grouped as multi-flat buildings with riser, multi-flat buildings with common meter room, single-family houses and high-rise buildings [48]. The differences in wiring topology of different buildings also cause diversity in signal attenuation and interferences in the PLC channel [49].

4.3.2 Underground cables characteristics

Underground cables are composed of one or more conductors that are covered with proper insulation materials, and are enclosed by a sheath. These cables present several advantages such as less chance of damage from storm and/or lightning, more reliable among faults and voltage drop issues and low maintenance costs. Nonetheless, main disadvantages of these cables are high installation cost and insulation problems in high voltage transmission. Thermal capacity that limits overloading in long-distance applications is an important characteristic of underground cables. Several disruptive effects such as noise, electromagnetic interference (EMI), impedance mismatch, and inter-symbol interference (ISI) may affect these cables because of the multi-branched structure of underground wiring. In addition, heat dispersion is a critical problem for underground cables and it may cause performance degradation on power transmission [1, 59, 60]. Thermoplastic (i.e., Polyvinyl Chloride (PVC)) or thermosetting (i.e., Ethylene-Propylene Rubber (EPR), Cross-Linked Polyethylene (XLPE)) type materials are generally preferred as insulation material for underground cables. Even though capacitive coupling among ground and power cables causes performance degradation on NB-PLC technologies operating under 1 MHz, reliable communication over underground cables can be performed up to 300 m distance [61, 62]. In addition, shielding of underground cables according to electromagnetic interferences can decrease noise issues of broadband applications. Several underground cable examples employed underground network are listed in Table 4.3.

4.3.3 Overhead power-lines characteristics

The use of overhead cables for power transmission over long distances provides two important advantages. The first one is related to installation costs whereas the second is related to transmission requirements of electricity at high voltage levels due to the economic reasons. These cables are widely utilized to deliver or distribute electrical power in modern power systems. The service quality of

TABLE 4.3 Several practical underground cable examples utilized in PLC applications

No.	Cable length	Cable type	Cross sectional area	Standard specification	Voltage/frequency range	Application type
1	90–210 m	J1VV [63]	2.5–10 mm^2	IEC 60502-1	0.6–1 kV/Up to 110 kHz	Lighting application
2	15 m	NYY [64]	35 mm^2	IEC 60502 DIN VDE 0276-603	0.6–1 kV/1–30 MHz	AMR applications
3	(a) 122 m (b) 1 km	(a) AMCMK [65] (b) AXPK-PLUS [66]	(a) 21 mm^2 (b) 16 mm^2	(a) SFS4880 (b) IEC 60,502–1, CENELEC HD 603, HD 60364	0.6–1 kV/150–490 kHz	LV distribution networks
4	100 m	GKN [67]	150 mm^2	HD 603; IEC 60228	1 kV/1–30 MHz	LV distribution networks
5	500 m	(a) PA [68]	150 mm^2	(a) NFC32-321	(a) 0.6–1 kV/30 Hz-100 MHz	LV distribution networks

Continued

TABLE 4.3 Several practical underground cable examples utilized in PLC applications—cont'd

No.	Cable length	Cable type	Cross sectional area	Standard specification	Voltage/frequency range	Application type
		(b) PB [69] (c) XLPE [70]		(b) IEC 60502-1, IEC 60245, EN 50264 (c) TICW/02-2009	(b) 3.6–6 kV/30 Hz-100 MHz (c) 0.6–1 kV/30 Hz-100 MHz	
6	500 m	(a) NA2XSYR [71] (b) 3x120 AL [72]	50–150 mm^2	(a) IEC60502-2, BS 6622, HD 620 (b) AS/NZS 1429.1	8.7–15 kV/1–20 MHz	MV distribution networks

overhead lines are closely depend on the structural design of the line. Aluminum, galvanized steel copper, steel-cored aluminum and cadmium copper are widely preferred as conductor materials in production of overhead cables. The overhead cables may be exposed to several important problems such as weather conditions (storm, lightning, etc.), voltage drop, unbalanced load issues, impedance problems in networks, grounding problems and so on. These problems may lead to crucial interferences for both NB and BB communication systems. In addition, coupling schemes, characteristic features and connection types of employed cables and branch types can affect performance of PLC technologies operating over overhead transmission lines. Various overhead cable examples utilized in the transmission and distribution networks are listed in Table 4.4.

4.4 PLC regulations and standards

This section introduces recent developed standards of HDR NB-PLC systems for SG communications. These standards are considered as the second-generation standards while the first generation ones covering LDR NB-PLC systems were widely utilized by utilities. LonWorks (ISO/IEC 14908-3), KNX (ISO/IEC 14543-4-5), CEBus (CEA-600.31), X10, Insteon, IEC 61334-5-1/2/4, AMR Turtle and TWACS are good examples for the first generation standards developed for the UNB-PLC systems [1, 17].

Several organizations and industrial alliances initiated studies for defining new standards for next generation PLC systems that operate under 500 kHz frequencies with high data rates. These PLC systems (referred as HDR NB-PLC systems) utilize multicarrier communication methods as in the BB-PLC systems. The IEEE 1901 and ITU-T G.hn standards have been designed for these HDR NB-PLC systems. There exist four private ITU-T recommendations and IEEE 1901.2 standard for NB-PLC systems that are published at around the 2013. These standards can be summarized as follows.

- **ITU-T G.9901 "Narrowband OFDM Power Line Communication Transceivers-Power Spectral Density (PSD) Specification"** [80]**:** This recommendation defines several control parameters for spectral contents, PSD related definitions (i.e., PSD mask requirements, PSD transmit levels, PSD measurement methods). In addition, ITU-T G.9901 integrates PHY and data link layer (DLL) specifications of ITU-T G.9902 (G.hnem), ITU-T G.9903 (G3-PLC) and ITU-T G.9904 (PRIME).
- **ITU-T G.9902 "Narrowband OFDM Power Line Communication Transceivers for ITU-T G.hnem Networks"** [81]**:** This recommendation covers PHY and DLL specifications for NB-PLC systems operating over AC and DC electric power lines under 500 kHz frequencies. It is generally developed based on the G.9955 and G.9956 materials, and ITU-T G.9902 regularly references to recommendation G.9901.

TABLE 4.4 Several practical overhead cable examples utilized in PLC applications

No.	Cable length	Cable type	Cross sectional area	Standard specification	Voltage/ frequency range	Application type
1	20 km	ACSR [73]	150 mm^2	IEC 61089, ASTM B232, BS 215, DIN 48204	11–20 kV/3–95 kHz	Power distribution
2	7.46 m	XVB-F2 [74]	70 mm^2	TS HD 603 S1, NBN IEC 502-NAD	0.6–1 kV/2–24 MHz	Power distribution
3	11 m	FG7OR [75]	1.5–300 mm^2	CEI-UNEL 35375, CEI 20-13, CEI 20-22 II	0.6–1 kV/ 500 kHz-30 MHz	Power, control and signaling applications

4	40 m	YJLV [76]	240 mm^2	GB12706	8.7–10 kV/ 40 kHz–1.5 MHz	Power distribution
5	–	CYKY [77]	10–240 mm^2	IEC 60332-1; VDE 0482 T332-1-2, TP-KK-133-01	0.6–1 kV/10 kHz- 10 MHz	Power distribution
6	240 m	EPR [78]	240 mm^2	ICEAS-94-649	11–20 kV/3– 148.5 kHz	Power distribution
7	2.52 km	AAC [79]	185 mm^2	IEC 61089	11–20 kV/3– 148.5 kHz	Power distribution

- **ITU-T G.9903 "Narrowband OFDM Power Line Communication Transceivers for G3-PLC Networks"** [82]**:** This recommendation covers PHY and DLL specifications for G3-PLC systems operating over AC and DC electric power lines under 500 kHz frequencies. It is developed based on the G.9955 and G.9956 materials (especially using Annexes A and D of G.9955 and Annex A of G.9956). Similar to the ITU-T G.9902, ITU-T G.9903 regularly references to recommendation G.9901.
- **ITU-T G.9904 "Narrowband OFDM Power Line Communication Transceivers for PRIME Networks"** [83]**:** This recommendation covers PHY and DLL specifications for PRIME NB-PLC systems operating over AC and DC electric power lines in the CENELEC-A band. It is developed based on the G.9955, G.9956 and G.9956 Amd1 materials (especially using Annex B of G.9955, Annex B of G.9956 and G.9956 Amd1). Similar to the ITU-T G.9902 and ITU-T G.9903, ITU-T G.9904 regularly references to recommendation G.9901.
- **IEEE 1901.2 "IEEE Standard for Low-Frequency (less than 500 kHz) Narrowband Power Line Communications for Smart Grid Applications"** [84]**:** This standard that is developed by the IEEE 1901.2 Working Group includes PHY layer, MAC layer, coexistence and EMC requirements for NB-PLC systems operating over AC, DC or non-energized electric power lines under 500 kHz frequencies.

Even though PHY layer structure of these standards is similar, the upper layers have important differences when compared to each other. The following subsections introduce details of these standards.

4.4.1 ITU-T G.9902 G.hnem standard

This standard is developed for SG applications, plug-in electrical vehicles (PEV), smart metering and energy management applications. In addition to supporting the Internet Protocol v6 (IPv6) protocol, ITU-T G.9902 G.hnem standard [81] provides several specifications for PHY layer and DLL layer. In addition, this standard can operate with other network protocols by employing suitable sublayers. It has been specifically defined for the NB-PLC systems exploiting OFDM, and it is able to support both indoor and outdoor communications such as indoor and outdoor communications over LV lines, indoor and outdoor communications over MV lines, and among MV and LV power lines over transformers in urban and rural areas, and vice versa. The reference model of this standard [81] is illustrated in Fig. 4.5 (ITU 5.12/5.13). This reference model is composed of three reference points that are application interface (A), physical medium-independent interface (PMI) and medium dependent interface (MDI). The layers upper the A reference point are outside the scope of ITU-T G.9902.

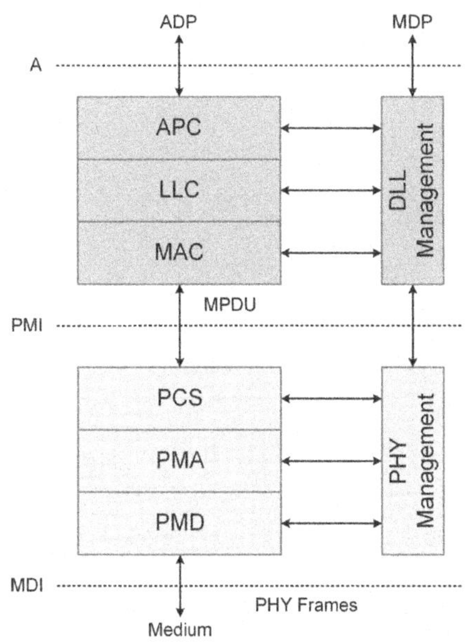

FIG. 4.5 Reference model of ITU-T Rec. G.9902 (Gh.nem). *(From Narrowband orthogonal frequency division multiplexing power line communication transceivers for ITU-T G.hnem networks, ITU-T Rec. G.9902, October 2012. [Online] Available: http://www.itu.int/rec/T-REC-G.9902 (Fig. 5-12), with permission.)*

The A reference point is a transition point among network layer and DLL layer, and its structure may change according to the selected network layer such as IPv6, Ethernet, and so on. Application protocol convergence (APC), which is a sublayer, is responsible to provide an interface for network layer. The logical link control (LLC) sublayer manages all connections with other nodes by taking into account the quality of service (QoS) requirements of each connection. The last sublayer of DLL is the MAC that organizes the access to the medium by using CSMA/CA algorithm. The PMI reference point is a transition stage between DLL and PHY layer. The PHY layer contains three sublayers called physical coding sublayer (PCS), physical medium attachment sublayer (PMA) and physical medium dependent sublayer (PMD). The PCS ensures bit-rate adaptation among PHY and MAC layers in addition to controlling and managing several processes in the PHY layer. The PMA sublayer maintains forward error correction (FEC) encoding and interleaving processes in the PHY layer. The last sublayer performs modulation and demodulation processes in this layer.

On the other hand, this standard also describes a management interface employed by upper layers to manage transactions of PHY and DLL layers. The ITU-T Rec. G.9902 (G.hnem) network topology is established in a logical domain structure in which each domain has specific sets of nodes related to the

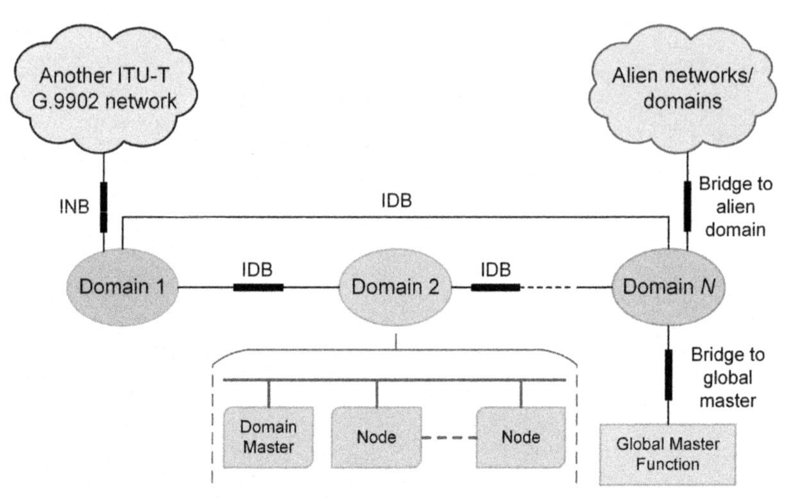

FIG. 4.6 Generic network architecture model of ITU-T Rec. G.9902 (Gh.nem). *(From Narrowband orthogonal frequency division multiplexing power line communication transceivers for ITU-T G.hnem networks, ITU-T Rec. G.9902, October 2012. [Online] Available: http://www.itu.int/rec/T-REC-G.9902 (Fig. 5-1), with permission.)*

domain. Each registered node in a domain is identified by its domain ID and node ID. Each domain has a domain master (DM) that is responsible for managing the operations of all other nodes. The DM also performs admission, resignation and other domain-wide operations. In the event of the node serving as DM fails, the DM mission is delivered to another node of the domain. Inter-domain bridges (IDB) provide the connection between different domains in the same network. It is worth noting that G.hnem domains can also connect to non-G.hnem domains that are called "alien" domains. A G.hnem network containing several interconnected domains are depicted in Fig. 4.6.

4.4.1.1 PHY layer

Even though the G.9902 can support different bandplans, it should necessarily support at least one bandplan. The CENELEC bands in the range of 3–148.5 kHz are considered as three bandplans: CENELEC-A (35.9375–90.625 kHz), CENELEC-B (98.4375–120.3125 kHz) and CENELEC-C/D (125–143.75 kHz). Similarly, the FCC band in the range of 9–490 kHz is taken into account as three bandplans that are FCC (34.375–478.125 kHz), FCC-1 (34.375–137.5 kHz) and FCC-2 (150–478.125 kHz). On the other hand, the ARIB bandplan employs 154.7–403.1 kHz frequency range.

The PHY transmitter of ITU-T G.9902 is based on the G3-PLC. This transmitter structure employs Reed-Solomon (RS) and convolutional coding based FEC systems. While the inner convolutional coding utilizes ½ rate and 7 constraint length, the outer RS encoder uses input blocks up to 239 bytes. An interleaving scheme with two-dimensional characteristic is exploited to prevent the occurrence of burst errors that may be arisen from time and frequency dependent noises. The mapping scheme of the ITU-T G.9902 is generally based on coherent quadrature amplitude modulation (QAM) modulations with 1, 2, 3 or 4 bits per carrier, and channel estimation process is coped with the channel estimation symbols located to the PHY frames.

4.4.1.2 MAC layer

The ITU-T G.9902 uses prioritized contention-based medium access that is a CSMA/CA algorithm with four priority levels. While the three lower levels (level 2, 1 and 0) are employed for user data frames, the highest priority (level 3) is exploited for beacons and emergency data transfers. In order to present more dynamic network management, management communications are performed via priority level 2 that is the highest level of application data priority. There is also an optional and synchronized medium access opportunity. If it is activated, the DM synchronously transmits network beacons. Each inter-beacon interval may contain a series of *contention-based* and *contention-free periods* as in IEEE 802.15.4 standard [85]. Since the MAC layer structure of the ITU-T G.9902 is similar to the IEEE 1901.2 standard, both layer 2 and layer 3 routing are also supported.

4.4.2 ITU G.9903 G3-PLC standard

In August 2009, Maxim Integrated Products, Inc. (United States) published the G3-PLC specifications as an open specification to meet the requirements of Electricité Réseau Distribuion France (ERDF) in the SG applications such as smart metering, AMI, home automation, HAN applications and further SG applications. In order to develop and introduce specifications, certification process and technology, several companies in the areas of utilities, system integrators, meter manufacturers and silicon vendors with the leadership of ERDF founded G3-PLC Alliance in 2011. The alliance standardized the specifications as ITU-T Rec. G.9903 at the end of 2012, and several updates were released in 2014, 2015 and 2017, respectively. This standard characterizes PHY, MAC and adaptation layers to authorize IP-based communication over LV and MV power grids. In order to cope with more destructive noise scenarios, this standard offers a variety of advanced properties such as adaptive tone mapping, two-dimensional interleaving and robust mode. In addition, it can enable communications between LV and MV power systems, and vice versa. Furthermore, the MAC layer structure developed for the IEEE 802.15.4 standard is adapted to work on the PHY layer. While IPv6 over Low power Wireless Personal Area Network (6LoWPAN) adaptation layer is also enabled to send IPv6 packets over power line channels, the layer 2.5 mesh routing protocol is enabled to detect best path among remote network nodes. Main technical parameters of ITU-T G.9903 G3-PLC standard are given in Table 4.5.

TABLE 4.5 Some characteristics of the ITU G.9903 G3-PLC standard

Frequency bands	ARIB (from 154.7 to 403.1 kHz) FCC-1 (from 154.6875 to 487.5 kHz) FCC-1.a (from 154.6875 to 262.5 kHz) FCC-1.b (from 304.687 to 487.5 kHz) CENELEC A (from 35.938 to 90.625 kHz)
Mapping scheme	DBPSK, DQPSK, or D8PSK mapping scheme based OFDM (BPSK, QPSK, 8-PSK or 16-QAM may be optionally utilized)
Data rate	Maximum 300 Kbps
Data link layer	IEEE 802.15.4 based MAC sublayer and IETF RFC 4944 based adaptation sublayer
Channel access	CSMA/CA with a random back-off time
Convergence layer	IPv6 6LoWPAN
Security	EAP-PSK, AES-128 key and CCM encryption

4.4.2.1 PHY layer

ITU G.9903 standard based systems can be arranged to support the FCC, CENELEC, ARIB bands in the frequency range between 10 and 490 kHz. Several PHY parameters of this standard for these bands are listed in the Table 4.6. Even though the differential modulation schemes are specified in the G3-PLC, differential encoding is preferred in the time direction. In addition, 16-ary QAM scheme can be optionally employed in the standard. The FEC of the standard is based on the combination of convolutional codes (CC) with ½ rate and RS code with (255,239) code length. Either four-times or six-times repetition is also exploited in robust transmission mode (called ROBO mode). Data rates of PHY layer may typically change from 4 to 200 Kbps [82].

The block diagram of PHY layer transmitter is illustrated in Fig. 4.7. The standard has three different operation modes that are called *normal*, *robust* and *super robust* modes. The super robust mode is defined for extremely destructive channel environments. As mentioned before, the two-dimensional interleaver is employed to protect systems against burst errors that may be occurred by time and frequency dependent noises. Different modulation schemes and optional coherent modulation techniques such as BPSK, QPSK, 8PSK and 16QAM can be selected as mapping method in this standard. In addition, there is an opportunity to deactivate subcarriers with low SNR for improving total transmission reliability.

TABLE 4.6 PHY layer parameters of the ITU G.9903 G3-PLC standard

Parameter	CENELEC A/B	FCC/ARIB
Sampling frequency (kHz)	400	1200
FFT size	256	256
Subcarrier spacing (kHz)	400/256 = 1.5625	1200/256 = 4.6875
Number of subcarriers	36 (CENELEC A) 16 (CENELEC B)	72 (FCC) 54 (ARIB)
Guard interval (samples/μs)	30 (CENELEC A) 75 (CENELEC B)	30 (FCC) 25 (ARIB)
Cyclic prefix (samples/μs)	22 (CENELEC A) 55 (CENELEC B)	22 (FCC) 18.33 (ARIB)
FCH modulation	DBPSK	BPSK
Interleaver	Per Reed-Solomon block	Per Reed-Solomon block
Preamble length (ms)	6.080	2.027
FCH length (symbols) of the data frames	13 (CENELEC A) 30 (CENELEC B)	12 (FCC) 16 (ARIB)
Max payload length (symbols)	252	511
FEC FCH (data frames)	Super robust mode: convolutional code (with ½ rate and 7 constraint length) + repetition coding by 6	
FEC payload	Normal mode: Reed-Solomon (with 255 block length, 239 message length and 8 correctable symbol errors) and convolutional code (with ½ rate and 7 constraint length) Robust mode: Reed-Solomon (with 255 block length, 245 message length and 8 correctable symbol errors) and convolutional code (with ½ rate and 7 constraint length) and repetition coding by 4	

4.4.2.2 MAC sublayer

The MAC layer of the G3-PLC was developed in accordance with IEEE 802.15.4-2006, and 6LoWPAN is employed to enable IPv6 in the standard. The channel access is achieved by means of CSMA/CA mechanism with a random back-off time. The random back-off time structure assists to decrease collision probability. If a node desires to transmit data frames, it has to wait for the

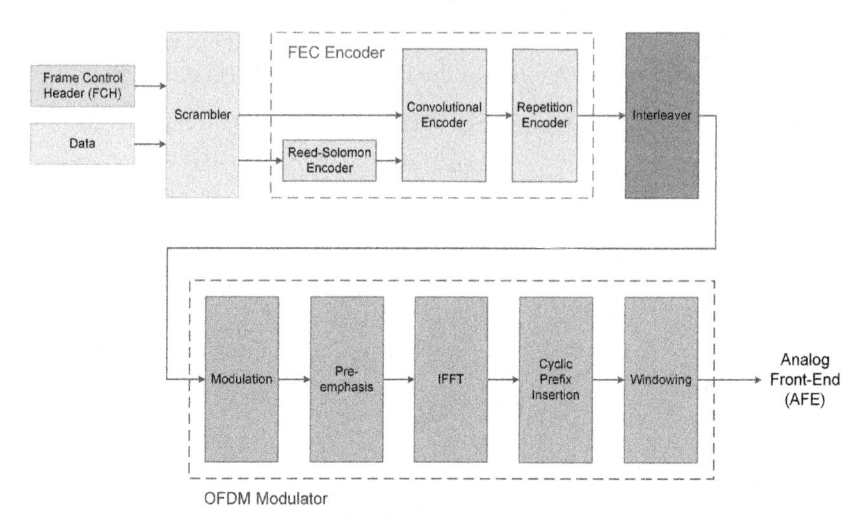

FIG. 4.7 PHY layer transmitter of ITU-T Rec. G.9903 (G3-PLC). *(From Narrowband orthogonal frequency division multiplexing power line communication transceivers for G3-PLC networks, ITU-T Rec. G.9903, August 2017. [Online] Available: http://www.itu.int/rec/T-REC-G.9903 (Fig. 7-1), with permission.)*

beginning of next contention period (CP) considering the message priority, and then random period starts. In the event of the channel is detected idle following the random back-off time, the node can convey its data. On the other hand, if the channel is busy following the random back off, the device has to wait for the next CP with respect to message priority, and then new random period is initiated for channel access process. The random back-off mechanism expands the time over which stations wishes to send through an additive decrease multiplicative increase (ADMI) back-off mechanism.

Two priority levels (high and normal) and two contention time windows that are depicted in Fig. 4.8 are enabled to define channel access prioritization. Three different slots are identified during the CP. The first slot is called contention-free slot (CFS) that is employed to convey next segments of MAC frame without exploiting back-off mechanism to avoid possible interruption from other nodes. The other slots are related to high-priority contention window (HPCW) and normal-priority contention window (NPCW). The messages are assigned by MAC layer into the one of these contention windows with respect to their priority.

In the standard, cyclic redundancy check (CRC) and automatic repeat request (ARQ) mechanism are used to provide delivery and integrity of messages. If a node obtains a MAC frame, it firstly controls the CRC of the message and then accepts the frame in the event of an accurate reception. After the accurate reception of a frame, an acknowledge (ACK) frame is transmitted. The accuracy of the received frame is analyzed by controlling the CRC. When a corruption is found, the receiver unit transmits a negative ACK (NACK) frame. In

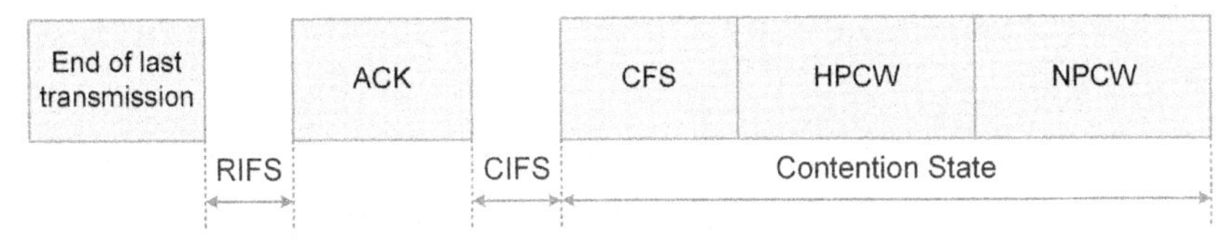

FIG. 4.8 Priority contention windows in ITU-T Rec. G.9903 (G3-PLC). *(From Narrowband orthogonal frequency division multiplexing power line communication transceivers for G3-PLC networks, ITU-T Rec. G.9903, August 2017. [Online] Available: http://www.itu.int/rec/T-REC-G.9903 (Fig. 9-4), with permission.)*

addition, the packet originator looks forward to a certain time (macAckWait-Duration) for the reception of an acknowledgement. In the lack of an acknowledgement or the reception of an NACK, it is assumed that the transmission was unsuccessful and retransmission process is performed in the appropriate contention period. The "macMaxFrameRetries" MAC parameter defines the maximum number of retransmissions.

Security and data protection structure of the G3-PLC is developed according to the MAC security suite of IEEE 802.15.4 clause 7.6 that employs an encryption mechanism depending symmetric keys ensured by higher layers. All nodes of a Personal Area Network (PAN) distributes a group master key (GMK) that is employed to encrypt all MAC data messages in the event of security activated. The GMK is shared with all nodes during the bootstrap process.

4.4.2.3 Adaptation layer

In the ITU G.9903 G3-PLC Standard, the 6LoWPAN has been selected to adapt IPv6 into the PLC. It also combines the layer 2 routing, fragmentation, header compression and security. Furthermore, it can supply a network protocol usage for transport protocols such as TCP, UDP and ICMPv6.

4.4.3 ITU-T G.9904 PRIME standard

In 2009, PRIME (PoweRline Intelligent Metering Evolution) Alliance was founded for the purpose of improving and employing open and standardized solutions for smart metering and SG applications under the leadership of Spanish Iberdrola [86]. The PRIME Alliance aims to develop specifications for OFDM based NB-PLC systems operating in the CENELEC-A band. The fundamental application area of the PRIME is the communications among the smart meters located in the LV power lines and the base node that is generally located on the data concentrators established in transformer stations. In 2012, PRIME version 1.3.6 was standardized by ITU with title of ITU-T Rec. G.9904 (PRIME) [83].

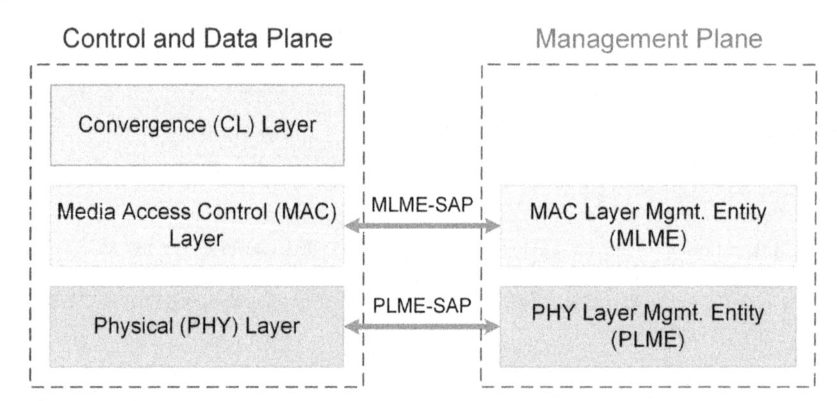

FIG. 4.9 Reference protocol layer model for ITU-T Rec. G.9904. *(From Narrowband orthogonal frequency division multiplexing power line communication transceivers for PRIME networks, ITU-T Rec. G.9904, October 2012. [Online] Available: http://www.itu.int/rec/T-REC-G.9904 (Fig. 4-1), with permission.)*

The reference model of PRIME standard is depicted in Fig. 4.9. The convergence layer (CL) manages classifying the traffic that is associated with the appropriate MAC connection. In addition, this layer also exploits the mapping of all traffic to be covered by MAC service data units (MSDUs). Several Service Specific Convergence Sublayers (SSCSs) are created to host several types of traffic in the MSDUs. The MAC layer ensures important transactions such as fundamental MAC operations, connection construction and maintenance, bandwidth allocation and topology resolution. On the other hand, the PHY layer sends and receives the MPDUs among neighbor nodes by employing OFDM.

Typically, a PRIME system consists of several subnetworks where each subnetwork is characterized in the scope of a transformer station. A subnetwork can be modeled as a tree structure that contains two kinds of nodes called "base node" and "service nodes". The base node is also called as "master node" that administers resources and connections of subnetworks through beacon signal transmitted periodically. In addition, the base node has a charge of PLC channel access, and each network can only contain one base node. While the base nodes are considered as the root of tree structure, the service nodes are considered as leaves or branch points of the tree structure. The service nodes are firstly in disconnected mode and should perform a registration process to be a part of the network. They take over two duties. One of them is to keep connectivity to the other nodes in the network while the second is the switching the data to expand connectivity.

4.4.3.1 PHY layer

The PHY layer of this standard is based on the OFDM method and serves in CENELEC-A band covered between 3 and 95 kHz. A typical block diagram of the PHY layer transmitter is illustrated in Fig. 4.10. The existing spectrum

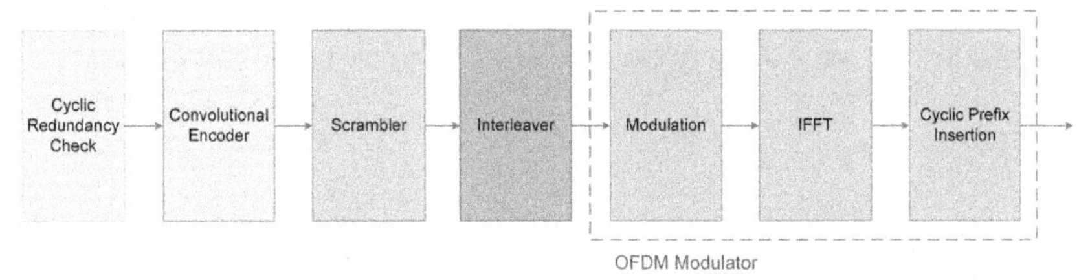

FIG. 4.10 Physical layer transmitter model of ITU-T Rec. G.9904. *(From Narrowband orthogonal frequency division multiplexing power line communication transceivers for PRIME networks, ITU-T Rec. G.9904, October 2012. [Online] Available: http://www.itu.int/rec/T-REC-G.9904 (Fig. 7-1), with permission.)*

is assigned to subcarriers where the first subcarrier is centered at 42 kHz and the last one is at 89 kHz. The sampling frequency is set to $f_S = 250$ kHz while the FFT size is chosen as 512. Thus, the space among OFDM carriers is 488 Hz. In addition, 96 subcarriers and 1 pilot subcarrier are used in the PHY transmitters of this standard. Furthermore, the transmitters contain a CRC detection code that are exploited to confirm wholeness of the MAC service data unit (SDU). The FEC may be optionally performed by employing CC codes with rate $R = 1/2$ and constraint length $K = 7$. After the optional FEC encoder, the transmitter contains a mandatory scrambler that performs random selections for bit streams in order to decrease peak values at the IFFT output. Afterwards, the data are fed to a block interleaver for randomizing the circumstances of bit errors before decoding. Three differential modulation schemes such as differential binary, quaternary, and eight-ary PSK (DBPSK, DQPSK, and D8PSK) can be used in this standard. Therefore, PHY layer data rate may be between 21.4 and 128.6 Kbps. The cyclic prefix composed of 48 samples is periodically included into the OFDM symbols at the IFFT output. It is worth noting that the standard presents an opportunity for the use of ARQ in order to overcome unpredictable impulsive noise cases.

The structure of a PHY layer frame is shown in Fig. 4.11 where a chirp signal is employed as preamble with duration of 2048 μs. In preamble, a constant envelope signal is exploited in lieu of OFDM symbols for maximizing the energy. In addition, the use of preamble helps to provide frame detection and automatic control gain adjustment in an efficient way. After the preamble, one header with two OFDM symbols is included to the frame in which the header is modulated by DBPSK scheme to transmit data related to payload length, modulation, FEC, MAC header and CRC.

The OFDM symbols of PHY header contain 84 data subcarriers and 13 pilot subcarriers that are included to the header for generating phase reference needed for channel estimation process. Only one carrier is exploited as pilot subcarrier in the time of payload where 96 subcarriers are employed for data transmission. The main parameters of PHY layer of PRIME standard are listed in Table 4.7.

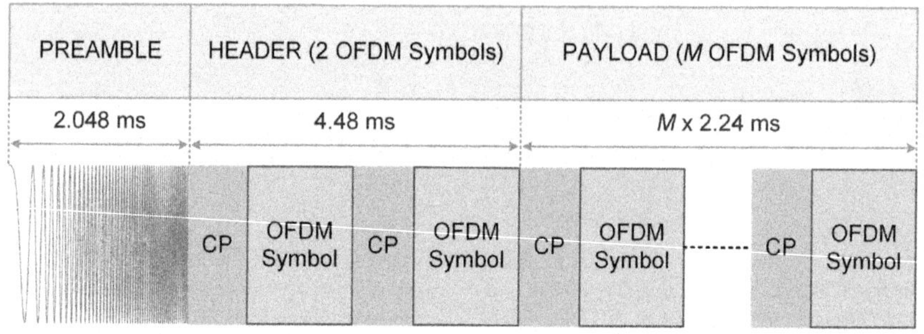

FIG. 4.11 A PHY frame example for ITU-T Rec. G.9904. *(From Narrowband orthogonal frequency division multiplexing power line communication transceivers for PRIME networks, ITU-T Rec. G.9904, October 2012. [Online] Available: http://www.itu.int/rec/T-REC-G.9904 (Fig. 7-2), with permission.)*

TABLE 4.7 PHY layer parameters of the ITU G.9904 PRIME standard

Parameter	Value
Sampling frequency (kHz)	250
FFT size	512
Carrier spacing (Hz)	≈488
Number of subcarriers	97
Used frequency band (kHz)	42–89
Pilot carrier number in the PHY header	13
Cyclic prefix (μs)	192
Mapping schemes	DBPSK, DQPSK, D8PSK
Forward error correction	Convolutional code (with ½ rate and 7 constraint length)
Interleaver	Per symbol
Preamble length (ms)	2.048
Header length (ms)	4.48
Max payload length (ms)	63 × 2.24 = 141.12

4.4.3.2 MAC layer

The MAC frames are used by the G.9904 devices for accessing the channel, which are defined depending on the mixed time intervals. A typical MAC frame structure is illustrated in Fig. 4.12. The channel access method in MAC layer is characterized employing both CSMA/CA and time-division multiplexing (TDM). The nodes on a network are able to access channel through shared

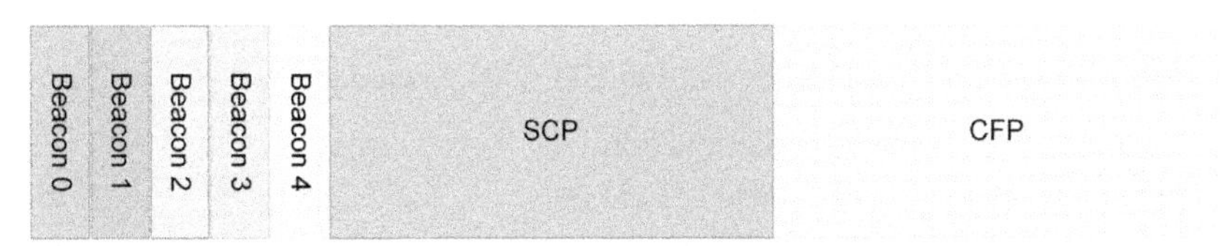

FIG. 4.12 A MAC frame structure example for ITU-T Rec. G.9904. *(From Narrowband orthogonal frequency division multiplexing power line communication transceivers for PRIME networks, ITU-T Rec. G.9904, October 2012. [Online] Available: http://www.itu.int/rec/T-REC-G.9904 (Fig. 8-7), with permission.)*

contention period (SCP) or contention-free period (CFP). Devices need to demand for allocation from base node in the CFP period. The base node either approves or rejects the request according to the channel status. In the SCP period, there is no an agreement requirement. Nonetheless, devices should consider the timing limits of the SCP period. Typically, a MAC frame consists of one or more beacons, one SCP and one or no CFP.

The ARQ is optionally provided in MAC layer of the standard in order to accomplish a safe data transmission. Furthermore, the standard supports two types of security operations. While the first profile security is performed in the upper layers, the second one is achieved by using 128 bit Advanced Encryption Standard (AES) encryption.

4.4.3.3 Convergence layers

The CL provides an adaptation layer between upper protocol layers and MAC layer. The main function of the CL is the acquiring traffic from upper supported layers and summarizing them to the SDUs. The CL is composed of two sublayers that are Common Part Convergence Sublayer (CPCS) and Service Specific Convergence Sublayer (SSCS). While the CPCS supports many general services, the SSCS supports certain services. The standard defines four SSCS that are Internet Protocol version 4 (IPv4), IPv6, IEC 61334-4-32 and NULL SSCS. The IEC 61334-4-32 is the most widely used SSCS method for the AMI applications. The most important service of CPCS is the segmentation and reassembly (SAR).

4.4.4 The IEEE 1901.2 standard

The IEEE Standards Association confirmed IEEE P1901.2 standard in 2013 [84]. The IEEE P1901.2 standard describes the PHY and MAC layers for HDR NB-PLC systems working on the frequencies below 500 kHz. This standard is generally depending on the ITU-T Rec. G.9903 (G3-PLC). In addition, this standard contains several annexes where a channel model is suggested in Annex D, and an EMC measurement method with specific limits is presented in Annex E.

TABLE 4.8 Some characteristics of the IEEE 1901.2 standard

Frequency bands	ARIB (from 155 to 403 kHz) CENELEC A (from 35 to 91 kHz) CENELEC B (from 98 to 122 kHz) FCC (from 10 to 487.5 kHz) FCC above CENELEC (from 155 to 488 kHz)
Mapping scheme	DBPSK, DQPSK, or D8PSK mapping scheme based OFDM (BPSK, QPSK, 8-PSK or 16-QAM may be optionally utilized)
Data rate	Maximum 300 Kbps
Data link layer	IEEE 802.15.4 based MAC sublayer and IETF RFC 4944 based adaptation sublayer
Channel access	CSMA/CA with a random back-off time
Convergence layer	IPv6 6LoWPAN
Security	AES-128 key in CCM encryption mode

Since the IEEE 1901.2 PLC standard was developed based on the G3-PLC standard, there exist many similarities among these standards. Nonetheless, the main difference between IEEE 1901.2 and G3-PLC is shown in MAC technology where the MAC technology of IEEE 2901.2 supports both route-over and mesh-under routing methods, and it can support more IPv6 functionalities. Main features of the IEEE 1901.2 standard are listed in Table 4.8.

The reference model of this standard contains data and management planes where data plane is composed of PHY and DLL layers that contains only the MAC sublayer. The management transactions are maintained by MAC sublayer management entity (MLME) and PHY layer management entity (PLME) in the standard.

4.4.4.1 PHY layer

The PHY layer of the standard is composed of OFDM containing channel-coding methods. According to the employed frequencies, minimum sampling frequency is $f_S = 1.2$ MHz while maximum number of carriers is $N_{carrier} = 128$. In addition, maximum PHY data rate of IEEE P1901.2 standard reaches up to 500 kbps. Since the PHY layer signal processing is very similar with ITU-T G.9903 (G3-PLC), only differences will be introduced as follows [30, 61].

- The frame control header (FCH) bits are mapped through BPSK for all bands.

- There are two options for number of preamble symbols, which are 9.5 and 13.5.
- Two different guard interval (GI) may be used in this standard. The first one is called standard GI with 30 samples that is the same as used in G3-PLC. The second is optional one with 52 samples, which can be exploited in FCC and ARIB bandplans.
- A super robust mode is available for the payload where FEC is performed by using RS code, CC and repetition code. This mode is optional for CENELEC bands while mandatory for others.

4.4.4.2 MAC layer

Similar to the ITU-T Rec. G.9903 (G3-PLC), the IEEE 1902.1 MAC layer employs services of IEEE 802.15.4-2006. Both standards use the same definitions for medium access, interframe spacing structure, ARQ procedure, tone map negotiation and segmentation of frames. The one remarkable difference among these standards is related to the optional multi-tone mask mode where it can operate in both beacon-enabled and non-beacon enabled modes [30, 61, 87]. The network model of the standard can be reduced to two devices as,

- A PAN coordinator that is the network coordinator can be taken into account as a master node.
- An IEEE 1901.2 device that includes an application of IEEE 1901.2-2013 MAC.

In addition, the network can be established in star, tree or mesh topology. Nonetheless, each network should have one IEEE 1901.2 device to serve as the network coordinator. The access to physical medium is realized via CSMA/CA protocol.

4.5 EMC regulations for PLC systems

The PLC system operates together, not in an isolated environment. Especially, the power lines are principally constructed to feed electrical loads, devices and home appliances instead of transmitting PLC signals. Thus, it is evident that PLC devices should obey regulations that aim to control level of signals connected to the power lines. In addition, PLC systems should comply with radiated emission limits [28, 88]. Since EMC regulations aim to exist together with other communication systems, they restrict frequency band and signal levels inserted to power lines by the PLC systems. This aim can be accomplished by restricting both radiated and conducted emissions. Radiated emissions occur owing to the common mode (CM) currents that is also called as antenna currents. Differential mode currents occur because of the fact that the PLC signals are injected between two power lines. At the same time, CM currents also emerge owing to the irregularities of the network [30, 89]. While the CM currents flow in one direction, differential mode currents flow in opposite directions. Therefore, the CM currents may cause important EMI problems. The level of radiated EMI generated by the CM may be crucial if the length of radiating source is in the level that can be compared with the wavelength of the transmitted signal. Therefore, radiated emissions are critical only for the BB-PLC systems.

4.5.1 NB PLC regulations

As stated in Section 4.2, the NB-PLC systems exploit frequencies between 3 and 500 kHz. The European Norm (EN) 50065 that encouraged the advance in both LDR and HDR NB-PLC was released by CENELEC in 1991 [29]. The EN 50056 characterizes transmission limits and procedures for measurements in the range of 3–148.5 kHz frequencies, and it also defines four CENELEC bands. Frequency ranges and transmission limits of CENELEC bands are illustrated in Fig. 4.13 according to the EN 50056. The transmission limits should be taken into account by considering measurement procedures explained in the EN 50065. While the CENELEC-A band is assigned for utilities, the others are assigned for consumers. In addition, devices to be operated in CENELEC-C band have to support CSMA/CA protocol that necessitates transmission and carrier sensing process in the range of 131.5 and 133.5 kHz, 125 ms minimum pausing period and 1 s maximum channel holding features [87, 90].

The EMC regulations for PLC emissions is arranged by the FCC Code of Federal Regulations, Title 47, Part 15 (47 CFR §15) in the United States [31]. While Section 15.3 clause (t) of [31] identifies "power line carrier" systems, Section 15.113 points out that power utilities can exploit 9–490 kHz band for PLC systems without protection and interference. In Japan, the frequencies between 10 and 450 kHz are assigned for PLC transmission by the ARIB. This standard also explains transmission limits and related measurement methods [91].

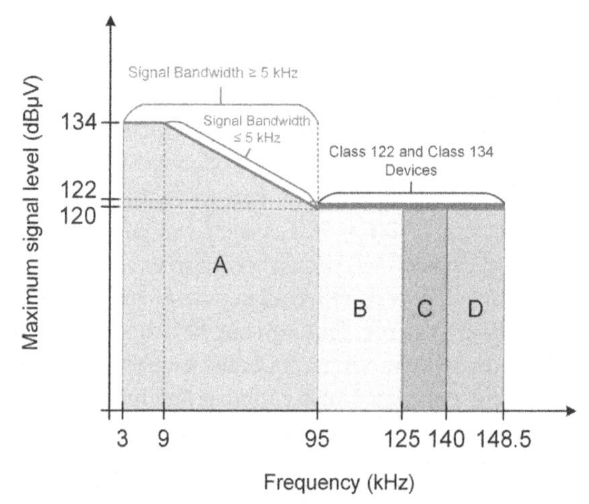

FIG. 4.13 Signal transmission levels in CENELEC bands.

4.5.2 BB PLC regulations

In Europe, EN 50561-1 standard [92] was accepted in 2012, which describes transmission limits and measurement methods for indoor BB-PLC systems operating over LV power networks in the frequency range 1.6065–30 MHz. This standard separately considers AC mains power, PLC ports, and telecommunication networks. Following the EN 50561-1, the EN 50561-3 is released for frequencies between 30 and 87.5 MHz in 2016 [93]. The EN 50561-1 requires a cognitive frequency exclusion operation that constitutes several notches in the pass band if a broadcast service is determined. The EN 50561-2 aims to present specifications for outdoor BB-PLC but it is not yet completed.

In the United States, the Code of Federal Regulations, Title 47, Part 15, arranges both indoor and outdoor BB-PLC in the range of 1.705–80 MHz. In addition, it specifies limits for both conducted and radiated emissions. The particular radiation limits are presented for PLC systems in two parts. The first one covers frequencies between 1.705 and 30 MHz while the second one is for the frequencies between 30 and 88 MHz. The conversion of radiation limits into the PSD values is performed through a coupling factor that may depend on many parameters such as wiring, topology, and characteristics of power grid.

The BB PLC systems in Japan are only applicable for in-home applications. Therefore, the BB-PLC systems are designed to operate frequencies between 2 and 30 MHz in Japan [94]. On the other hand, emission limits of common mode currents are defined in two parts where the first one is valid for 2–15 MHz and the second one is for 15–30 MHz bands.

4.6 Case study PLC applications for smart grids

In order to provide case study examples of PLC applications, this section presents several outdoor PLC applications performed for remote monitoring of renewable energy sources (RESs). The section introduces two PLC applications reported by the authors of the book [5, 8]. The observing, metering and management processes are essential for renewable energy systems similar to the SG applications of conventional power grids. In the first case study, remote monitoring requirements of the RESs are investigated for solar microgrid systems [8]. The aim of this case study is to eliminate monitoring costs since the power lines are not only used to deliver electricity, but also are exploited to transmit several parameters of loads located at the back-end of the modeled solar microgrid. The block diagram of designed microgrid structure containing generation, energy conversion, transmission, distribution and monitoring sections is depicted in Fig. 4.14. The DC-AC energy conversion system of this model is composed of solar power plant and multilevel inverter (MLI) system. While the solar plant contains three separate solar plants with perturb and observe maximum power point tracking (P&O MPPT) system, the multilevel

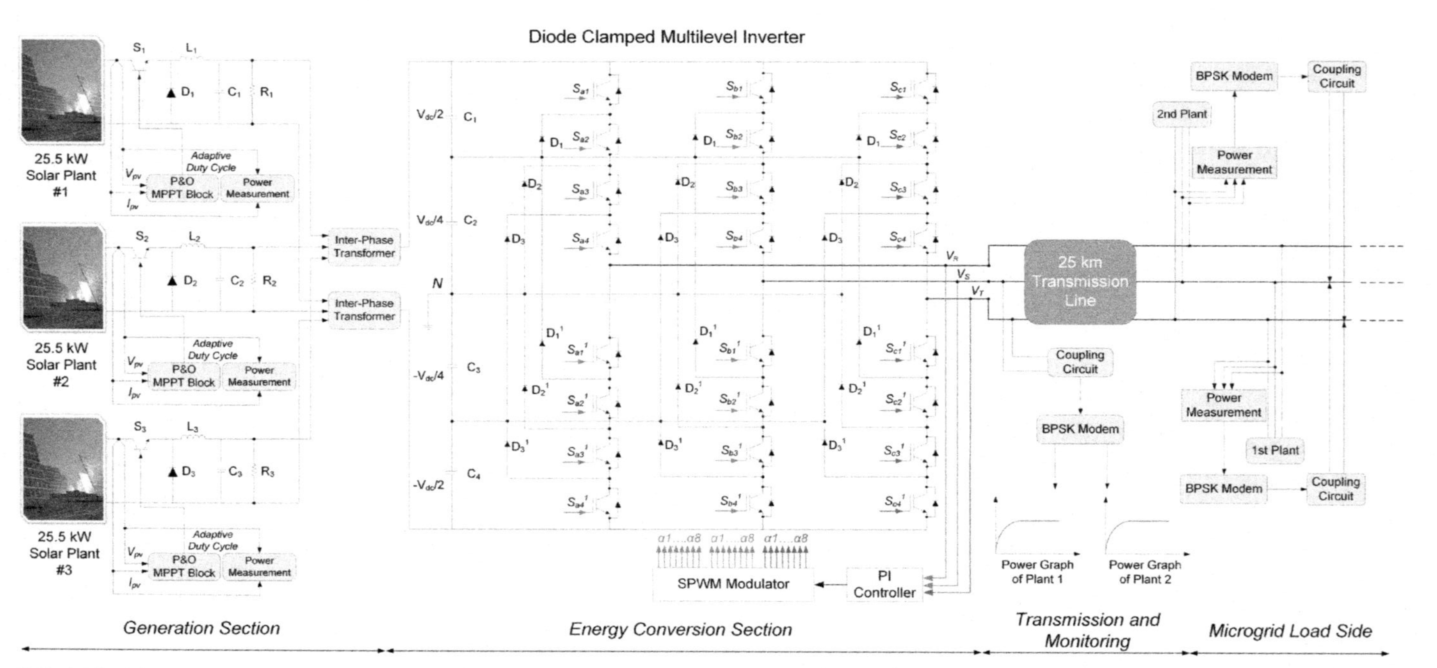

FIG. 4.14 The block diagram of the designed solar microgrid system.

inverter is employed to generate three-phase AC line voltages. PI regulator and phase disposition sinusoidal pulse width modulation (PD-SPWM) algorithm are employed to manage the MLI. Each solar plant contains 150 photovoltaic (PV) modules where each PV module with 170 W power are designed with respect to Sharp NE-170UC1 PV panel [95]. The designed buck converters to feed diode clamped MLI system regulate the generated electricity by the solar plants.

The buck converters are managed by employing P&O MPPT algorithm that provides to obtain maximum power and regulated DC output voltage. The transmission and distribution section of designed microgrid is modeled by using a 25 km transmission line with realistic impedance values, and the parameters of utilized transmission line are listed in Table 4.9. This transmission line is also used as a PLC channel environment. The distribution line part of the design is created with two load plants whose power rates are 1500 and 2500 W. The PLC communication infrastructure of the designed remote monitoring system is based on designed BPSK modems that are placed in different locations. The power consumption rates of the loads are measured and are transmitted by employing the designed modems.

The designed solar microgrid system contains three-phase PI section transmission line whose channel characteristics are firstly determined by performed simulation studies since it will be utilized as a communication medium as well as power transmission. As stated earlier, two different loads that are formed as 1500 and 2500 W rated powers are taken into account in this microgrid. Each load is observed by modeled BPSK modems operating on different frequencies to transmit measured parameters. Different carrier configurations are performed to modulate multichannel input data where the carrier frequencies of modems are selected as 6 and 8 kHz. There are two different reasons related to employing modems with dissimilar carrier frequencies. While the first reason is regarding adjacent channel interference problem, the other is related to decreasing corruptive effects of power lines. It is worth noting that these frequencies can be rearranged if the number of loads increases. Fig. 4.15A depicts block diagram of the BPSK modem while modulator and demodulator structures of

TABLE 4.9 Parameters of employed transmission line

Unit	Value
Length	25 km
Frequency	50 Hz
Resistance	0.2568 Ω/km
Inductance	4×10^{-7}H/km
Capacitance	8.6×10^{-9} F/km

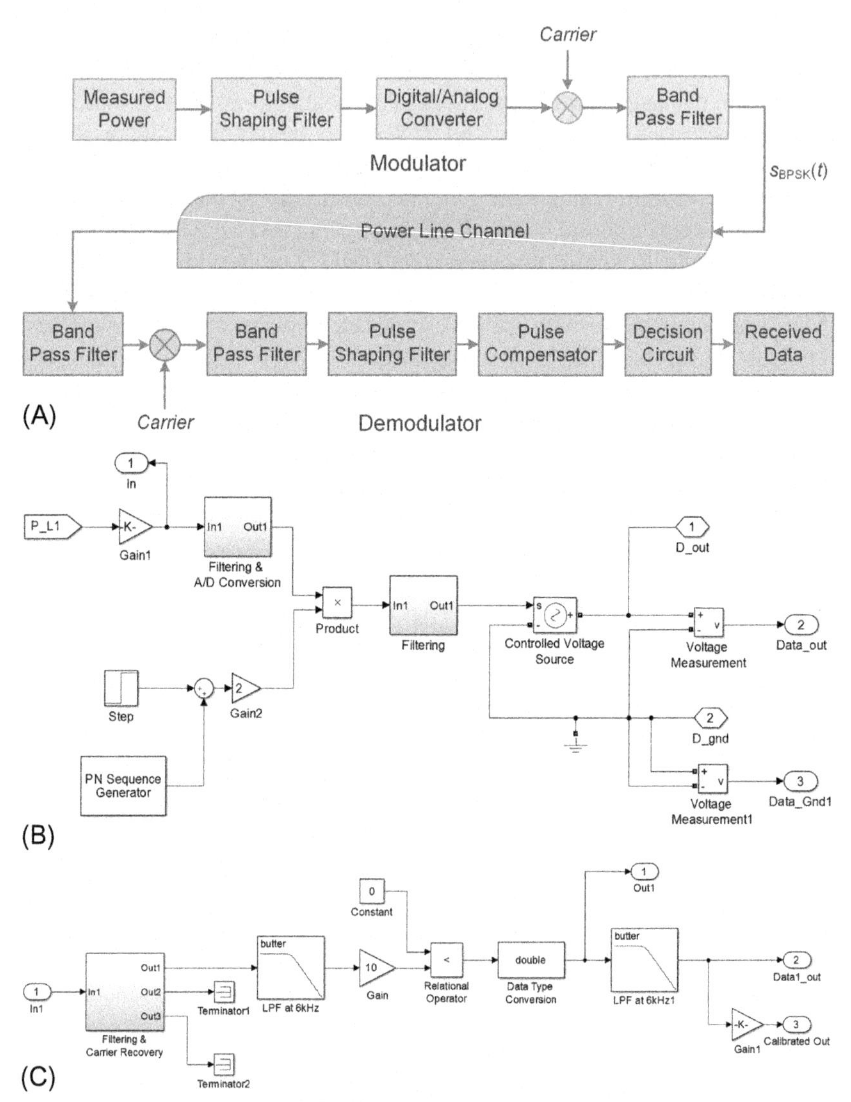

FIG. 4.15 PLC infrastructure of designed remote monitoring system; (A) Block diagram of PLC modem based on BPSK, (B) BPSK modulator in Matlab/Simulink, (C) BPSK demodulator in Matlab/Simulink.

the modem in Matlab/Simulink are illustrated for one channel in Fig. 4.15B and C, respectively.

Fig. 4.16A shows line voltage total harmonic distortion (THD) analysis result for unloaded case of the MLI. The THD ratio for this case is calculated as 0.04% in which all baseband and carrier harmonics are nearly removed. The THD of line voltage analysis for fully loaded case is depicted in Fig. 4.16B where the MLI is loaded

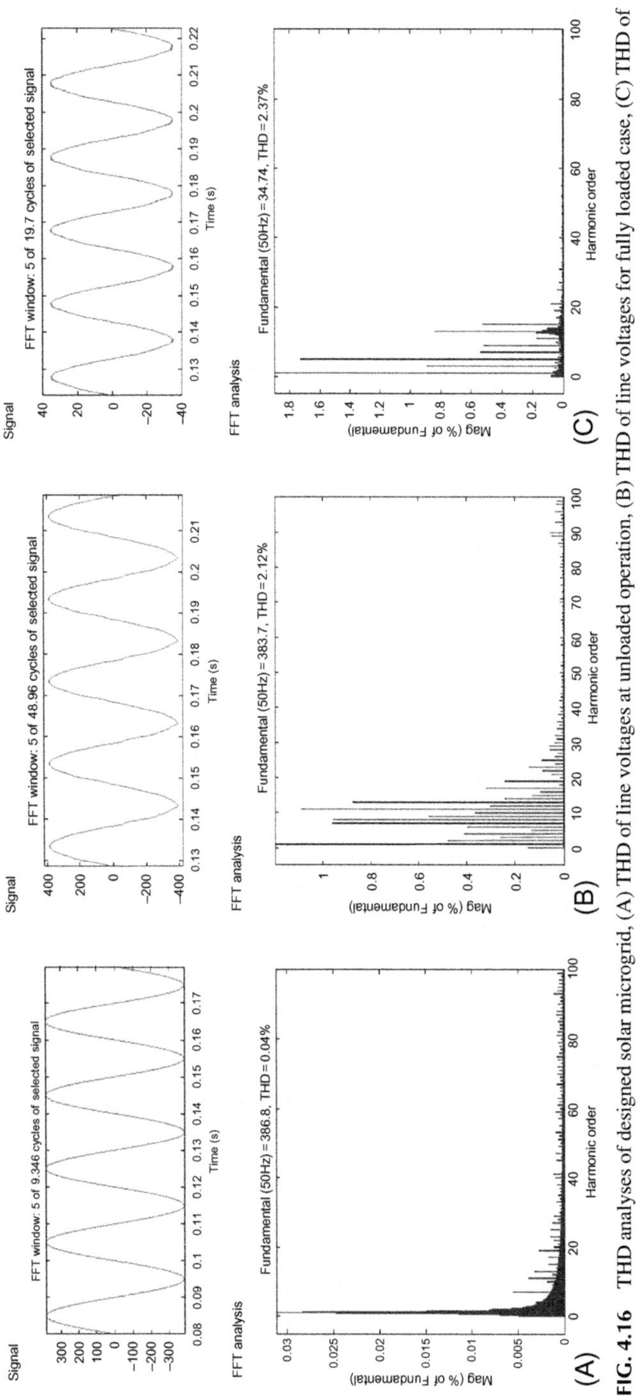

FIG. 4.16 THD analyses of designed solar microgrid, (A) THD of line voltages at unloaded operation, (B) THD of line voltages for fully loaded case, (C) THD of line currents when loaded with 13 kW.

through 75 kW resistive and inductive load. The fifth, seventh, ninth and eleventh harmonics, which are supposed as the most important harmonic orders, are observed lower than 1% for this case. The third analysis shows that the THD of line currents is calculated as 2.37% as can be seen from Fig. 4.16C when the MLI is loaded with 13 kW resistive and inductive load.

Acquired performance analysis results for the observed loads are illustrated in Figs. 4.17 and 4.18, respectively. Each load plant comprises a BPSK modem that is able to quantize applied power consumption rate to its input, and modulates this data to convey over transmission lines. The first axis of Figs. 4.17 and 4.18 shows the measured power consumptions of loads while the second axis depicts modulating data that are obtained by attenuating the measured power data at a rate of 1:1000. The first plant is modulated via a carrier signal with 8 kHz frequency while the second is modulated via a carrier signal with 6 kHz frequency. The filtering analyses are conducted for both different filter order and cut-off frequency values where 400 Hz with second order, 200 Hz with second order and 50 Hz with fourth order cases are taken into account as can be seen from the results.

When the presented analysis results are considered, it is evident that the designed remote monitoring system for solar microgrids can be exploited for observing the RESs located in different places in an efficient way. Moreover, it is important to note that this monitoring system eliminates installation costs, due to using power lines as a communication medium, when compared to other monitoring systems based on SCADA, wireless communications or in any Ethernet-based systems. Besides, this system can be simply upgraded for increased number of the load plants.

The second case study focuses on a multi-channel PLC infrastructure that is particularly modeled for monitoring a hybrid RESs constituted with solar and wind sources. The wind turbine structure of the modeled system is based on a permanent magnet synchronous generator (PMSG). Six-pulse uncontrolled rectifier and DC-DC buck converter are used to connect wind energy conversion system (WECS) to the DC bus. The solar plant of the designed test bed is created with eight PV panels that are similar panels explained in the first case study. The employed buck converter in the design is managed with P&O MPPT algorithm while the inverter part of the design is modeled with SPWM method. The designed system aims to carry several measurement data obtained from the hybrid RESs. The voltage, current and power rates of each energy plant are separately measured and are conveyed to monitoring center thanks to the designed PLC modems that are based on QPSK mapping scheme. A modulated signal that is composed of six-channel measurement data is conveyed over transmission lines to the energy monitoring center. Transmission line length is adjusted as 25 km with realistic parameters given in Table 4.9. The schematic diagram of the modeled hybrid RES system is shown in Fig. 4.19. This system is modeled with a DC coupled configuration where several RESs may be connected to a DC bus over DC-DC converters or AC-DC rectifiers.

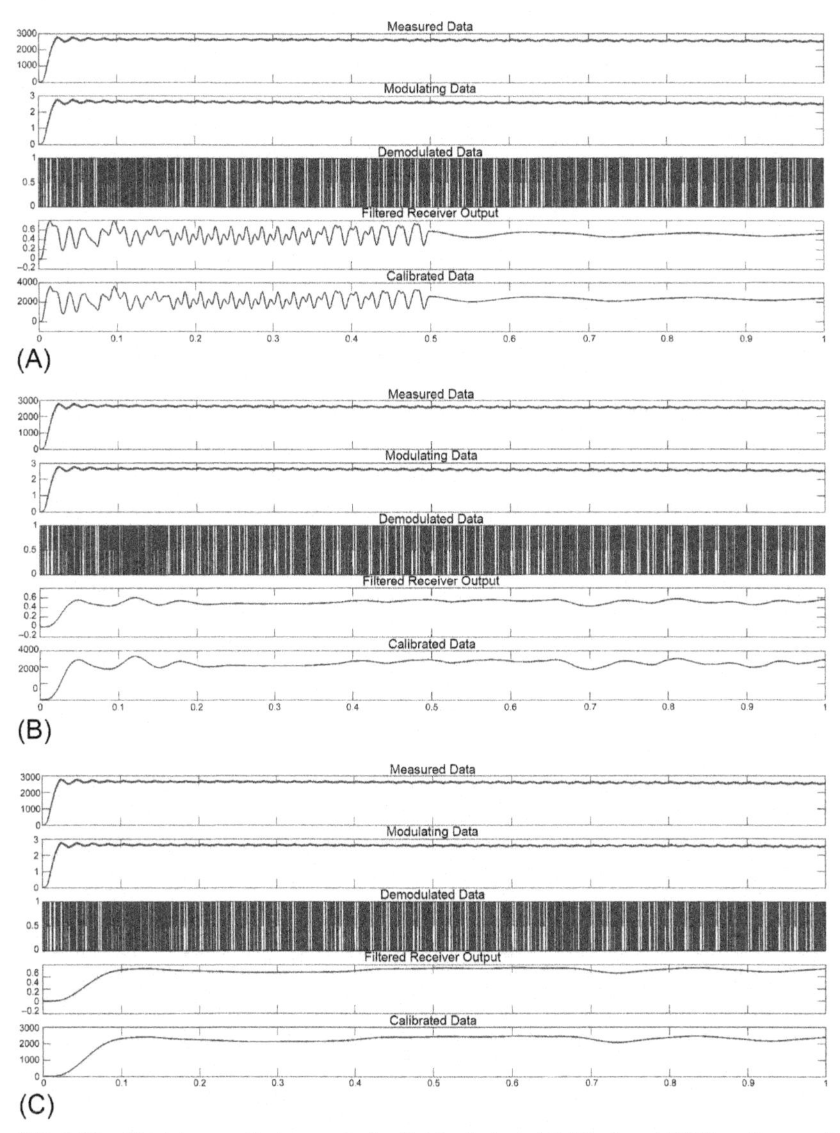

FIG. 4.17 Obtaining monitoring results for first load plant, (A) filtering at 400 Hz with second order, (B) filtering at 200 Hz with second order, (C) filtering at 50 Hz with fourth order.

The schematic and circuit diagram of the designed systems is illustrated in Fig. 4.20 that can be classified into four categories as WECS, solar plant, DC-AC conversion system and PLC system. While the PLC section is depicted with blue boxes, the others are from top to below, respectively.

The wind turbine of the modeled system is established by considering technical properties of Hummer wind turbine with 3 kW power, and cut-off speed value is

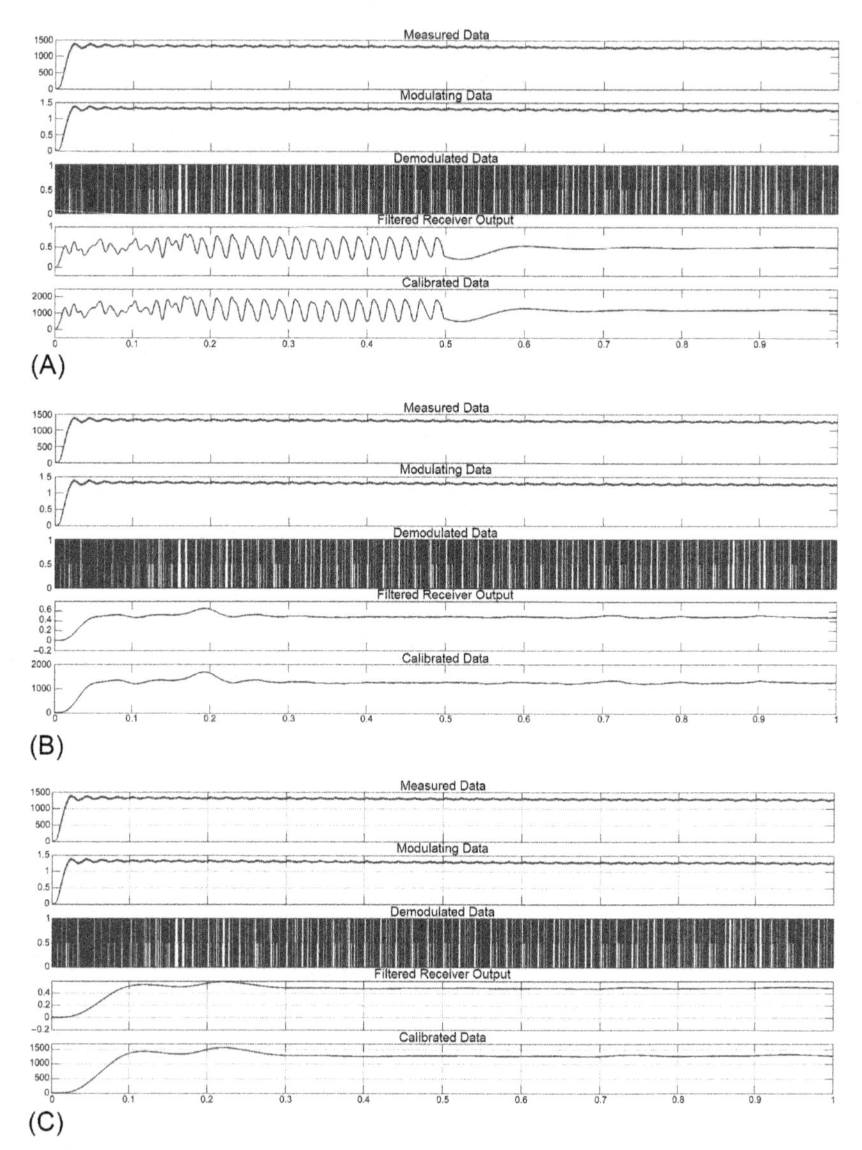

FIG. 4.18 Obtaining monitoring results for second load plant, (A) filtering at 400 Hz with second order, (B) filtering at 200 Hz with second order, (C) filtering at 50 Hz with fourth order.

adjusted as 20 m/s according to the technical parameters of this turbine model. The outputs of WECS in terms of voltage, current and power signals are illustrated in Fig. 4.21 in which the output voltage is fixed by the used PI controller in a very short time as much as 0.04 s. The DC output voltage of buck converter is fed to the DC coupling bus over inductive loads as can be seen from the Fig. 4.20.

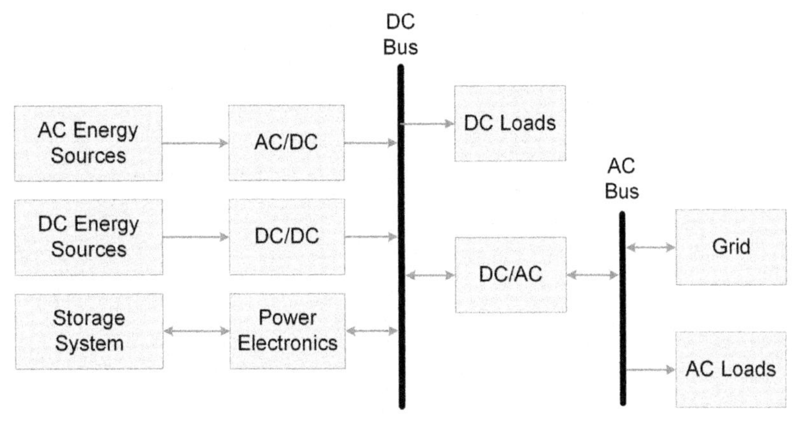

FIG. 4.19 Schematic diagram of designed hybrid RES system with DC coupling.

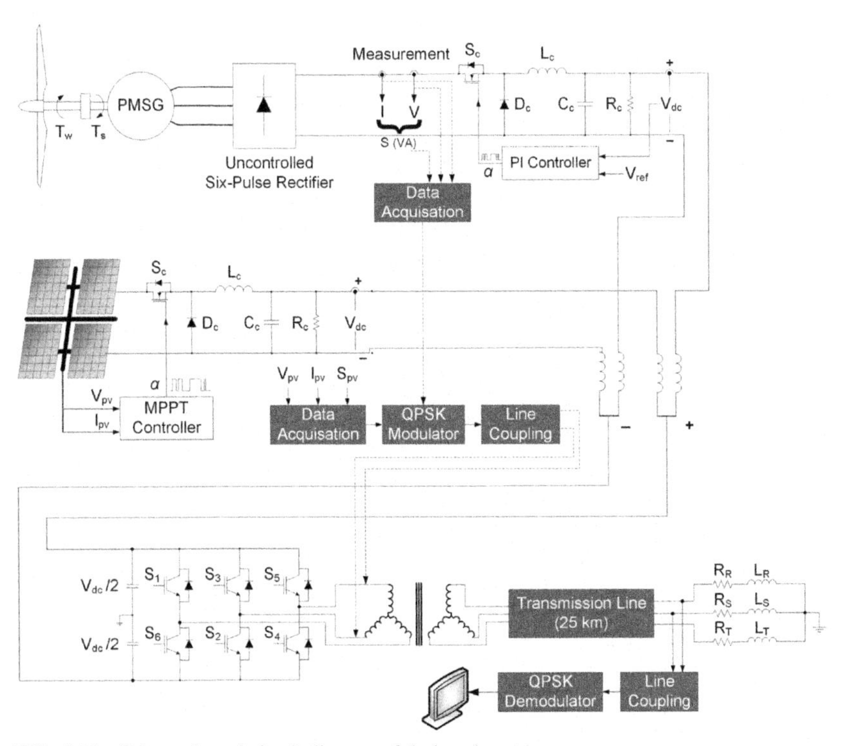

FIG. 4.20 Schematic and circuit diagram of designed system.

The characteristic analyses of PV modules are tested by changing irradiation level in the range of 200 and 1000 W/m^2. The acquired curves are also confirmed by comparing to the reference model of Sharp that is the same PV model

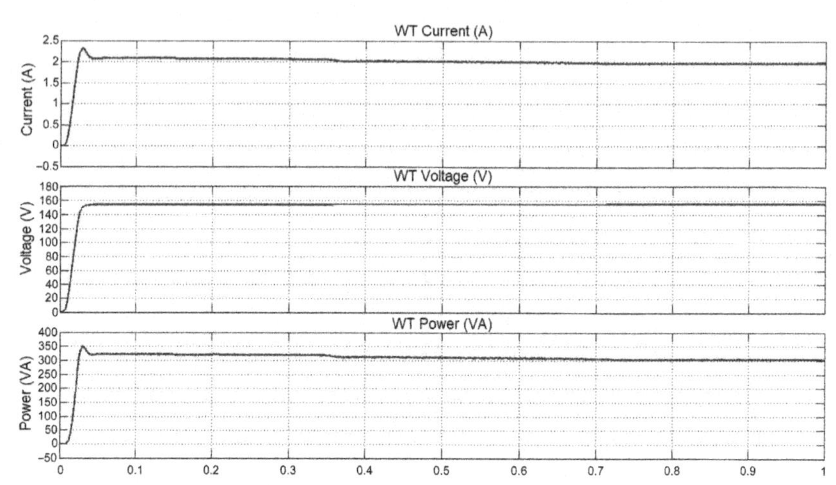

FIG. 4.21 Obtained output waveforms for the WECS.

exploited in the first case study [95]. The power output of solar plant versus changing irradiation values is shown in Fig. 4.22. Even though the changing irradiation leads to fluctuations on the PV voltage output, the designed buck converter overcomes this issue with respect to employed MPPT algorithm and the output can be fixed to desired power level.

The separate voltage outputs of WECS and solar plant are combined via DC bus system and it is fed to the designed inverter at 24 V constant value,

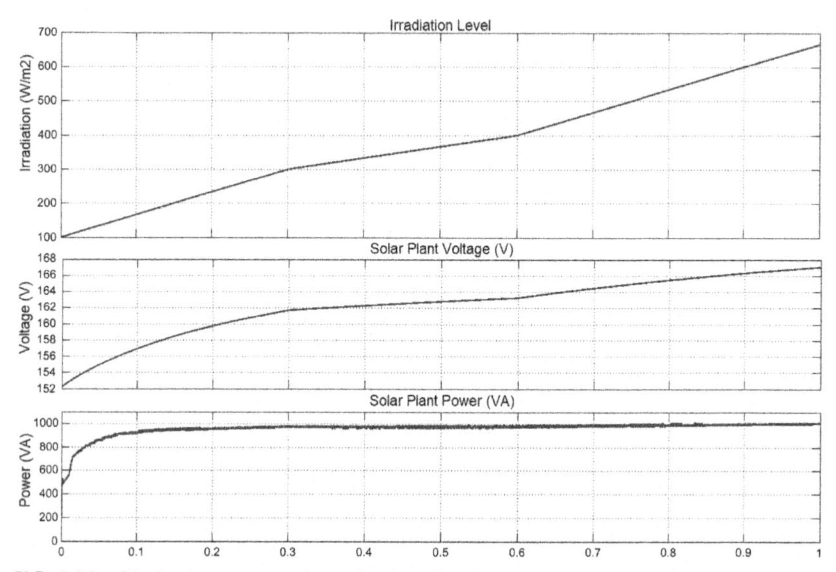

FIG. 4.22 Obtained output waveforms for the solar plant.

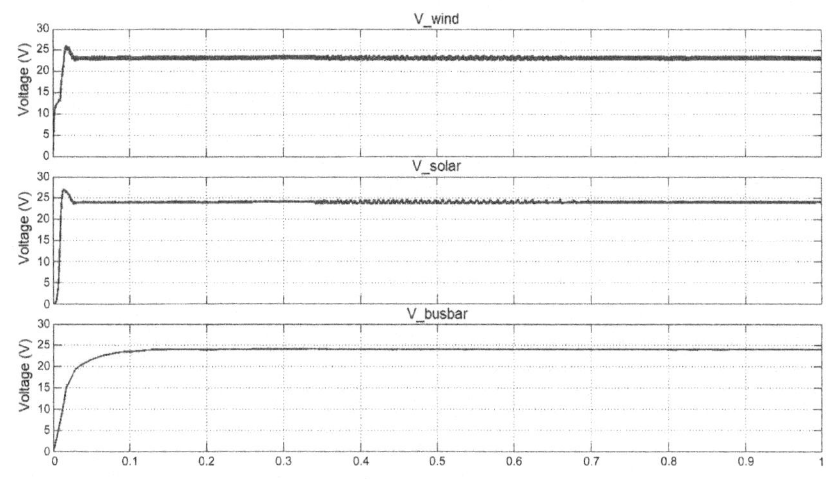

FIG. 4.23 Output voltage waveforms of WECS, solar plant, and DC bus.

which is illustrated in Fig. 4.23. Afterwards, the DC bus voltage is fed to the full bridge inverter to convert DC voltages into the AC voltages. In this case study, DC bus voltage is converted to 380 V three-phase line voltages over a wye-wye connected three-phase transformer as depicted in Fig. 4.24. In addition, the designed full bridge inverter is controlled by 5 kHz switching frequency signals.

The PLC infrastructure of this case study intends to transmit measured current, voltage and power data of WECS and solar plant to energy monitoring center as stated earlier. After the values to be transmitted are measured, they are converted to quantized data in QPSK modulator system. Transmitted and received measurement data are shown in Fig. 4.25. The first and second

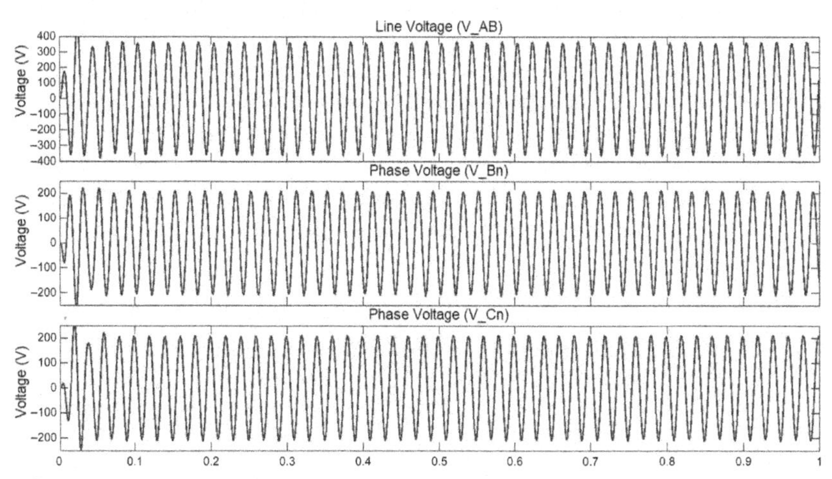

FIG. 4.24 Line voltage (380 V) and phase voltages (220 V) generated by full bridge inverter.

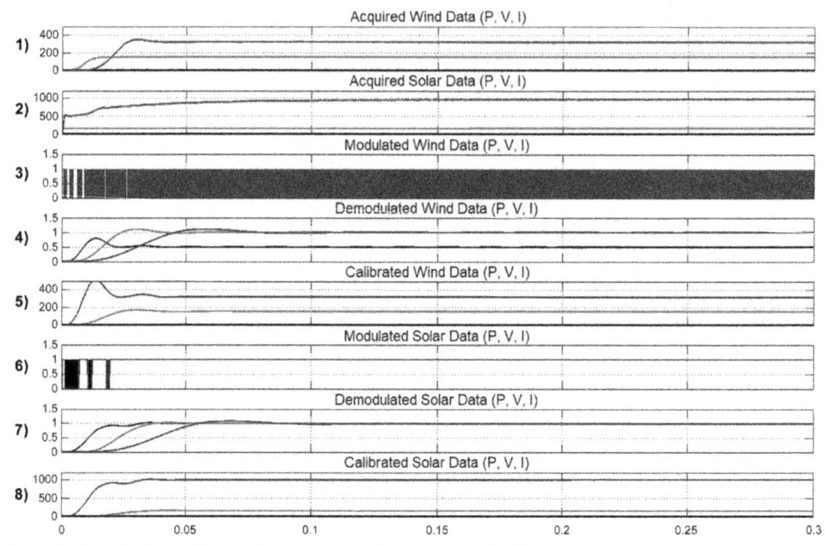

FIG. 4.25 Performed waveform analyses by designed PLC modems for measured, modulated, demodulated and calibrated data.

waveforms stand for the measured voltage, current and power values of WECS and solar plants, respectively. Modulated wind measurement, demodulated wind measurement and calibrated wind measurement are shown in third, fourth and fifth waveform of Fig. 4.25, respectively. In addition, these transactions are also realized for the solar plant system in a similar way. Modulated measurement of solar plant, demodulated measurement of solar plant and calibrated measurement of solar plant are illustrated in sixth, seventh and eighth waveform of Fig. 4.25, respectively. By employing various carriers with different frequencies, modulated data of solar plant are combined with modulated WECS data.

The presented analyses for the second case study clearly show that designed monitoring system can be efficiently used to track different parameters of several energy sources. Furthermore, the presented multi-channel PLC system can be simply arranged to monitor much more parameters of energy generation and distribution systems. Moreover, this system can be also exploited as a control system by sending any control data from grid side to energy source side.

References

[1] Y. Kabalci, A survey on smart metering and smart grid communication. Renew. Sust. Energ. Rev. 57 (2016) 302–318, https://doi.org/10.1016/j.rser.2015.12.114.

[2] E. Kabalci, Y. Kabalci, I. Develi, Modelling and analysis of a power line communication system with QPSK modem for renewable smart grids. Int. J. Electr. Power Energy Syst. 34 (2012) 19–28, https://doi.org/10.1016/j.ijepes.2011.08.021.

[3] I. Develi, Y. Kabalci, A. Basturk, Performance of LDPC coded image transmission over realistic PLC channels for smart grid applications, Int. J. Electr. Power Energy Syst. 62 (2014) 549–555.
[4] I. Develi, Y. Kabalci, Highly reliable LDPC coded data transfer in home networks by using Canete's PLC channel model. Int. J. Electr. Power Energy Syst. 62 (2014) 912–918, https://doi.org/10.1016/j.ijepes.2014.05.051.
[5] E. Kabalci, Y. Kabalci, Multi-channel power line communication system design for hybrid renewables, in: Power Engineering, Energy and Electrical Drives (POWERENG), 2013 Fourth International Conference on, IEEE, 2013, , pp. 563–568.
[6] E. Kabalci, Y. Kabalci, A measurement and power line communication system design for renewable smart grids. Meas. Sci. Rev. 13 (2013) 248–252, https://doi.org/10.2478/msr-2013-0037.
[7] I. Develi, Y. Kabalci, A. Basturk, Artificial bee colony optimization for modelling of indoor PLC channels: a case study from Turkey. Electr. Power Syst. Res. 127 (2015) 73–79, https://doi.org/10.1016/j.epsr.2015.05.021.
[8] Y. Kabalci, E. Kabalci, Modeling and analysis of a smart grid monitoring system for renewable energy sources. Sol. Energy 153 (2017) 262–275, https://doi.org/10.1016/j.solener.2017.05.063.
[9] I. Develi, Y. Kabalci, Analysis of the use of different decoding schemes in LDPC coded OFDM systems over indoor PLC channels, Elektronika IR Elektrotechnika 20 (2014) 76–79.
[10] Y. Kabalci, I. Develi, E. Kabalci, LDPC coded OFDM systems over broadband indoor power line channels: a performance analysis, in: Power Engineering, Energy and Electrical Drives (POWERENG), 2013 Fourth International Conference on, IEEE, 2013, , pp. 1581–1585.
[11] J. Routin, C.E.L. Brown, Improvements in and Relating to Electricity Meters, Patent GB189 724 833(1898).
[12] C.H. Thordarson, Electric Central Station Recoding Mechanism for Meters, U.S. Patent US 784 712(1905).
[13] S. Remseier, H. Spiess, Making power lines sing, ABB Rev. 2 (2006).
[14] M. Schwartz, Carrier-wave telephony over power lines: early history [history of communications], IEEE Commun. Mag. 47 (2009) 14–18.
[15] D. Dzung, I. Berganza, A. Sendin, Evolution of Powerline Communications for Smart Distribution: From Ripple Control to OFDM. IEEE, 2011, pp. 474–478, https://doi.org/10.1109/ISPLC.2011.5764444.
[16] K. Dostert, Powerline Communications, Prentice Hall PTR, Upper Saddle River, NJ, 2001.
[17] S. Galli, A. Scaglione, Z. Wang, For the grid and through the grid: the role of power line communications in the smart grid. Proc. IEEE 99 (2011) 998–1027, https://doi.org/10.1109/JPROC.2011.2109670.
[18] M.M. Rahman, C.S. Hong, S. Lee, J. Lee, M.A. Razzaque, J.H. Kim, Medium access control for power line communications: an overview of the IEEE 1901 and ITU-T G.hn standards. IEEE Commun. Mag. 49 (2011) 183–191, https://doi.org/10.1109/MCOM.2011.5784004.
[19] J. Brown, J.Y. Khan, Key performance aspects of an LTE FDD based smart grid communications network. Comput. Commun. 36 (2013) 551–561, https://doi.org/10.1016/j.comcom.2012.12.007.
[20] F. Nouvel, P. Tanguy, S. Pillement, H.M. Pham, Experiments of in-vehicle power line communications. in: M. Almeida (Ed.), Advances in Vehicular Networking Technologies, InTech, 2011https://doi.org/10.5772/14258.
[21] I.S. Stievano, F.G. Canavero, V. Dafinescu, Power Line Communication Channel Modeling for in-Vehicle Applications. IEEE, 2012, pp. 376–379, https://doi.org/10.1109/I2MTC.2012.6229152.

[22] V. Degardin, I. Junqua, M. Lienard, P. Degauque, S. Bertuol, Theoretical approach to the feasibility of power-line communication in aircrafts. IEEE Trans. Veh. Technol. 62 (2013) 1362–1366, https://doi.org/10.1109/TVT.2012.2228245.

[23] S. Galli, T. Banwell, D. Waring, Power Line Based LAN on Board the NASA Space Shuttle. IEEE, 2004, pp. 970–974, https://doi.org/10.1109/VETECS.2004.1388975.

[24] S. Barmada, L. Bellanti, M. Raugi, M. Tucci, Analysis of power-line communication channels in ships. IEEE Trans. Veh. Technol. 59 (2010) 3161–3170, https://doi.org/10.1109/TVT.2010.2052474.

[25] SAE International, Broadband PLC Communication for Plug-in Electric Vehicles, SAE International, 2014. https://www.sae.org/standards/content/j2931/4_201410/.

[26] ISO, Road Vehicles—Vehicle to Grid Communication Interface—Part 3: Physical and Data Link Layer Requirements, https://www.iso.org/standard/59675.html, 2015.

[27] P. Karols, K. Dostert, G. Griepentrog, S. Huettinger, Mass transit power traction networks as communication channels. IEEE J. Sel. Areas Commun. 24 (2006) 1339–1350, https://doi.org/10.1109/JSAC.2006.874410.

[28] L. Lampe, L.T. Berger, Power line communications. in: Academic Press Library in Mobile and Wireless Communications, Elsevier, 2016, , pp. 621–659, https://doi.org/10.1016/B978-0-12-398281-0.00016-8.

[29] C. EN50065, Signalling on Low-Voltage Electrical Installations in the Frequency Range 3 kHz to 148,5 kHz—Part 1: General Requirements, Frequency Bands and Electromagnetic Disturbances, European Committee for Electrotechnical Standardization (CENELEC), Bruxelles, Belgium. Standard EN 50065-1:2011(2011).

[30] J.A. Cortes, J.M. Idiago, Smart metering systems based on power line communications. in: E. Kabalci, Y. Kabalci (Eds.), Smart Grids and Their Communication Systems, Springer Nature, Singapore, 2019, , pp. 1–43, https://doi.org/10.1007/978-981-13-1768-2_4.

[31] FCC, Title 47 of the Code of Federal Regulations (CFR), Tech. Rep. 47 CFR §15, U.S. Federal Communications Commission (FCC),. (n.d.). https://goo.gl/cN4JMC.

[32] I.O. for Standartization, Information Technology-Home Electronic System (HES) Architecture-Part 3-5: Media and Media Dependent Layers Powerline for Network Based Control of HES Class 1, International Standard ISO/IEC 14543-3-5, ISO, 2007.

[33] A.N.S.I.I. Association (ANSI/EIA), Control Network Power Line (PL) Channel Specification, ANSI/CEA-709.2-A, ANSI September(2006).

[34] I.O. for Standartization, Interconnection of Information Technology Equipment-Control Network Protocol-Part 3: Power Line Channel Specification, International Standard ISO/IEC 14908–3, ISO. Revision Vol. 11(2011).

[35] Y. Kabalci, Communication methods for smart buildings and nearly zero-energy buildings. in: N. Bizon, N. Mahdavi Tabatabaei, F. Blaabjerg, E. Kurt (Eds.), Energy Harvesting and Energy Efficiency, Springer International Publishing, Cham, 2017, , pp. 459–489, https://doi.org/10.1007/978-3-319-49875-1_16.

[36] D. Clark, Powerline communications: finally ready for prime time? IEEE Internet Comput. 2 (1998) 10–11.

[37] H. Meng, S. Chen, Y. Guan, C. Law, P. So, E. Gunawan, T. Lie, Modeling of transfer characteristics for the broadband power line communication channel, IEEE Trans. Power Delivery 19 (2004) 1057–1064.

[38] S. Galli, T. Banwell, A novel approach to the modeling of the indoor power line channel-part II: transfer function and its properties, IEEE Trans. Power Delivery 20 (2005) 1869–1878.

[39] S. Barmada, A. Musolino, M. Raugi, Innovative model for time-varying power line communication channel response evaluation, IEEE J. Sel. Areas Commun. 24 (2006) 1317–1326.

[40] D. Sabolic, A. Bazant, R. Malaric, Signal propagation modeling in power-line communication networks, IEEE Trans. Power Delivery 20 (2005) 2429–2436.

[41] T. Sartenaer, P. Delogne, Deterministic modeling of the (shielded) outdoor power line channel based on the multiconductor transmission line equations, IEEE J. Sel. Areas Commun. 24 (2006) 1277–1291.

[42] F.J. Canete, J.A. Cortes, L. Diez, J.T. Entrambasaguas, A channel model proposal for indoor power line communications, IEEE Commun. Mag. 49 (2011) 166–174.

[43] C. Hensen, W. Schulz, Time dependence of the channel characteristics of low voltage power-lines and its effects on hardware implementation, AEU Int. J. Electron. Commun. 54 (2000) 23–32.

[44] H. Philipps, Modelling of powerline communication channels, in: Proc. 3rd Int'l. Symp. Power-Line Commun. and its Applications, 1999, , pp. 14–21.

[45] M. Zimmermann, K. Dostert, A multipath model for the powerline channel, IEEE Trans. Commun. 50 (2002) 553–559.

[46] M. Tlich, A. Zeddam, F. Moulin, F. Gauthier, Indoor power-line communications channel characterization up to 100 MHz—part I: one-parameter deterministic model, IEEE Trans. Power Delivery 23 (2008) 1392–1401.

[47] M. Tlich, A. Zeddam, F. Moulin, F. Gauthier, Indoor power-line communications channel characterization up to 100 MHz—part II: time-frequency analysis, IEEE Trans. Power Delivery 23 (2008) 1402–1409.

[48] A. Rubinstein, F. Rachidi, M. Rubinstein, EMC guidelines, in: The OPERA Consortium, IST Integrated Project Deliverable D9v1. 1, IST Integrated Project No. 026920, 2008.

[49] A. Vukicevic, Electromagnetic Compatibility of Power Line Communication Systems, Ph.D. thesisEcole Polytechnique Federale de Lausanne (EPFL), Lausanne, 2008.

[50] B. Varadarajan, I.H. Kim, A. Dabak, D. Rieken, G. Gregg, Empirical measurements of the low-frequency power-line communications channel in rural North America, in: Power Line Communications and Its Applications (ISPLC), 2011 IEEE International Symposium on, IEEE, 2011, , pp. 463–467.

[51] K. Razazian, A. Kamalizad, M. Umari, Q. Qu, V. Loginov, M. Navid, G3-PLC field trials in US distribution grid: initial results and requirements, in: Power Line Communications and Its Applications (ISPLC), 2011 IEEE International Symposium on, IEEE, 2011, , pp. 153–158.

[52] K. Razazian, J. Yazdani, Utilizing beyond CENELEC standards for smart grid technology, in: Innovative Smart Grid Technologies (ISGT Europe), 2011 2nd IEEE PES International Conference and Exhibition on, IEEE, 2011, , pp. 1–6.

[53] M. Gotz, M. Rapp, K. Dostert, Power line channel characteristics and their effect on communication system design, IEEE Commun. Mag. 42 (2004) 78–86.

[54] D. Anastasiadou, T. Antonakopoulos, Multipath characterization of indoor power-line networks, IEEE Trans. Power Delivery 20 (2005) 90–99.

[55] X. Ding, J. Meng, Characterization and modeling of indoor power-line communication channels, in: Proceedings of the 2nd Canadian Solar Buildings Conference, Calgary, 2007, , pp. 1–7.

[56] X. Ding, J. Meng, Channel estimation and simulation of an indoor power-line network via a recursive time-domain solution, IEEE Trans. Power Delivery 24 (2009) 144.

[57] P. Meier, M. Bittner, H. Widmer, J. Bermudez, A. Vukicevic, M. Rubinstein, F. Rachidi, M. Babic, J.S. Miravalles, Pathloss as a function of frequency, distance and network topology for various LV and MV European powerline networks, in: The OPERA Consortium, Project Deliverable, EC/IST FP6 Project, 2005D5v0.9.

[58] A. Sendin, I. Peña, P. Angueira, Strategies for power line communications smart metering network deployment, Energies 7 (2014) 2377–2420.

[59] E. Marthe, F. Issa, F. Rachidi, Analysis of power line communication networks using a new approach based on an efficient measurement technique, in: Electromagnetic Compatibility, 2003 IEEE International Symposium On, IEEE, 2003, , pp. 367–371.

[60] L. Peretto, R. Tinarelli, A. Bauer, S. Pugliese, Fault location in underground power networks: a case study, in: Innovative Smart Grid Technologies (ISGT), 2011 IEEE PES, IEEE, 2011, , pp. 1–6.

[61] S. Galli, T. Lys, Next generation narrowband (under 500 kHz) power line communications (PLC) standards, China Commun. 12 (2015) 1–8.

[62] I.H. Kim, W. Kim, B. Park, H. Yoo, Channel measurements and field tests of narrowband power line communication over Korean underground LV power lines, in: Power Line Communications and its Applications (ISPLC), 2014 18th IEEE International Symposium On, IEEE, 2014, , pp. 132–137.

[63] J1VV, Low Voltage Lightning Cable, Available from: https://goo.gl/Yv1XBq, 2018.

[64] Klaus Faber AG, NYY-J Power Cable, Available from: https://goo.gl/fg24Ua, 2018.

[65] Kajote Oy, AMCMK Lov Voltage Power Cable, (n.d.). Available from: https://goo.gl/ASRDgo.

[66] Draka Keila Cables, AXPK-PLUS 0,6/1 kV Power Cable, Available from: https://goo.gl/siEzqi, 2018.

[67] Nexans Switzerland Ltd, Low Voltage Cable Type GKN Alu, Available from: https://goo.gl/kAHvEE, 2018.

[68] Nexans Switzerland Ltd, NF C 32–321 Cables According to U-1000 AR2V, Available from: https://goo.gl/vGMmnx, 2018.

[69] Wind Cluster Denmark, WILBAwind MC 0.6/1kV, CU, Available from: https://goo.gl/ewAB62, 2018.

[70] XLPE Power Cable, 0.6/1 kV Low Voltage, Available from: https://goo.gl/4Qck3R, 2018.

[71] Seval Kablo, NA2XSYR MV Power Cable, Available from: https://goo.gl/PGTM1U, 2018.

[72] NKT cables GmbH, 3x120 AL MV Power Cable, Available from: https://goo.gl/hJ9E5z, 2018.

[73] Cangneng Electric Power Equipment Co., Ltd., ACSR Conductor for Overhead Power Transmission Lines, Available from: https://goo.gl/gP3hVW, 2018.

[74] Seval Kablo, XVB-F2 Single Core Overhead Power Cable, Available from: https://goo.gl/uVFAMZ, 2018.

[75] La Triveneta Cavi, FG7OR Power Cable, Available from: https://goo.gl/oFQyPC, 2018.

[76] Nexans (Yanggu) New Rihui Cables, YJLV MV Power Cable, Available from: https://goo.gl/nXekAp, 2018.

[77] NKT cables GmbH, CYKY Power Cable, Available from: https://goo.gl/i7N5K5, 2018.

[78] American Wire Group, EPR MV Power Cable, Available from: https://goo.gl/G5pceA, 2018.

[79] Oznur Kablo, Stranded Aluminum Conductors for Aerial Lines (AAC), Available from: https://goo.gl/YgNf9V, 2018.

[80] ITU-T, Narrowband Orthogonal Frequency Division Multiplexing Power Line Communication Transceivers—Power Spectral Density Specification, https://goo.gl/dHhTSh, 2012.

[81] ITU-T, Narrowband Orthogonal Frequency Division Multiplexing Power Line Communication Transceivers for ITU-T G.hnem Networks, https://goo.gl/ejZRgA, 2012.

[82] ITU-T, Narrowband Orthogonal Frequency Division Multiplexing Power Line Communication Transceivers for G3-PLC Networks, https://goo.gl/pVJHYj, 2013.

[83] ITU-T, Narrowband Orthogonal Frequency Division Multiplexing Power Line Communication Transceivers for PRIME Networks, https://goo.gl/vReXnh, 2012.

[84] IEEE Standards Association, IEEE 1901.2-2013 for Low-Frequency (Less Than 500 kHz) Narrowband Power Line Communications for Smart Grid Applications, https://goo.gl/EYqxPJ, 2013.

[85] IEEE Standards Association, others, Wireless Medium Access Control (MAC) and Physical Layer (PHY) Specifications for Low-Rate Wireless Personal Area Networks (WPANs), IEEE Std 802.15.4-2006, IEEE, 2006.

[86] P.R.I.M.E. Alliance, PRIME Alliance, PRIME – a future proven global reality, beyond metering, (2018) https://goo.gl/YALF5q.

[87] I. Berganza, G. Bumiller, A. Dabak, R. Lehnert, A. Mengi, A. Sendin, PLC for smart grid. in: L. Lampe, A.M. Tonello, T.G. Swart (Eds.), Power Line Communications, John Wiley & Sons, Ltd, Chichester, 2016, , pp. 509–561, https://doi.org/10.1002/9781118676684.ch9.

[88] H. Hirsch, M. Koch, N. Weling, A. Zeddam, Electromagnetic compatibility, in: Power Line Communications: Principles, Standards and Applications from Multimedia to Smart Grid, 2016, , pp. 178–222.

[89] A. Sugiura, Y. Kami, Generation and propagation of common-mode currents in a balanced two-conductor line, IEEE Trans. Electromagn. Compat. 54 (2012) 466–473.

[90] M. Girotto, A.M. Tonello, EMC regulations and spectral constraints for multicarrier modulation in PLC, IEEE Access 5 (2017) 4954–4966.

[91] A. of R.I. and Businesses (ARIB), Power Line Communication Equipment (10kHz–450kHz), Standard STD-T84, https://goo.gl/u1KBjS, 2002.

[92] C. EN50561, Power Line Communication Apparatus Used in Low-Voltage Installations—Radio Disturbance Characteristics—Limits and Methods of Measurement—Part 1: Apparatus for in-Home Use, European Committee for Electrotechnical Standardization (CENELEC), Bruxelles, Belgium, 2013. Standard EN 50561-1:2013 https://goo.gl/27wBd3.

[93] C. EN50561-3, Power Line Communication Apparatus Used in Low-Voltage Installations—Radio Disturbance Characteristics—Limits and Methods of Measurement—Part 3: Apparatus Operating Above 30 MHz, European Committee for Electrotechnical Standardization (CENELEC), Bruxelles, Belgium, 2016. Standard EN 50561–3:2016 https://goo.gl/VLyrud.

[94] ITU, The Impact of Power Line High Data Rate Telecommunication Systems on Radiocommunication Systems Below 470 MHz, (2013).

[95] Sharp, USA, Sharp NE-170UC1 Multipurpose Module, https://goo.gl/s8JyDJ, 2008. Accessed 30 July 2018.

Chapter 5

Emerging wireless communication technologies for smart grid applications

Chapter outline

5.1 Introduction

Wireless sensor networks (WSNs) have attracted considerable attention over the last decade because of their cost advantages on the practical implementations. A general WSN scheme is composed of wireless nodes that may be a combination of individual devices. The wireless nodes may be linked to each other to transfer data packets in which every node can be connected to one or more sensors for monitoring environmental variables and various physical quantities. The quantities to be perceived by sensors may be temperature, infrared, sound, movement, magnetism, vibration, pressure and light. In general, the WSN nodes contact with each other by exploiting wireless communications. The

From Smart Grid to Internet of Energy. https://doi.org/10.1016/B978-0-12-819710-3.00005-3

IEEE 802.15.4, which is formed to empower short-range applications with both low power consumption and low data speeds, is the most commonly preferred standard for the WSN systems. The well-known application areas of the WSNs are smart homes, smart cities, smart grids (SGs), smart environments, environmental surveillance, internet of things (IoT), cloud computing, vehicular ad hoc networks (VANETs), machine-to-machine (M2M) communication, cyber physical systems (CPSs), healthcare monitoring and military investigation [1–7]. The WSN systems may be used in healthcare systems for observing status of patients whereas they may be utilized in environmental surveillance systems to monitor real-time ecological conditions. On the other hand, they can be used to manage home appliances through mobile devices (e.g., smart phones, tablets) remotely. Another crucial application area of the WSNs is in the SGs that are assumed as the transformation of the traditional power grid systems. A typical SG system, which includes billions of sensors and smart meters (SMs), aims to present improvement on power quality and more efficiency against to power consumption demands. In the SG systems, there is no requirement for human intervention to manage generation, transmission and distribution stages of power systems since various powerful services are provided by the WSNs. Traditional power systems are mostly based on the use of wired networks. Hence, only certain regions can be covered owing to the limitations of the network scheme. Otherwise, the WSNs can be utilized in all places without any limitation. By perceiving several critical characteristics of power units such as vibration, pressure, light, temperature and sound, the WSN nodes can additionally inform the control and management center of power systems before a fault occurs. Thus, crucial problems that may give rise to significant issues affecting power quality of users and efficiency of the power generation systems can be prevented through measures taken by the service providers. Furthermore, the WSNs utilized in the SGs can monitor transmission and distribution layers of the power systems, which present advantage to efficiently handle demand management requirements in utility grids. The energy theft, fault detection and insulation breakdown issues can be also prevented by adapting WSN applications into the SGs, which improves the management and operating characteristics of the SGs compared to traditional power grids [8].

The SGs associate information and communication technology (ICT) with conventional power grids in order to provide more efficient, secure and flexible power grid system where the observation and management requirements of energy generation systems and customer demands can be simply handled [9]. The fundamental components of the SGs are the SMs and sensors that are spread over the entire network to enable peak value detection of electricity utilization for the purpose of demand response management (DRM). In addition, security is so crucial concern for the SGs and smart environments such as smart homes, smart cities. Hence, the utilization of secure communication systems among system components and customers should be ensured in these environments. Nonetheless, real-time monitoring and device management can be realized by using secure access gateways in home area networks (HANs) [10]. The WSNs can also present practical solutions with low cost as well wide coverage advantage. These advantages make them more prominent technology than that of other technologies. On the other hand, it is worth noting that low power consumption is a vital requirement of massive networks operating long-term.

In order to enable a communication opportunity with low power consumption, low cost and wide coverage properties for the WSNs, IEEE 802.15.4 standard has been improved by the IEEE 802.15 working group. Physical (PHY) and media access control (MAC) layers of this standard exploit exclusive scheduling protocols for sleep/wake up modes to accomplish energy-efficient data transmission. Furthermore, different network schemes (e.g., peer-to-peer (P2P), cluster-tree and star topology) are supported by this standard. Due to these distinguished characteristics, IEEE 802.15.4 standard has been considered as one of the most crucial nominees for the WSNs and wireless control networks [11–14].

Another potential wireless technology for SG systems is based on the IEEE 802.11 standard that is commonly exploited standard in wireless local area networks (WLANs). This standard defines various essential functionalities related to PHY layer, MAC layer, protocols, authentication, quality of service (QoS) and so on. From the concept of SG applications, IEEE 802.11 based technologies (e.g., Wireless Fidelity (Wi-Fi)) that can provide efficient performance in both shared spectrum and noisy channels are crucial for Smart Energy Profile 2.0 (SEP 2.0). It is also important to note that this technology can support many application scenarios and IP based protocols.

Cellular communication systems are also considered as one of the key technologies for SG applications such as smart metering, phasor measurement unit, smart energy management, and remote monitoring. Cellular communication systems can support the requirements of SG applications such as reliability, security, latency, and overall performance. Furthermore, several service providers and operators are currently introducing solutions and services for SG applications. In Fig. 5.1, different wireless communication technologies are compared in terms of data rate and coverage. Detailed information regarding these communication technologies are given by the following sections of this chapter.

5.2 Short-range communication technologies

5.2.1 IEEE 802.15.1 and Bluetooth

Bluetooth is a widespread wireless communication technology developed for enabling short-range communication. Ericsson designed this technology in 1994 to wirelessly connect mobile phones and other mobile devices and enabling communication between these devices. Then, a group of companies formed the Bluetooth Special Interest Group (SIG) in 1998. The IEEE approved

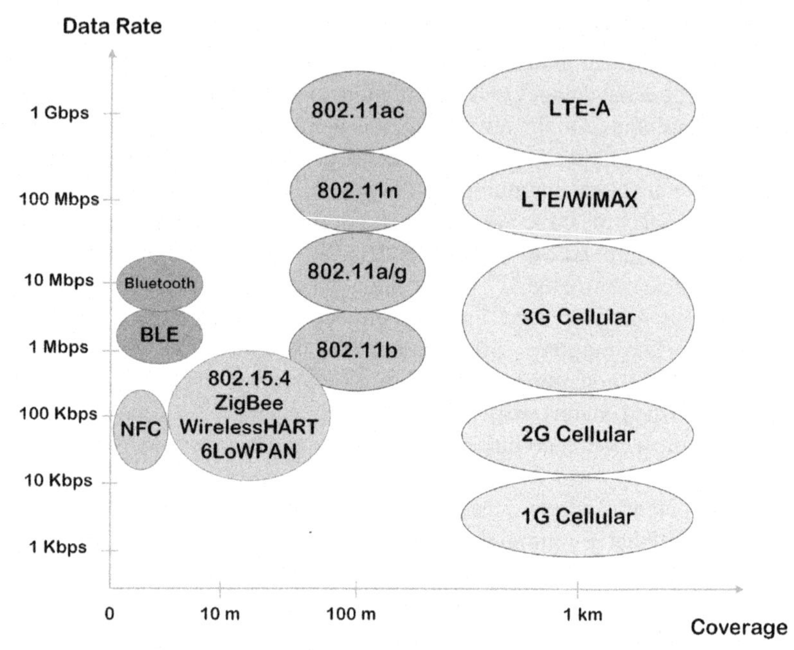

FIG. 5.1 Comparison of emerging wireless communication technologies.

the 802.15.1 specification as the standard of Bluetooth in 2002. The PHY and MAC layers of seven-layer Open Systems Interconnection (OSI) model are characterized by the standard to ensure wireless communication with low power consumption. The technology is principally developed for providing wireless data transfer over short ranges. Bluetooth technology serves on the industrial, scientific and medical (ISM) band (in 2400–2480 MHz) by utilizing frequency hopping spread spectrum (FHSS) method and it transmits data by dividing into the packets. The early versions of the technology provided 1 Mbps data rate whereas Bluetooth 3.0/high-speed version (Bluetooth 3.0 + HS) theoretically presented data transfer speed up to 24 Mbps. Two different topologies are employed in the technology that are called as Piconet and Scatternet. Wireless Personal Area Network (WPAN) may typically form a Piconet where one mobile device behaves as a master and the other devices act as the slaves. A Scatternet scheme is composed of two or more Piconets. On the other hand, a master that manages the packet exchange process by generating clocks can connect up to seven slaves in a Piconet topology to enable communications among the master and slaves. A device can change its master/slave status if the device has a requirement to join more than one Piconet.

An improved version of the technology called as Bluetooth Low Energy (BLE) or Bluetooth 4.0 has been presented with the aim of providing lower-power consumption and lower-latency features while protecting communication range.

This new version of the technology optimizes the protocol to significantly reduce power consumption when compared with earlier ones. Thus, the batteries of BLE enabled devices can supply energy for longer periods of time. In 2016, Bluetooth SIG introduced Bluetooth 5 whose main new feature is related to IoT technology. In addition, the Bluetooth 5.0 presents improvements on connectionless services, and some important improvements provided by this version of the technology can be listed as [15].

- Slot availability mask (SAM)
- 2 Msym/s PHY for low energy (LE)
- LE long range
- High duty cycle non-connectable advertising
- LE advertising extensions
- LE channel selection algorithm #2

Since its low power consumption and rapid data transmission characteristics, this technology may be preferred for communications between smart home appliances, energy management system and SM.

5.2.2 Near-field communication

Near-Field Communication (NFC) that is a set of very short-range radio transmission standards provides data exchange opportunity among electronic devices in a short range as much as 4 cm (1.6 in.). Data transmission via this method can be performed by bringing devices close enough or contacting them physically. In addition, communication can be accomplished between an NFC device and a tag that is an unpowered NFC chip. This technology works at 13.56 MHz frequency that is an unlicensed band, and employs inductive coupling. The NFC Forum collects several vendors, manufacturers and service providers, and carries out the studies to ensure interoperability. There exist three different operation method for the NFC systems that are called as NFC card emulation method, P2P method, and reader/writer method. In the first method, the NFC acts similar to a radio-frequency identification (RFID) system. Therefore, the NFC system in this operating mode can be regarded as an extension of RFID technology. The P2P mode authorizes NFC devices to transmit data among them in an ad hoc fashion. The last mode allows NFC devices to read and write information on tags.

Even though the technology was actually improved by Sony and NXP Semiconductors, there exist various proposed standards for NFC technology. The ISO/IEC 18092 is the international standard characterized by NFC Forum that is created by Sony, Nokia and Philips in 2004. In addition, the NFC Forum supports NFC, and approves compliance of devices. Also, Sony developed FeliCa and it complies with Japanese Industrial Standard (JIS) X6319-4. The FeliCa actually is a contactless RFID smart card system widely employed for payments in Japan. NXP Semiconductors developed MIFARE that is based on the

ISO/IEC 14443 standards for Type A. ISO/IEC 21481 standard defines communication mode selection mechanism for devices maintaining ISO/IEC 18092, ISO/IEC 14443 or ISO/IEC 15693. Current and predicted applications of the NFC technology cover contactless payment processes, identification, access control, and automation. Fast development of smart devices also provides novel services for the NFC technology.

5.3 Communication technologies for low-rate WPANS

5.3.1 IEEE 802.15.4 standard

In 2003, IEEE 802.15.4 standard has been introduced to meet demands for communication technologies that are able to provide low-power consumption feature. The IEEE 802.15.4 working group (WG) developed this standard as the first standard for low-rate WPANs (LR-WPANs). The main purpose of the 802.15.4 is to put forward a new infrastructure for wireless network applications that has several critical characteristics such as low complexity, energy saving mode and low cost [16]. The standard defines both PHY and MAC layers as the base layers of the protocol. The fundamental signal processing transactions such as modulation, channel selection, energy management and data transmission are performed in the PHY layer whereas frame management and collision prevention processes are realized in the MAC layer. The upper layers of the protocol stack are specially defined by the technologies such as ZigBee, Wireless Highway Addressable Remote Transducer (WirelessHART) and IPv6 over Low power Wireless Personal Area Networks (6LoWPAN).

In the IEEE 802.15.4 standard, there exist different network node possibilities that may be full-function device (FFD) or reduced-function device (RFD). The first one contains nodes that can entirely ensure the standard due to the fact that the nodes can cover all features of the network. Hence, the FFDs in any network scheme may act as an end-device, a local coordinator, or a personal area network (PAN) coordinator. It is worth noting that at least one FFD should be employed in an 802.15.4 network. The RFD node type only acts an end-device and forms basic nodes that have some capabilities of the network due to memory and processing scarcities. Since the RFDs behave as end-devices in the network, they cannot transmit messages to their ultimate targets. In addition, a local coordinator, which should be associated with a PAN coordinator or a previously related local coordinator, provides synchronization owing to beacon transmissions.

As can be seen from Fig. 5.2, the IEEE 802.15.4 standard can support several node topology schemes called star, mesh (P2P) and cluster-tree. In a star topology, all devices can communicate with a PAN coordinator by using master/slave network model. One of the FFDs can take over the coordinator role of the network, and the remain nodes (other FFDs and RFDs) can only convey with the network coordinator. The nodes may create more complicated network scheme

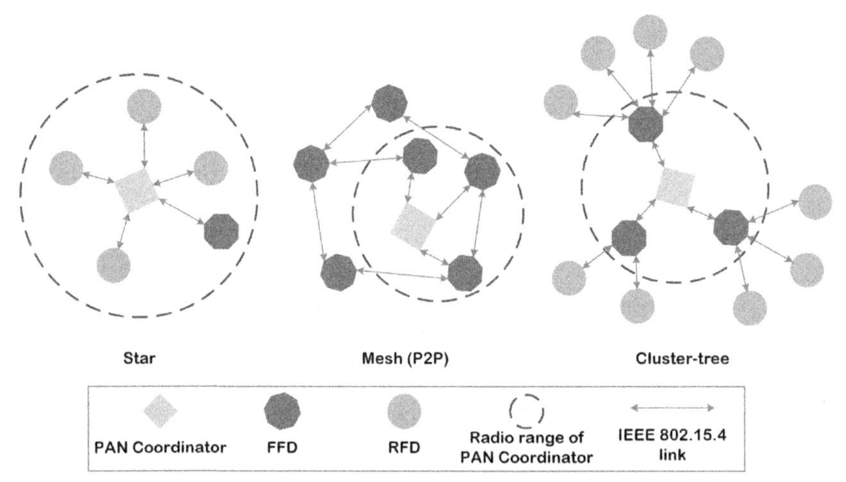

FIG. 5.2 Node topologies employed in IEEE 802.15.4 based WSNs.

in mesh topology in which whole devices of the network can communicate with each other. Nonetheless, the devices aiming to communicate with others should primarily contact with the PAN coordinator. The cluster-tree network scheme is considered as an exclusive version of mesh (P2P) network topology in which a large number of devices is the RFDs. Typically, communication range of the IEEE 802.15.4 changes from 10 to 75 m because of the low-powered transmission characteristics. On the other hand, the coverage of the network may be extended thanks to multi-hop and cooperative networking where some nodes in the network can serve as a relay as in the cluster-tree topology [17].

5.3.1.1 PHY layer of IEEE 802.15.4 standard

The PHY layer of the standard is responsible for data processing transactions regarding transmission and receiving like modulation, spreading and channel selection [18]. Although the PHY layer defines several transmission mediums in various bands, the most utilized operation bands in applications are 868 MHz, 915 MHz, and 2.4 GHz frequency bands. The ISM band (2.4 GHz frequency band) is the most widely employed unlicensed frequency band all over the world. Channel allocation of these frequency bands in the standard is illustrated in Fig. 5.3. As the 868 MHz frequency channel is utilized through one channel with 20 Kbps data rate in Europe, the 915 MHz frequency channel is employed through 10 channel with 40 Kbps data rate in North America. The frequencies between 2.4 and 2.483 GHz are employed by the last band to present 16 channels with 250 Kbps data rate. Each channel space is specified as 2 MHz in the 915 MHz frequency band whereas channel space is determined as 5 MHz in the 2.4 GHz frequency band as can be seen from the Fig. 5.3. In addition, other PHY layer specifications of the standard is summarized in Table 5.1. In addition, the

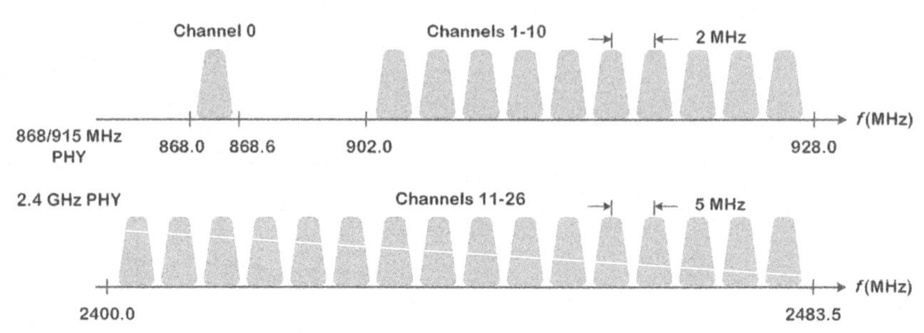

FIG. 5.3 Channel allocation types for 802.15.4 based WSNs.

TABLE 5.1 Several PHY layer specifications of IEEE 802.15.4 standard

Frequency range (MHz)	Spreading parameters		Data parameters		
	Chip rate (kchips/s)	*Modulation*	*Bit rate (Kbps)*	*Symbol rate (ksymbol/s)*	*Symbols*
868–868.6	300	BPSK	20	20	Binary
868–868.6	400	ASK	250	12.5	20-bit PSSS
868–868.6	400	O-QPSK	100	25	16-ary quasi-orthogonal
902–928	600	BPSK	40	40	Binary
902–928	1600	ASK	250	50	5-bit PSSS
902–928	1000	O-QPSK	250	62.5	16-ary quasi-orthogonal
2450–2483.5	2000	O-QPSK	250	62.5	16-ary quasi-orthogonal

minimum transmission power level of a transmitter is specified as −3 dBm (approximately 0.5 mW) by the standard. Nevertheless, transmitters should hold their power levels as low as possible to avoid interference to other communication systems. Therefore, local regulators specify the maximum transmission power level.

As in other wireless systems, noise and interferences significantly affect the 802.15.4 signals while they are transmitting over the wireless environments. This issue may cause performance degradation in communication systems. The wireless communication systems compatible with the IEEE 802.15.4

standard exploit spread spectrum (SS) methods such as direct sequence SS (DSSS) and parallel sequence SS (PSSS) to provide reliability to interference and noise effects. In addition, these methods provide some advantages in terms of interoperability. For example, IEEE 802.11 WLAN systems also serve in the ISM frequency band similar to the IEEE 802.15.4 systems. Nevertheless, the effects of WLAN systems on the IEEE 802.15.4 systems are taken into account as a wideband interference effect because the systems exploit different spreading codes and SS methods. If the Bluetooth (IEEE 802.15.1) systems in the ISM band are taken into account, the IEEE 802.15.1 systems behave as a narrowband interference source for other wireless systems because of the narrow-band structure of the IEEE 802.15.1 systems. Thus, the SS methods employed in the IEEE 802.15.4 based wireless communication systems provide durability to these destructive effects.

The PHY layer of the standard contains a typical packet scheme to characterize a general MAC interface. This packet is also defined as PHY protocol data unit (PPDU) that comprises four main sections called preamble, packet delimiter, PHY header, and PHY service data unit (PSDU). A typical packet structure exploited in the PHY layer of the standard is depicted in Fig. 5.4.

The PHY layer of the standard supports various significant features such as activation/deactivation of transceivers, link quality indication (LQI), energy detection (ED), channel frequency selection and clear channel assessment (CCA). These characteristics are summarized below.

- *Activation/Deactivation of Transceivers:* The transceivers compatible with the IEEE 802.15.4 standard should support different operation modes including transmitting, receiving and sleeping. The MAC layer controls activation or deactivation of the transceivers. The turnaround time that defines the required time among transmission and reception processes is determined to be less than 12 symbol periods.
- *Energy Detection (ED):* The ED parameter is related to power level prediction of received signal. In the ED process, there is no requirement for additional transactions such as demodulation, decoding or signal identification. A period of eight symbols is specified as the ED duration in the IEEE 802.15.4 standard. The acquired ED result is used on the network layer

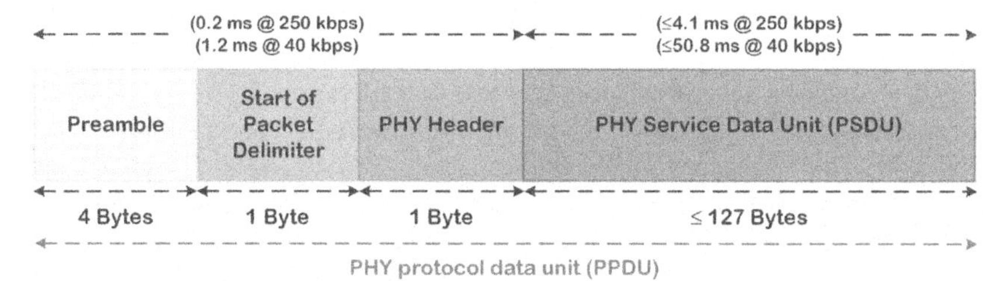

FIG. 5.4 PHY packet structure of IEEE 802.15.4 standard.

for various processes such as channel selection algorithm and CCA. The zero value of detected ED means that the received signal power is 10 dB above the receiver sensitivity level.

- *Link Quality Indication (LQI):* The LQI determines a quality measure for data packets and it is considered for every received data packet. While this transaction can be accomplished by using various techniques, the most widely preferred techniques are ED method, signal-to-noise ratio (SNR) estimation or the combination of these two methods. It is worth noting that the IEEE 802.15.4 standard does not specify how the LQI parameter will be used neither on the network layer nor on the application layer. The minimum and maximum values of this parameter show the quality range depend on the detectable signals by receivers. Furthermore, it is also desirable that distribution of the LQI values is uniform within the range.
- *Clear Channel Assessment (CCA):* The CCA transaction can be performed over several methods that are ED, carrier sense, a combination of ED and carrier sense methods, and *ALOHA*. In the first method (in ED), the channel is considered as busy if an energy level is detected larger than the predetermined threshold. In the second method, the CCA reports an occupied channel when it detects a modulated signal including spreading characteristic of the standard. In addition, it is worth noting that the energy level of the perceived signal is ignored in the carrier sense method. In third method, the CCA reports an occupied channel if it perceives a modulated signal having spreading characteristics and energy level is detected larger than the predetermined threshold. In fourth mode (in *ALOHA*), the CCA will ever report a free channel.
- *Channel Frequency Selection:* Even though there exist 27 wireless channel alternatives in the standard, only a part of channels is allowed to be used by a network. Thus, PHY layer should be able to set its radio dynamically.

5.3.1.2 MAC layer of IEEE 802.15.4 standard

The MAC layer comprises two services called MAC data service and MAC management service. The former service empowers MAC protocol data units (MPDUs) for receiving and/or transmitting data across the PHY data service. The major features of the MAC layer can be listed as guaranteed time slot (GTS) management, beacon management, frame validation, acknowledged frame delivery, channel access, association, and disassociation. In addition, the layer presents several chances to carry out appropriate security schemes for different type applications. There are two channel access methods in the IEEE 802.15.4 standard, which are named as *beacon-enabled method* and *non-beacon method*. If unslotted carrier sense multiple access with collision avoidance (CSMA/CA) is employed, the standard requires the use of non-beacon mode. In the event of the slotted CSMA/CA is employed, the standard requires the use of beacon-enabled mode where PAN coordinator periodically conveys beacon frames to whole end-devices existing in the network. The beacons are responsible for three significant purposes that can be sorted as ensuring synchronization among devices, identifying PAN infrastructure and defining structure of superframes. A superframe is employed in beacon-enabled mode to manage communication in the wireless channel. The duration between two beacons is defined as a beacon interval (BI) that is composed of an active period and an optional inactive period. The nodes can preserve their power sources by activating sleeping mode (low-power mode) during the inactive period. The active period of the superframes is called as superframe duration (SD). Every SD is composed of 16 time periods with equal lengths. Furthermore, Contention Access Period (CAP) and Contention Free Period (CFP) sections are also defined in the active periods. The CFP that is managed by a PAN coordinator is generally exploited in low-latency applications, and contains up to seven GTSs. The nodes can contact with each other by employing a slotted CSMA/CA or *ALOHA* in the CAP. The operational modes of IEEE 802.15.4 MAC layer are illustrated in Fig. 5.5.

A PAN coordinator describes the superframe scheme which is defined on the basis of *macBeaconOrder* (BO) and *macSuperframeOrder* (SO) values. The BO defines the time period whereas the SO identifies the active period and beacon frame. A general superframe scheme is illustrated in Fig. 5.6 where the beacon-enabled mode is taken into account. The relationship between BI and BO can be given as follows.

$$BI = aBaseSuperframeDuration \times 2^{BO} \text{ symbols, } for\, 0 \leq BO \leq 14 \qquad (5.1)$$

In the event of the value of BO is 15, the value of SO is ignored and beacon frames are not sent unless a special request is available. Furthermore, the relationship between SO and SD can be given as follows.

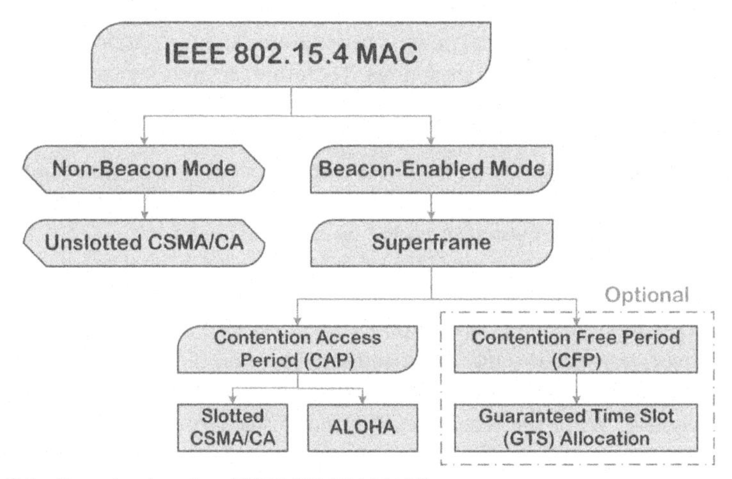

FIG. 5.5 Operational modes of IEEE 802.15.4 MAC layer.

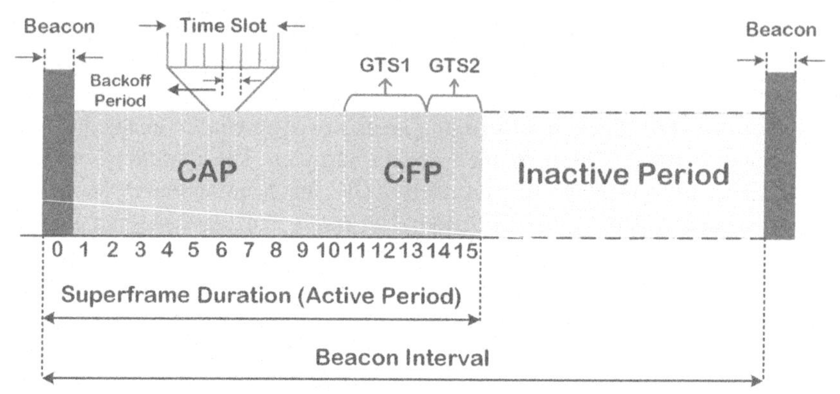

FIG. 5.6 Superframe structure for IEEE 802.15.4 standard.

$$SD = aBaseSuperframeDuration \times 2^{SO} \text{ symbols}, \ for\ 0 \leq SO \leq BO \leq 14 \quad (5.2)$$

In the event of the value of SO is 15, the superframe will not carry on the active mode after the beacon. When the value of the BO is 15, the value of SO will be ignored.

The MAC layer responsibilities can be given as follows.

- *Forming of Beacons for Coordinator:* A coordinator can select its operating mode as beacon-enabled mode or non-beacon mode. The coordinator runs by using superframe structure in the beacon-enabled mode where the number of superframes is constrained by network beacons. Every active part of a superframe contains 16 equally spaced slots (*aNumSuperframeSlots*). Furthermore, coordinators periodically send network beacons in order to synchronize devices connected to the network.
- *Supplying Synchronization among PAN Coordinator and End-Device:* The synchronization is a critical process to determine status of devices in the network and energy saving transactions. Therefore, an end-device operating in beacon-enabled mode should follow beacons for presenting synchronization to the PAN coordinator.
- *Association and Disassociation:* In star and mesh network types, automatic setup and self-configuration features are enabled through association and disassociation information provided by the MAC layer.
- *Channel Access Process* via *CSMA/CA Technique:* Channel access process in the IEEE 802.15.4 standard is accomplished by employing CSMA/CA method similar to other popular protocols developed for wireless networks. However, request-to-send (RTS) and clear-to-send (CTS) mechanisms are not utilized in this standard.
- *GTS System:* While the beacon-enabled mode is active, coordinator can commit some parts of the efficient superframe into a device. These parts of the superframe are called GTSs and they also include CFP of the superframe.

- *Providing Safe Connections between MAC Entities:* The MAC layer utilizes various transactions such as CSMA/CA, re-transmission, frame acknowledgement and CRC data verification in order to advance connection reliability between MAC entities.

5.3.2 IEEE 802.15.4g (smart utility network, SUN)

The IEEE 802.15.4g standard (Smart Utility Network, SUN) presents amendments to the IEEE 802.15.4 standard to ensure a global standard that simplifies large-scale process control applications such as SG networks capable of supporting large, geographically heterogeneous networks with minimum infrastructure and potentially millions of fixed endpoints [19]. This global standard harmonizes mapping schemes, data rates, power levels, PHY layer, and other technical features. It is mainly developed for outdoor communication applications that function in the unlicensed frequency bands covering the 700 MHz to 1 GHz band and 2.4 GHz ISM band. The date rate of the IEEE 802.15.4g may change from 5 to 400 Kbps. This standard uses time division multiple access (TDMA) based frequency-hopping MAC different from the IEEE 802.15.4 standard, which can be considered as coordinated sampled listening (CSL). The CSL authorizes receivers to regularly sample the channel for arriving transmission data at low duty cycles. The transmitter and receiver devices can be synchronized to diminish transmit overhead. The mesh topology is utilized by the standard to realize practical applications in the SG systems since this network topology can provide robust access from/to the SMs at an acceptable deployment cost. The 802.15.4e MAC sublayer has improved the MAC layer of IEEE 802.15.4. The combination of 802.15.4g PHY and 802.15.4e MAC sublayer provides energy-efficient and robust solutions for the IEEE 802.15.4g standard. In addition, the PHY layer of the SUN provides interoperable communication systems among SMs and SG devices.

The Wireless Smart Utility Network (Wi-SUN) is an international interoperable standard that is developed in accordance with the IEEE 802.15.4g. Wi-SUN Alliance [20] is a consortium of several companies that aims to form a global organization for driving the adoption of interoperable SUNs. The Alliance performs comprehensive tests to indicate interoperability in sub-GHz frequency bands for wide-coverage markets including Japan, North America, Australia/New Zealand, and Latin America markets.

5.3.3 IEEE 802.15.4k—(low energy critical infrastructure monitoring, LECIM*)*

The IEEE 802.15.4k standard was started by characterizing PHY and MAC layer for low-energy critical infrastructure monitoring (LECIM) in 2013 [21]. The LECIM networks can be utilized for several monitoring applications covering industrial metering systems that have wide coverage, reliable, secure, and low-energy consumption features. Especially, the LECIM networks aim to serve for outdoor applications containing thousands of nodes with low data rate. Some application areas of the LECIM networks are related to SG, smart metering, smart city, IoT applications, transportation, industrial monitoring, agriculture, and so on [22]. The major properties of the LECIM networks are minimum infrastructure requirement with star network topology, wide coverage areas with high receiver sensitivity and low power consumption [21].

In order to support LECIM applications, there exist two different PHY layer specifications [23–25]. While one of them is based on frequency shift keying (FSK) modulation, the other is based on the DSSS method. The DSSS based PHY layer of the IEEE 802.15.4k defined for a star topology can reach to -148 dBm receiver sensitivity by employing 32,768 (2^{15}) spreading factor. Thus, the coverage area of the LECIM may be widen to tens of kilometers in rural areas and several kilometers in urban areas. The data rate of DSSS based LECIM networks may change according to utilized frequency band and region. There are different frequency bands for the LECIM networks. For instance, 400 and 917 MHz frequencies are used in South Korea, 470 and 780 MHz frequency bands are employed in China, 868 MHz is exploited in Europe, 902 MHz is utilized in North America, 920 MHz is used in Japan and 2.4 GHz ISM band is available all over the world. On the other hand, the FSK based PHY layer of LECIM systems does not aim to present a multi-rate PHY [25]. This type of the PHY layer can operate in the frequencies of 863, 915, 920, and 2450 MHz by taking into account different mapping and channel parameters.

5.3.4 ZigBee

There exist new wireless technologies that have been developed based on the IEEE 802.15.4 standard. As stated earlier, the standard merely defines PHY and MAC layer specifications whereas the technologies such as ZigBee [26], WirelessHART [27], ISA100.11a [28], 6LoWPAN [29] and 6TiSCH [30] identify the upper layers separately. The ZigBee Alliance has developed ZigBee technology as an extensive wireless protocol for applications with low-power consumption. This technology can support control and monitoring applications with low cost, and ZigBee transceivers can be simply embedded in many devices. In addition, the ZigBee systems can be operated via small batteries over several years due to low power consumption feature of the technology. Therefore, the use of this technology prevents the need for frequent replacement of device batteries. The ZigBee Alliance has identified network layer and application layer above the IEEE 802.15.4 MAC layer. The network layer of ZigBee handles the routing processes whereas the application layer accomplishes suitable frameworks for several application types. The technology can operate on either 2.4 GHz or 868/915 MHz unlicensed ISM bands to empower network topologies based on the star, mesh, or cluster tree schemes in the WSNs. There exist 16 channels in the 2.4 GHz frequency band, and each of them is composed of 5 MHz bandwidth. Whereas the European 868 MHz frequency band presents maximum 20 Kbps data rates, the United States and Australian 915 MHz frequency band offer maximum 20 Kbps data rates. The other frequency band can present up to 250 Kbps data rates. Furthermore, this technology provides authentication, encryption and data routing capabilities. On the other hand, manufacturer-specific profiles and public profiles are supported by the ZigBee technology. Automation of home and buildings, telecommunication services, advanced metering applications, SEP 1.0 and SEP 2.0 can be considered as public profile examples.

A ZigBee network may be configured by using three different devices according to their roles that are called as end-device, router and coordinator. As a coordinator controls entire transactions of the network, a router transmits data among the nodes available in the network, and an end-device executes the commands of coordinator. It is worth noting that a router may serve more than one node in the network. Later the first version of ZigBee, development of ZigBee Pro was completed in 2007, and it became a widespread technology all over the world. This new version is able to provide easy network configuration, more flexible network architecture and lower energy consumption characteristics. Hence, the ZigBee Pro technology has been broadly utilized in numerous applications, especially in the control and monitoring applications.

5.3.5 WirelessHART

WirelessHART that is a developed version of the Highway Addressable Remote Transducer (HART) protocol is the IEEE 802.15.4 standard based wireless communication technology. In 2007, the HART Communication Foundation (HCF) introduced this technology as an open standard. In April 2010, International Electrotechnical Commission approved the technology with IEC 62591 as an international and industrial wireless communication standard. The WirelessHART technology exploits the channel-hopping technique in order to prevent interference, and uses transmission power adaptation. In addition, this technology promotes mesh network topology with time synchronization, self-organization and self-healing features. The main aim of this technology is to provide wireless communication opportunity with high reliability and security for industrial applications.

Field devices, mobile devices, network manager, access points and gateway are the main components of a WirelessHART system. The field devices are used at the industrial facilities to accomplish routing and data acquisition processes possible. Then, obtained data are transmitted to the gateways through access points. A network manager is utilized to organize and manage wireless communications among devices existing in the network. The WirelessHART technology is also able to support several network topologies such as star, mesh and combination of star and mesh topology. It is also important to note that this technology cannot contact with other networks configured according to different technologies unless a specific gateway is utilized among the networks.

5.3.6 ISA100.11a

International Society of Automation (ISA) developed the ISA100.11a in 2009 that is a wireless networking technology, and it is approved by the IEC with 62734 standard code in 2014. Different from the formerly proposed wireless networking technologies, the ISA100.11a aims to provide new perspectives in terms of connectivity, flexibility, reliability, coverage and security concepts. This technology is powerful against noise and interference effects that are widely available in industrial environments. Therefore, the ISA100.11a technology ensures robust, reliable and secure wireless communications for uncritical monitoring, controlling and management applications whose latency value may be high as much as several hundreds of milliseconds. In addition, it is able to function with other wireless technologies developed based on the IEEE 802.11 and IEEE 802.15 standards.

The network and transport layers of this technology are designed depend on the Internet Protocol Version 6 (IPv6), 6LoWPAN and User Datagram Protocol (UDP) standards whereas data link layer is developed individually in which graph routing, frequency hopping, and time-slotted TDMA transactions are accomplished [28]. There exist two major devices in the network established based on the ISA100.11a that are named field devices and infrastructure devices. The field devices are composed of data acquisition devices, routers and mobile devices whereas infrastructure devices comprise backbone routers, gateways and security devices [31, 32]. The backbone routers present some important advantages that may be sorted out as enhancing network reliability, developing throughput, reducing latency and reducing traffic problems. The comparison of protocol characteristics of IEEE 802.15.4 based emerging technologies is illustrated in Fig. 5.7. The layers of these technologies are abstracted by considering seven-layer OSI reference model where blanks shown in the figure refer that standards do not enclose interested protocol(s).

5.3.7 6LoWPAN

In 2005, a working group of the Internet Engineering Task Force (IETF) in the Internet area has aimed to develop a wireless communication technology enabling the IPv6 over IEEE 802.15.4 networks, and introduced 6LoWPAN technology. This technology intended to present a new concept in which networks based on the IEEE 802.15.4 standards are able to communicate with each other by means of the IP system directly. Therefore, 6LoWPAN technology ensures the interoperability of wireless communication technologies by integrating WSNs to the Internet. The 6LoWPAN technology determines some methods such as encapsulation and header compression that permit IPv6

	ZigBee Pro	WirelessHART	ISA 100.11a	6LoWPAN	6TiSCH
APP	Object Oriented Profile Protocol	Command Oriented HART Protocol	Object Oriented Native Protocol	Object Oriented Profiles (HTTP)	Object Oriented Apps
APS	Device Discovery Blinding		Basic and Smart Tunneling		IETF CoAP
Transport		Block Data Transfer Stream Transport	Optional Security Connectionless Service	UDP, TCP, ICMP	IETF RPL
Network	Addressing, Routing, Network Joining	Graphic/Source Superframe Transport	Addressing, Routing, Address Translation	IPv6, RPL and Adaption Layer	IETF 6LoWPAN
DLL		Routing, Slot Timing, TDMA/CSMA, Hopping	Routing, Slot Timing, TDMA/CSMA, Hopping		IETF 6P (6TiSCH)
MAC	IEEE 802.15.4 MAC	IEEE 802.15.4 MAC	IEEE 802.15.4 MAC	IEEE 802.15.4 MAC	IEEE 802.15.4 MAC TSCH
PHY	IEEE 802.15.4 PHY (2.4 GHz Radio Freq.)	IEEE 802.15.4 PHY (2.4 GHz Radio Freq.)	IEEE 802.15.4 PHY (2.4 GHz Radio Freq.)	IEEE 802.15.4 PHY (2.4 GHz Radio Freq.)	IEEE 802.15.4 PHY (2.4 GHz Radio Freq.)

FIG. 5.7 OSI reference model comparison of emerging communication technologies based on IEEE 802.15.4 standard (*APP*, Application Layer; *APS*, Application Support sublayer; *DLL*, Data Link Layer).

packets to be transmitted to and/or received from 802.15.4 based wireless networks. In addition, the 6LoWPAN is a promising technology for the IoT applications since it makes possible all capabilities of the IPv6 protocol on small-scale devices. The data payload length of IEEE 802.15.4 standard is 127 bytes whereas the standard IPv6 packet header size is 40 bytes. The packet size adaptation between IEEE 802.15.4 network and IPv6 network, and address resolution arranging among these networks should be provided by the 6LoWPAN technology. Therefore, an adaptation layer is characterized by the 6LoWPAN technology that is defined on the IEEE 802.15.4 MAC layer, and the adaptation layer intends to meet requirements of the IPv6.

Since the 6LoWPAN employs a new concept, this feature of technology makes it superior than that of the other low-powered WSN technologies based on the IEEE 802.15.4 standard. However, the technology is not yet as popular as the other ones. In the event of considering the progress of systems exploiting data packets, it is predicted that the 6LoWPAN will be a common wireless technology in a short while.

5.3.8 6TiSCH

The advances on time slotted channel hopping (TSCH), deterministic and synchronous multichannel extension (DSME) methods, which are specified by the IEEE 802.15.4e standard in 2012, encouraged the use of IPv6 protocol in industrial networks. IETF 6TiSCH working group adapted these improvements to develop a new wireless technology [30] that is called as 6TiSCH technology. It is a novel standard depend on IEEE 802.15.4 standard similar to ZigBee Pro, WirelessHART, 6LoWPAN, and ISA100.11a [31, 33, 34]. By connecting MAC sublayer with network layer, the 6TiSCH technology intends to provide wireless communications through IPv6 protocol over the TSCH. Moreover, this technology intends to develop a novel standardizing method for assisting various resource scheduling procedures. The deficiencies appeared in scheduling and managing processes, which was not accomplished by the IEEE 802.15.4e standard, are resolved by operation sublayer of the 6TiSCH technology. The main procedures utilized in IEEE802.15.4e TSCH such as the combining of channel hopping and time synchronization are very similar to procedures employed in ISA100.11a and WirelessHART technologies. Thus, it is predicted that the scheduling procedures to be exploited by the 6TiSCH technology may be compatible with these standards. In addition, there exist well-known wireless devices such as OpenWSN, Contiki and RIOT that promote the utilization of 6TiSCH in the WSNs [33, 34].

5.4 IEEE 802.11 based technologies for WLANS

The IEEE 802.11 based standards are the most widely utilized wireless standard for WLANs. One of the most popular standards based on the IEEE 802.11 is

Wi-Fi that is developed by Wi-Fi Alliance [35]. The Wi-Fi term has been created by the Wi-Fi Alliance to present more attractive name rather than the IEEE 802.11 technology. The Wi-Fi Alliance that is formed to promote the technology in 1999 is a non-profit international association of several carriers, vendors, service providers, manufacturers and other related companies. Furthermore, it also manages certification and interoperability issues. The Wi-Fi technology characterizes an upper layer protocol by using IP for enabling wireless communications over the Internet. Even though the Wi-Fi and IEEE 802.11 terms may be employed mutually, there exist minor differences among them. The IEEE 802.11 is a collection of standards, which identifies several fundamental functionalities such as PHY layer, MAC layer, protocols, authentication, QoS and so on. On the other hand, the Wi-Fi Alliance determines technical specifications by referring to IEEE 802.11 standards and manages certification process of devices that ensures interoperability. In addition, Wi-Fi certified devices are immediately able to contact with each other thanks to Wi-Fi Direct technology developed by the Alliance that is presently activated in many smart devices such as smartphones, tablets and notebooks. Furthermore, the Wi-Fi Alliance researches in several issues to provide performance improvements in terms of data rate, security, QoS and hardware based issues [36–38].

The family of IEEE 802.11 standards characterizes the PHY and MAC layers, and supports IP addressing. The most popular standards of IEEE 802.11 family are 802.11a, 802.11b, 802.11g, 802.11n, and recently released 802.11ah. IEEE 802.11 networks are intended to function in unlicensed frequencies and they are serving on the ISM bands by generally using 2.4 and 5 GHz carrier frequencies. In addition, below 1 GHz and above 60 GHz frequency bands are presently being taken into account for the PHY layer characterization. Several PHY layer specifications, FHSS, DSSS and OFDM methods are exploited by the IEEE 802.11 standards. The FHSS based systems allocate the 2.4 GHz band into sub-channels where each sub-channel has 1 MHz bandwidth. In addition, transmitter devices dynamically vary the channels by employing predefined sequence sets in the standard. An improved channel utilization method is the DSSS technique that is also exploited in Code Division Multiple Access (CDMA) systems. Firstly, data are multiplied with a chip sequence in this method and then, either differential binary phase shift keying (DBPSK) or differential quadrature phase shift keying (DQPSK) modulation scheme is used to convey data. Complementary code keying (CCK) modulation scheme may be used to advance data rates of the DSSS, which is called as high rate DSSS (HR/DSSS). In addition, there are some standards based on IEEE 802.11 employing OFDM method that is a combination of mapping and multiplexing method. This method uses many orthogonal sub-carriers in order to carry data.

A generalized MAC frame format of the standard is illustrated in Fig. 5.8. In addition to the MAC data frame, standard also utilizes RTS and CTS control frames.

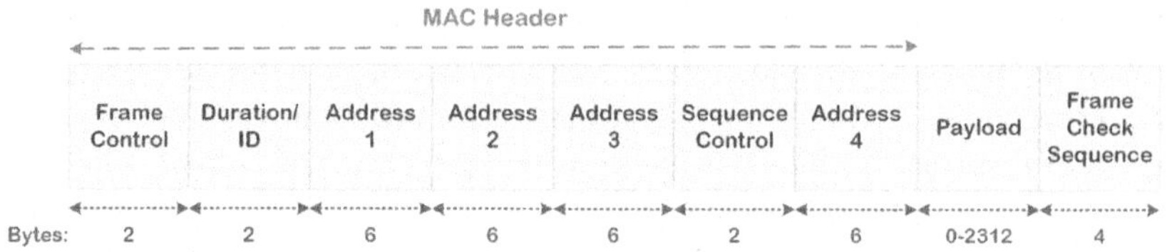

MAC Header

	Frame Control	Duration/ ID	Address 1	Address 2	Address 3	Sequence Control	Address 4	Payload	Frame Check Sequence
Bytes:	2	2	6	6	6	2	6	0-2312	4

FIG. 5.8 MAC frame structure of IEEE 802.11 standard.

IEEE 802.11a works on 5 GHz frequency by employing OFDM method with 52 subcarriers. Theoretically, the maximum data rate of this standard is 54 Mbps.

IEEE 802.11b works on 2.4 GHz frequency of the ISM band by using DSSS method. Data rate of this standard can reach up to 1 Mbps for outdoor environments while reachable data rate is around 11 Mbps for indoor environments. While the range for indoor is approximately 30–40 m, the range for outdoor is nearly 90–100 m [39–41].

IEEE 802.11g works on 2.4 GHz frequency of the ISM band by using either the DSSS or the OFDM method, and the standard is able to provide 54 Mbps data rates. In addition, this standard is compatible with devices based on the IEEE 802.11b standard [39–41].

IEEE 802.11n works on both 2.4 and 5 GHz frequency of the ISM band by using OFDM method. In addition, this standard can reach 600 Mbps data rates by supporting Multiple Input-Multiple Output (MIMO) scheme.

IEEE 802.11ad operates in millimeter wave band (60 GHz) since it intends to support ultra-high-throughput short-range communication. Its reachable data rate is around 6.76 Gbps.

IEEE 802.11af that re-characterizes PHY and MAC layers is also referred as White-Fi or Super Wi-Fi, and it was approved in 2014. The main aim of this standard is to enable WLAN in TV white spectrum (TVWS) between 54 and 790 MHz. This band was originally allocated for analog TV signals but it is not currently employed for transmission of analog TV signals due to digital transmission of TV signals.

IEEE 802.11ah that is an amendment of IEEE 802.11-2007 standard is published in 2017, and it is also called as Wi-Fi HaLow. This recent standard employs 900 MHz unlicensed bands in order to extend range of Wi-Fi networks when compared with traditional Wi-Fi networks functioning in the ISM bands with 2.4 and 5 GHz frequencies. The main advantages presented by the standard are its lower power consumption and longer coverage features. These features of the standard make it convenient for M2M, IoT and SG applications.

The PHY layer of this standard exploits OFDM method with 64 subcarriers and beamforming is supported for both multi user MIMO and single user systems. The developments on power saving mode are mainly realized in the MAC layer. Therefore, devices may be operated through batteries. The standard is intended to support star topology in infrastructure mode. Nonetheless, IEEE 802.11 based standards supporting mesh network scheme may be revised for supporting the IEEE 802.11ah to extend network with wider coverage. The CSMA/CA technology, which is an advanced version of CSMA technology employed in Ethernet, is exploited in MAC layer of this standard. Firstly, the CSMA/CA method senses the channel, and then data transmission is performed if the channel is idle. In the event of the channel is busy, station should wait for a while until continuing transmission completes, which is called as contention period.

IEEE 802.11e-2005 characterizes the MAC procedures to assist LAN applications with QoS requirements. The mentioned procedures are related to transmitting audio, voice and video over the WLANs. It provides remarkable improvements for delay constrained environments.

IEEE 802.11s defines several protocols for IEEE 802.11 stations to create self-configured multi-hop networks that are able to serve in both broadcast/multicast and unicast data transmission. This standard promotes Hybrid Wireless Mesh Protocol (HWMP) that is a routing protocol based on Ad hoc On-demand Distance Vector (AODV). This standard is generally employed in outdoor applications required wide coverage.

The coverage areas and operating frequencies of the IEEE 802.11 based standards are illustrated in Fig. 5.9 while technical specifications of them are listed in Table 5.2.

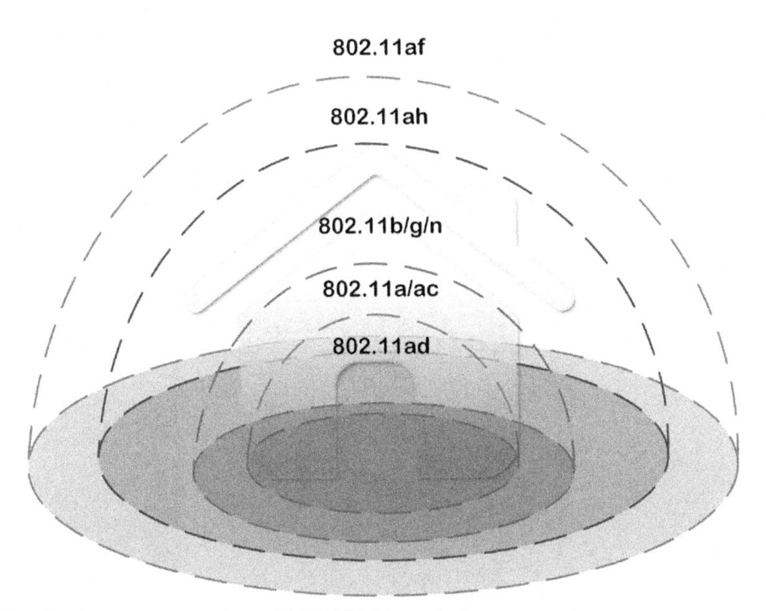

FIG. 5.9 Coverage comparison of IEEE 802.11 standards.

TABLE 5.2 Comparison of 802.11standards

	802.11								
	–	*b*	*a*	*g*	*n*	*ac*	*ad*	*af*	*ah*
Year adopted	1997	1999	1999	2003	2009	2014	2016	2014	2016
Frequency band	2.4 GHz	2.4 GHz	5 GHz	2.4 GHz	2.4/5 GHz	5 GHz	60 GHz	2.4/5 GHz	900 MHz
Bandwidth	20 MHz	20 MHz	20 MHz	20 MHz	20/40 MHz	20/40/80/160 MHz	2.16 GHz	6/7/8 MHz	1/2/4/8/16 MHz
Modulation scheme	BPSK to 256-QAM	BPSK to 256-QAM	BPSK to 256-QAM	BPSK to 256-QAM	BPSK to 256-QAM	BPSK to 256-QAM	BPSK to 64-QAM	BPSK to 256-QAM	BPSK to 256-QAM
Transmit scheme	FHSS, DSSS	DSSS, CCK	OFDM	DSSS, OFDM	OFDM	OFDM	SC, OFDM	SC, OFDM	SC, OFDM
Peak data rate	2 Mbps	11 Mbps	54 Mbps	54 Mbps	600 Mbps	6.93 Gbps	6.76 Gbps	26.7 Mbps	40 Mbps
Range	20 m	35 m	35 m	70 m	70 m	35 m	10 m	1 km	1 km
Maximum transmit power	100 mW	100 mW	100 mW	100 mW	100 mW	160 mW	10 mW	100 mW	100 mW

5.4.1 Network architectures of IEEE 802.11 standards

Network architectures of the IEEE 802.11 standards can be classified into two categories that are referred as infrastructure network architecture and ad hoc network architecture. In an IEEE 802.11 network, a device having the communication ability is called a station while a device establishing an infrastructure network is referred as an access point (AP). A connection between a station and an AP should be realized to enable communication in an infrastructure network. Hence, communication among the stations is generally relayed through the AP. Whole stations containing APs in the network establish a Basic Service Set (BSS). A Basic Service Set Identifier (BSSID) that identifies a MAC address of the AP in the network defines this service group. Each 802.11 network has a Service Set Identifier (SSID) that can be identified by using case-sensitive text strings as long as 32 characters. In addition, WLAN networks can be combined to extend range of network, which is referred as Extended Service Set (ESS). In ad hoc network architectures, Independent Basic Service Set (IBSS) authorizes the stations to enable P2P communication where the use of the APs does not require.

The RTS and CTS may optionally be employed to prevent frame collisions in IEEE 802.11 networks. The main aim of the use of RTS and CTS is to overcome hidden terminal problem (HTP) that is illustrated in Fig. 5.10. The HTP appears if there exists one AP (or station, Station 2) in the middle of two or more stations (Stations 1 and 3). In the event of the Stations 1 and 3 are remotely located, the stations are not able to connect with each other. However, they can contact with the Station 1. If the Stations 1 and 3 simultaneously send message to Station 2, a collision called HTP appears. This collision, the HTP, can be

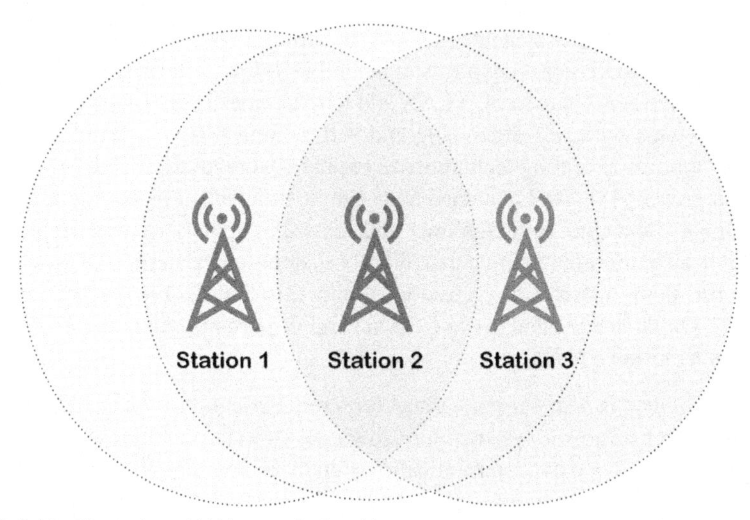

FIG. 5.10 Illustration of hidden terminal problem.

prevented by using the RTS and CTS mechanism. If the Station 1 intends to convey data, it transmits an RTS message to Station 2, and the Station 2 approves transmission by transmitting a CTS message that is sent to all stations within the coverage areas of the Station 2. After this, the stations except Station 1 and 2 avoid data transmission.

5.4.2 Application perspective of WLAN systems

From the concept of SG applications, Wi-Fi is an important technology for SEP 2.0. It is able to provide efficient performance in both shared spectrum and noisy channels. In addition, this technology can support many application scenarios and IP based protocols as mentioned before. It is worth noting that the Wi-Fi Alliance has characterized several specifications for certified devices that can fulfill requirements of the SEP 2.0. Since the Wi-Fi connectivity has been already covered all over the world, it can play a considerable role in popular application scenarios such as smart metering, smart energy management and smart home automation. This technology aims to provide services in home area networks (HANs), neighborhood area networks (NANs) and field area networks (FANs) of the SG systems. Using the Wi-Fi in these networks may also develop interoperability. On the other hand, there are some drawbacks of the Wi-Fi technology. Its high-power consumption has been considered as a disadvantage over a long time. The development of novel Wi-Fi chips with lower power consumption has presented new solutions to overcome this problem. In addition, these developed Wi-Fi chips are able to operate many years like ZigBee chips.

Ethernet based communication infrastructure for substation automation systems has been proposed by IEC 61850 standard. This standard based WLAN system is able to improve protection of distribution substations thanks to smart monitoring and control systems [42–44]. In addition, Wi-Fi systems including smart sensing and control devices can be employed in protection applications. Moreover, the combination of WLAN and wireless mesh concepts may provide robust systems with self-organizing and self-healing features. Wireless mesh exploits multi-hop routing techniques to extend coverage range with low transmission power [45]. Wi-Fi enabled SMs can be utilized for repeating information signals, and thus coverage area and network capacity may be improved [46]. It is also important to note that WLAN systems can also support these services for both distribution substations and Distributed Energy Resources (DERs). On the other hand, there exist several important problems for WLAN systems as follows [47].

- Interference is very crucial issue for communication signals. Electrical equipment of power systems may adversely affect communication signals or even may wipe out communication signals completely. In addition, other wireless devices may also generate interference effects, which will cause negative influences on communication signals.

- Wireless devices for Industrial applications are not ready yet. Nonetheless, new developments on error correcting codes and protocols have improved the stability of wireless communication systems.

Therefore, Wi-Fi technology is appropriate for applications that require low data rates and environments with low interference. Channel assignment of Wi-Fi systems is variable in the world. As a result of this situation, overlap issue is occurred in the channels, in other words, limited number of devices are connected to the Wi-Fi network [48]. In addition, interference issue may also lead to security problems in Wi-Fi applications, especially in Wi-Fi based SG applications. Security infrastructure of the Wi-Fi systems is implemented by using Wi-Fi Protected Access (WPA) that is a security standard developed for devices equipped with wireless internet connections. The Wi-Fi Alliance improved the WPA to ensure advanced data encryption and better user authentication than that of Wired Equivalent Privacy (WEP) that is the fundamental security standard of the Wi-Fi technology. The enhanced version of the WPA is WPA2 that is approved by the IEEE in 2004. While the WPA is based on Extensible Authentication Protocol (EAP) and Advanced Encryption Standard (AES), the WPA2 employs Counter Mode Cipher Block Chaining Message Authentication Code Protocol (CCMP) that is an AES-based encryption method. The Wi-Fi Alliance introduced WPA3 as a new security standard in January 2018. This security standard employs 128-bit data encryption in WPA3-Personal mode while 192-bit data encryption is used in WPA3-Enterprise mode. The WPA3 standard will reduce security problems occurred by weak passwords and will make easier configuration transactions of devices without having display interface.

5.5 Cellular communication technologies

Effective utilization of new ICTs has turned into a significant factor for developing global economy. One of the most critical factor for the universal ICT is wireless communications that are able to play a significant role on the advances of other sectors. European Mobile Observatory (EMO) has reported that mobile communication market has gained a great revenue level by outdistancing other major sectors [49, 50]. It is also clear that wireless and mobile communication technologies should universally continue to evolve for satisfying user demands. Recently, request of location and time independent communication has turned into one of the indispensable conditions for everyone, even for all smart things. In order to meet mentioned requests, the use of mobile communication technologies is highly suggested because of superiorities of these technologies.

When the development of cellular communication systems is taken into account from first generation (1G) to fourth generation (4G), it is obvious that each generation aims to offer more reliable communication system by eliminating deficiencies of former communication generation. These deficiencies are firmly related to coverage, spectral efficiency, data rate and mobility [51, 52]. The 1G cellular systems, which were designed on the basis of narrowband and analogue systems, were introduced in the early years of the 1980s. Advanced Mobile Phone System (AMPS), Total Access Communication System (TACS) and Nordic Mobile Telephone (NMT) were the most popular standards utilized in the 1G cellular systems. These standards exploited frequency-division multiple access (FDMA) technique, and typically provided 2.4 Kbps data rates. Weak spectral efficiency and security issues were the major problems of the 1G systems. Later, second generation (2G) cellular systems were launched at the beginning of 1990s. In order to provide enhanced capacity and wider coverage areas, 2G cellular systems used digital systems different from the 1G systems that were based on the analogue systems. The 2G systems not only advanced voice communication but also enabled text message opportunity for the users. These systems employed TDMA or CDMA method to present higher data rates than 1G systems, as much as 64 kbps data rates. Global System for Mobile communication (GSM), digital AMPS (D-AMPS), Personal Digital Cellular (PDC) and CDMA One (IS-95) were the most popular standards utilized in the 2G cellular systems. Even though 2G systems have provided important developments compared to 1G systems, reachable data rates of 2G systems were still limited that could not present sufficient speeds for the users.

Third generation (3G) mobile communication system, which was the first international standard introduced by International Telecommunication Union (ITU), aimed to provide significant improvements than previous generations. The major innovation of 3G cellular systems was their data based characteristics. Therefore, 3G systems could reach to two Mbps data rates by employing IP. In addition, new services such as multimedia messaging, online TV and video calling have been enabled for the users. Wideband CDMA (W-CDMA) technology could support frequency division duplex (FDD) and time division duplex (TDD) modes in 3G networks. International Mobile Telecommunications-2000 (IMT-2000), Universal Mobile Telecommunications Systems (UMTS), and CDMA 2000 were the most popular technologies utilized in the 3G cellular systems. After these standards, High Speed Uplink/Downlink Packet Access (HSUPA/HSDPA) and Evolution-Data Optimized (EVDO) technologies were introduced as 3.5G cellular technologies that could provide data rates as much as 30 Mbps [52–54]. The evolution of the cellular communication systems is illustrated in Fig. 5.11.

The ITU has defined main requirements and several specifications of the 4G communication systems which aim to provide 100 Mbps data rates for high-mobility users and 1 Gbps data speeds for low-mobility users. This generation of cellular communication systems can present wireless communications with high-speed by exploiting 20 MHz bandwidth when compared to former generations. There exist two popular technologies for 4G systems, which are Long-Term Evolution (LTE) and Worldwide Interoperability for Microwave Access (WiMAX). While the LTE is introduced by 3rd Generation Partnership Project

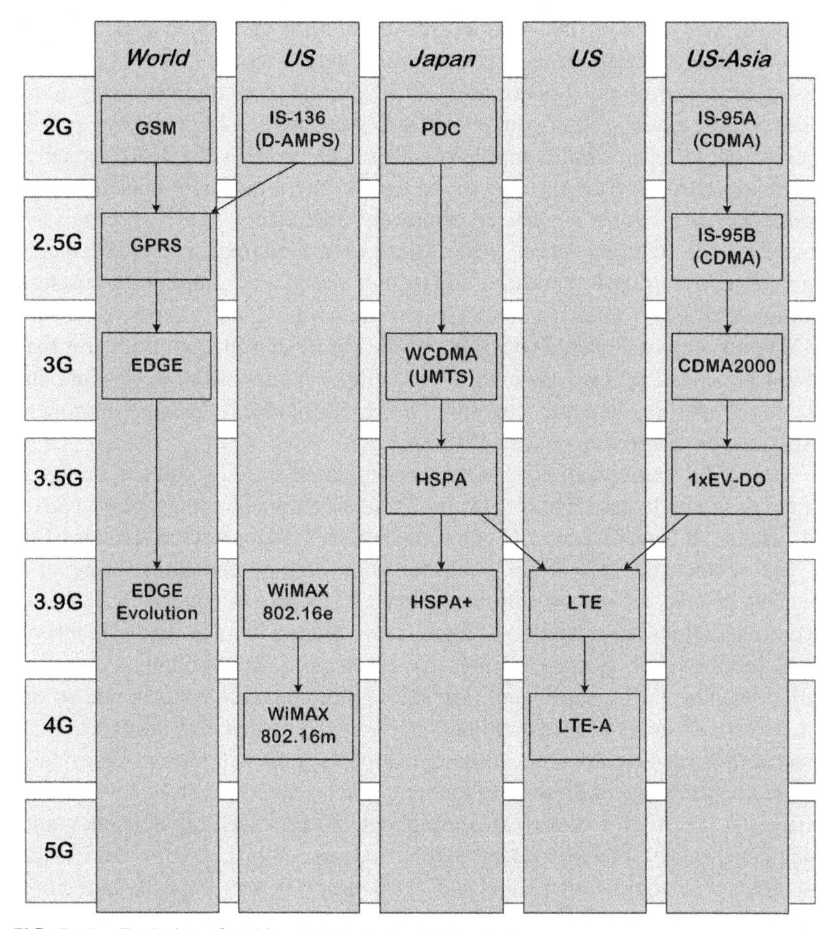

FIG. 5.11 Evolution of mobile communication technologies.

(3GPP), the WiMAX is proposed by the IEEE. The LTE technology utilizes orthogonal frequency division multiple access (OFDMA) for downlink connections and single carrier frequency division multiple access (SC-FDMA) for uplink connections. The WiMAX technology employs the OFDMA for both connection types [54].

Currently, mobile users are able to connect Internet with lower latency and higher data rates due to the fact that 4G communication systems are started to be employed worldwide. Recently, increasing popularity of smart devices (e.g., smart phones, netbooks, tablets etc.) and increasing internet access possibilities have caused remarkable changes in user habits. For instance, users have experienced watching high-definition videos, employing new and useful third party applications and social media platforms. As a result of the increased utilization of the Internet, service providers have encountered broader bandwidth demands. The 4G and first reported results of fifth generation (5G) cellular communication systems are showed that millimeter wave (mmWave) frequencies may present new opportunities for wireless communication systems because there exists a significant potential to attain higher data rates by exploiting existing idle bandwidths in mmWave frequencies [55–57]. Unfortunately, there exist various problems to be overcome by emerging communication standards. One of the most significant issues is insufficiency of RF spectrum for current mobile communication systems since ultra-high frequency (UHF) bands are thoroughly utilized. Another issue is high energy consumption of wireless technologies. For example, a cellular phone network typically needs about 40–50 MW power to maintain its operation [58]. The other important problems that should be solved by next generations of cellular communication systems are related to mobility, data rate, coverage, spectral efficiency, QoS and interoperability issues of heterogeneous networks.

Later 4G communication systems were released in 2011, researchers have started to research new technologies for cellular communication systems called 5G systems. It is waited that 5G communications systems will be standardized by 2020s. Recently, ITU-R has published some recommendations for general concepts of next generation cellular systems for 2020 and beyond [59]. These recommendations are related to utilization situations and main requirements of novel services such as wireless industry automation, SGs, augmented reality (AR), e-health, traffic safety and efficiency, remote tactile control, and so on [54, 59]. If new generation of cellular systems are compared with the 4G cellular systems, it is predicted that 5G communication systems will present significant developments. Some of these developments can be sorted out as 10-fold energy efficiency, 25 times average cell throughput, 10-fold spectral efficiency and data rates (i.e., 10 Gbps peak data rate for low-speed mobile systems and 1 Gbps for high-speed mobile systems), and 1000 times system capacity per km^2. Furthermore, this new generation system aims to support several application circumstances such as communication in high-speed vehicles, 4K video streaming without disconnection and so on, which could not be accomplished by the previous generations [50]. Various services and applications to be experienced by users through new generation cellular systems are illustrated in Fig. 5.12 [60]. There exist two essential trends directing expected scenarios. The first one is that next generation communication systems will gather all wireless systems under the same umbrella to enable data acquisition, observing and managing of devices. For example, SG systems will be more efficient and powerful since sensors, SMs and whole systems can communicate with each other over high-speed wireless connection. In addition, IoT and M2M services will be managed more effectively. The second expectation is related to big data that is a result of growing connected sensors, devices and novel services and applications such as remote health check, ultra-high-definition video streaming, AR and so on. Moreover, industry initiatives are determined major requirements of 5G cellular systems, which can be summarized as follows [51].

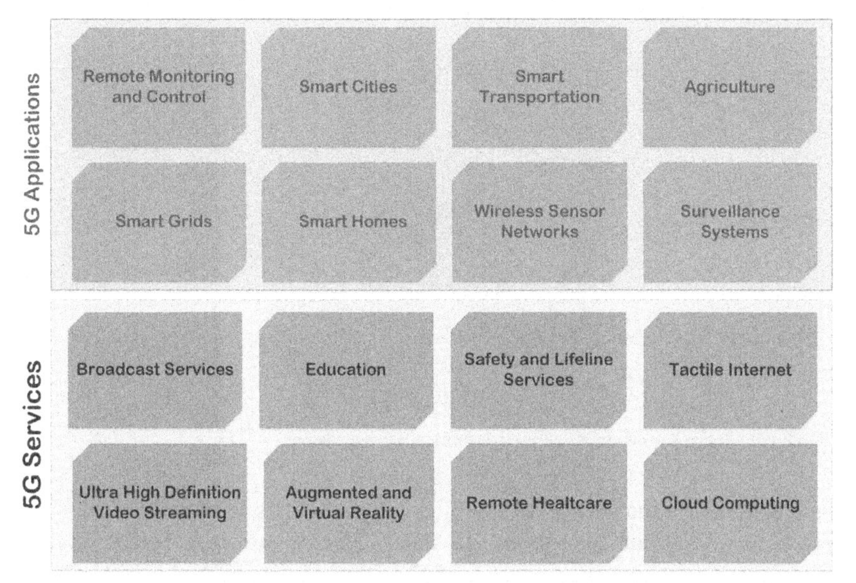

FIG. 5.12 Possible application and service samples for 5G cellular communication systems.

- Full coverage
- 99.999% availability
- 90% reduced energy consumption
- 1000-fold more bandwidth opportunity per unit area
- 1–10 Gbps data rates for entire components of the network
- Longer battery life as much as 10 years
- Providing service to about 100 times more devices
- 1 ms end-to-end (E2E) latency

The key requirements of new generation cellular systems are also summarized in Fig. 5.13. The 5G systems have to overcome forthcoming user traffics which will be larger and more complex than present networks. This problem is considered as one of the important challenges of dense network systems. In order to support growing user traffics, providing 1000x system capacity per km^2 than 4G systems is aimed in 5G systems. Another significant need is to provide higher data speed than that of the 4G networks. In the event of improvement of cloud systems and dense content applications are considered, the 5G cellular systems not only should ensure better quality of the user experience (QoE) but also should intent to present higher data rate services. The next generation wireless communication systems should allow plenty devices to be connected to the network in order to ensure better assistance [51, 60].

Cellular communication systems are regarded as one of the key technologies for SG applications such as smart metering, phasor measurement unit, smart energy management, and remote monitoring [61]. Reliability, security, latency, and entire performance requirements of SG applications can be supported by

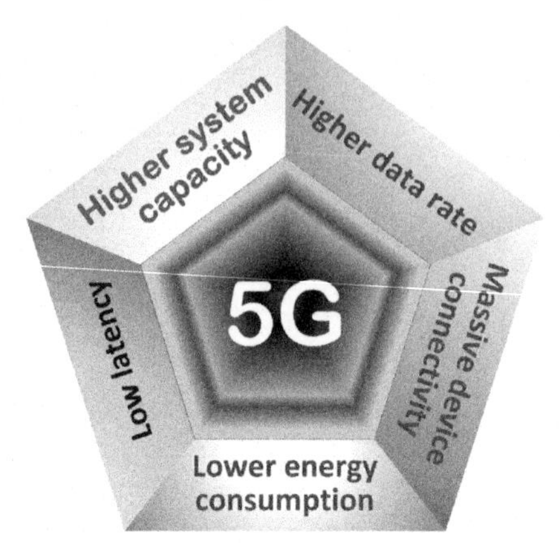

FIG. 5.13 The key aims of 5G cellular communication systems.

these communication networks. Several service providers and operators are currently presenting solutions and services for SG applications since there exist various reasons providing advantages for SG systems. In addition, this technology has been utilized by the utilities for monitoring purposes. There are transmission line monitoring devices that can transmit measurements and notification messages through the cellular communication systems. It is advantageous to employ cellular communication systems since the network is already in use and there is no installation cost. In addition, improved security mechanisms are used in these systems. On the other hand, bad system performance under emergency circumstances are regarded the disadvantage of the cellular communication systems.

5.6 IEEE 802.16/WiMAX

WiMAX technology, which is standardized thanks to IEEE 802.16 for enabling broadband wireless access in both mobile and non-mobile point-to-multipoint communications, is specifically developed for Wireless Metropolitan Area Network (WMAN) applications by providing high data rates as much as 70 Mbps within a large coverage area. Even though the primary aim of this technology is to accomplish worldwide interoperability for microwave access, the WiMAX technology is a good candidate to enable connection among data management points and SMs in SG systems. The IEEE 802.16 standard suggests utilizing 2–66 GHz frequency spectrum, and interoperability features of the technology has been specified by the WiMAX forum. The exploited spectrum of this technology can be classified in two groups according to line-of-sight (LOS) status. While the frequencies between 11 and 66 GHz are utilized for LOS conditions, the frequencies from 2 to 11 GHz are exploited for none-LOS (NLOS) cases. Also, the 2.3, 2.5, and 3.5 GHz licensed frequency bands have been allocated for mobile communications whereas 3.5 GHz licensed band and 5.8 GHz unlicensed band have been devoted for non-mobile communications. Typically, 25

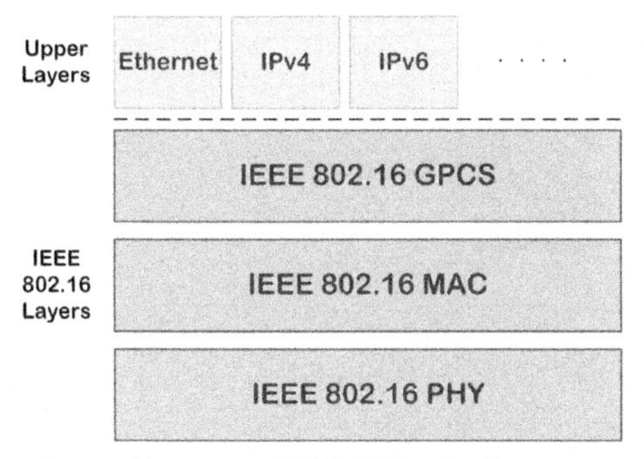

FIG. 5.14 GPCS based protocol structure of WiMAX technology.

or 28 MHz channel bandwidth is utilized in IEEE 802.16 standard where LOS is needed because of short wavelength.

The protocol structure of IEEE 802.16 standard that is illustrated in Fig. 5.14 contains PHY layer, MAC layer and a Generic Packet Convergence Sublayer (GPCS). The PHY layer of the technology is based on OFDMA technology. Theoretically, the technology can reach up to 70 Mbps data rates, and the range of WiMAX is approximately 50 km in non-mobile scenarios while it is about 5 km in mobile scenarios [44]. The improved versions of the technology can also support MIMO technology and beamforming methods. In order to ensure secure communication, WiMAX characterizes three steps that are authentication, key establishment, and data encryption. These steps are accomplished at the MAC layer through AES algorithm, Privacy Key Management protocol and EAP. In addition to the security transactions, the MAC layer of the technology can enable the use of scheduling mechanisms and sleep control mechanisms for energy saving processes. In addition, other important characteristics of the WiMAX are related to low latency, low operating costs, availability and scalability of traffic management systems.

Although this technology works like Wi-Fi technology for last mile access, it provides higher data rates, wider coverage and capacity than that of the Wi-Fi systems. In addition, WiMAX can present better connection quality when compared with Wi-Fi technology. In Wi-Fi systems, if there exist multiple users connected to an AP, the users have to compete with each other due to used contention-based CSMA mechanism for channel access. Therefore, each user may have different bandwidth in the Wi-Fi systems. On the other hand, WiMAX technology exploits a grant-request mechanism in MAC layer to allow data exchange. This characteristic of technology provides better use of radio resources. Furthermore, this technology can be employed to establish wide area wireless backhaul network that can function either in P2P or mesh modes. The backhaul-based WiMAX systems implemented in mesh mode can improve wireless coverage and they can ensure reduced deployment cost, reconfigurability and fast deployment.

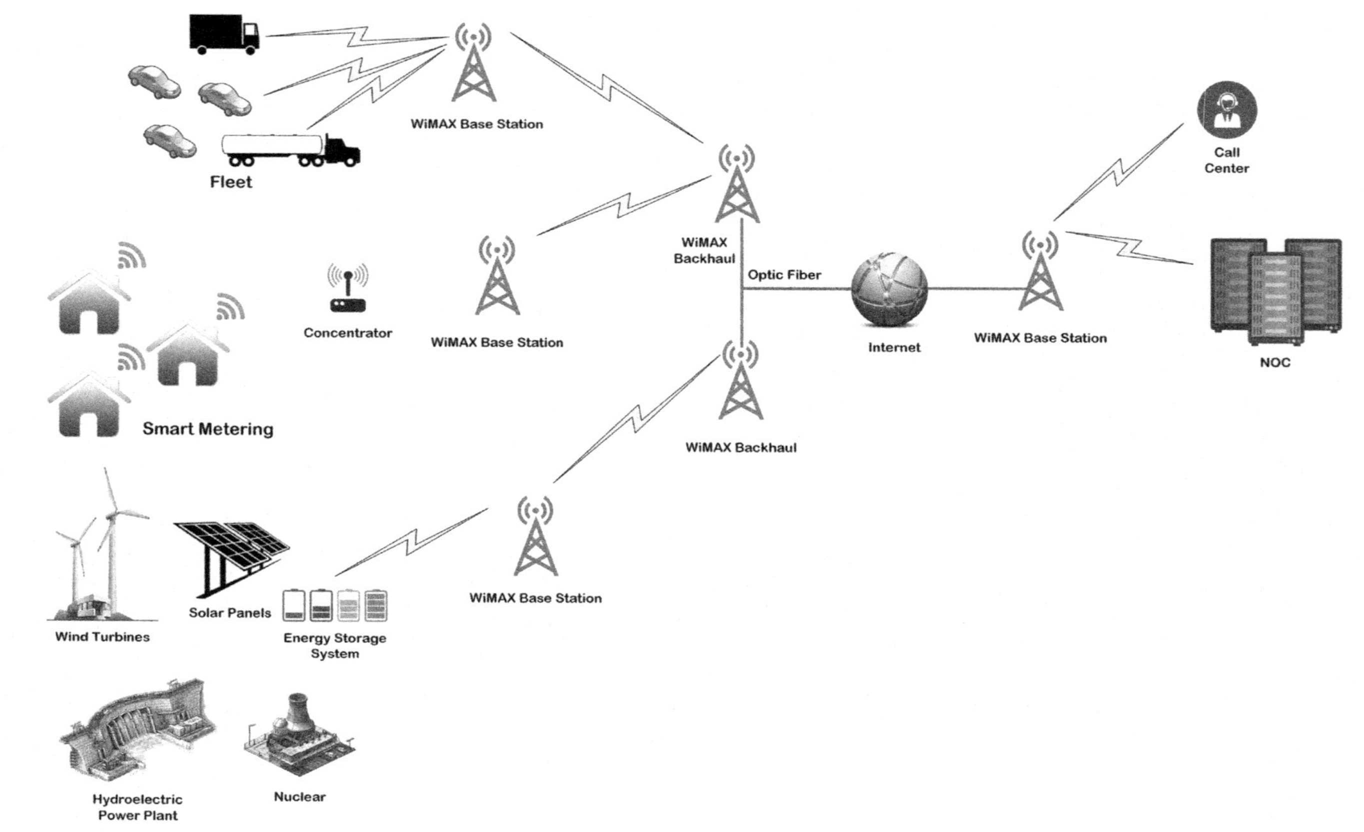

FIG. 5.15 Connectivity of WiMAX technology in smart grid metering applications.

Since the WiMAX technology was primarily designed for the WMANs, it can be regarded as a promising technology for SG WANs [45, 62]. This technology is regarded as a promising technology for Advanced Metering Infrastructure (AMI) applications because of low latency, lower costs, enhanced security and scalability advantages. If the AMI system is conducted via this technology, real time pricing services may effectively serve. In addition, this technology can be exploited for remote monitoring applications in SG systems. To sum up, the duty of this technology will change according to requirements, system background, infrastructures and environmental conditions. The deployment of WiMAX technology in the SG applications is illustrated in Fig. 5.15. The duties of the WiMAX technology in SG implementations can be classified into four groups as backhaul, last-mile connectivity, mobility and emergency [63]. In backhaul scenario, the technology can supply connections between base stations and network operating center (NOC) where the WiMAX ensures interoperability among several wired and wireless communication methods utilized in terminal devices. One of the most important advantage of this technology is mobility provided in the wide coverage areas. This advantage allows supporting mobile units and services at the same network infrastructure. Last-mile connectivity case is related to connection of technology directly to terminal devices available in the network, which is an important requirement for remote monitoring applications. When an emergency occurs, mobile base stations and related devices can be transferred to emergency area for establishing a tentative network that may exploit WiMAX or other communication technologies as backhaul.

References

[1] C. Sixto, L. Jorge, Scheduling the real-time transmission of periodic measurements in 802.15.4 wireless sensor network. Prog. Comput. Sci. 114 (2017) 499–506, https://doi.org/10.1016/j.procs.2017.09.015.

[2] Y. Zhan, Y. Xia, M. Anwar, GTS size adaptation algorithm for IEEE 802.15.4 wireless networks. Ad Hoc Netw. 37 (2016) 486–498, https://doi.org/10.1016/j.adhoc.2015.09.012.

[3] I.F. Akyildiz, W. Su, Y. Sankarasubramaniam, E. Cayirci, A survey on sensor networks. IEEE Commun. Mag. 40 (2002) 102–114, https://doi.org/10.1109/MCOM.2002.1024422.

[4] W.Z. Song, B. Shirazi, R. Huang, M. Xu, N. Peterson, R. LaHusen, J. Pallister, D. Dzurisin, S. Moran, M. Lisowski, S. Kedar, S. Chien, F. Webb, A. Kiely, J. Doubleday, A. Davies, D. Pieri, Optimized autonomous space in-situ sensor web for volcano monitoring. IEEE J. Sel. Top. Appl. Earth Obs. Remote Sens. 3 (2010) 541–546, https://doi.org/10.1109/JSTARS.2010.2066549.

[5] L.M.R. Peralta, B.A.I. Gouveia, D.J.G. de Sousa, C.d.S. Alves, Enabling museum's environmental monitorization based on low-cost WSNs. in: 2010 10th Annual International Conference on New Technologies of Distributed Systems (NOTERE), 2010 , pp. 227–234, https://doi.org/10.1109/NOTERE.2010.5536677.

[6] C. Gomez, J. Paradells, Wireless home automation networks: a survey of architectures and technologies. IEEE Commun. Mag. 48 (2010) 92–101, https://doi.org/10.1109/MCOM.2010.5473869.

[7] M.H. Rehmani, A.-S.K. Pathan (Eds.), Emerging Communication Technologies Based on Wireless Sensor Networks: Current Research and Future Applications, CRC Press, Boca Raton, FL, 2016.

[8] S.J. Isaac, G.P. Hancke, H. Madhoo, A. Khatri, A survey of wireless sensor network applications from a power utility's distribution perspective, in: AFRICON, 2011, 2011, pp. 1–5, https://doi.org/10.1109/AFRCON.2011.6072184.

[9] Z. Zhu, S. Lambotharan, W.H. Chin, Z. Fan, Overview of demand management in smart grid and enabling wireless communication technologies. IEEE Wirel. Commun. 19 (2012) 48–56, https://doi.org/10.1109/MWC.2012.6231159.

[10] T. Li, J. Ren, X. Tang, Secure wireless monitoring and control systems for smart grid and smart home. IEEE Wirel. Commun. 19 (2012) 66–73, https://doi.org/10.1109/MWC.2012.6231161.

[11] A. Nazim, D. Benot, T. Fabrice, Bandwidth and energy consumption tradeoff for IEEE 802.15.4 in multihop topologies. in: M. Matin (Ed.), Wireless Sensor Networks—Technology and Applications, InTech, 2012https://doi.org/10.5772/48295.

[12] W. Ye, J. Heidemann, D. Estrin, An energy-efficient MAC protocol for wireless sensor networks. in: Proceedings Twenty-First Annual Joint Conference of the IEEE Computer and Communications Societies, vol. 3, 2002, pp. 1567–1576, https://doi.org/10.1109/INFCOM.2002.1019408.

[13] W. Ye, F. Silva, J. Heidemann, Ultra-low duty cycle MAC with scheduled channel polling, in: Proceedings of the 4th International Conference on Embedded Networked Sensor Systems, ACM, 2006, pp. 321–334.

[14] F. Cuomo, E. Cipollone, A. Abbagnale, Performance analysis of IEEE 802.15.4 wireless sensor networks: an insight into the topology formation process. Comput. Netw. 53 (2009) 3057–3075, https://doi.org/10.1016/j.comnet.2009.07.016.

[15] Bluetooth, Bluetooth Core Specification V 5.0, https://goo.gl/Na3aco, 2016.

[16] S.S. Kulkarni, M. Arumugam, J. Zheng, M.J. Lee, S. Ci, H. Sharif, K. Nuli, P. Raviraj, Lower layer issues—MAC, scheduling, and transmission, in: S. Phoha, T. LaPorta, C. Griffin (Eds.), Sensor Network Operations, John Wiley & Sons, Inc, 2006, pp. 185–261, https://doi.org/10.1002/9780471784173.ch4.

[17] A. Willig, Placement of relayers in wireless industrial sensor networks: an approximation algorithm. in: 2014 IEEE Ninth International Conference on Intelligent Sensors, Sensor Networks and Information Processing (ISSNIP), 2014, pp. 1–6, https://doi.org/10.1109/ISSNIP.2014.6827685.

[18] IEEE, IEEE Standard for Low-Rate Wireless Networks, IEEE Std 802.15.4–2015 (Revision of IEEE Std 802.15.4–2011). (2016) pp. 1–709, https://doi.org/10.1109/IEEESTD.2016.7460875.

[19] IEEE, IEEE 802.15 WPAN Task Group 4g (TG4g) Smart Utility Networks, https://goo.gl/Bg35dt, 2019.

[20] Wi-SUN Alliance, https://goo.gl/45uAw9, 2019.

[21] IEEE, IEEE Draft Standard for Local and metropolitan area networks Part15.4, in: Low-Rate Wireless Personal Area Networks (WPANs) Amendment Physical Layer Specifications for Low Energy, Critical Infrastructure Monitoring Networks (LECIM), IEEE P802.15.4k/D5, 2013.

[22] U. Raza, P. Kulkarni, M. Sooriyabandara, Low power wide area networks: an overview. IEEE Commun. Surv. Tutorials 19 (2017) 855–873, https://doi.org/10.1109/COMST.2017.2652320.

[23] K.S. Kwak, B. Shen, Y. Jin, K.J. Kim, R. Yang, H. Lee, J. Huh, Legacy Based PHY Design for LECIM, IEEE P802.15 Working Group for Wireless Personal Area Networks (WPANs), https://goo.gl/qr2vWt, 2011.

[24] S. Dey, D. Howard, Y. Jin, S. Kato, K.S. Kwak, M.A. Ameen, X. Wang, S. Shimada, LECIM DSSS PHY Merger Update, IEEE P802.15 Working Group for Wireless Personal Area Networks (WPANs), https://goo.gl/AXrtD9, 2011.
[25] M. Brown, Preliminary Draft for IEEE 802.15.4k, IEEE P802.15 Working Group for Wireless Personal Area Networks (WPANs), https://goo.gl/2YZ1jX, 2012.
[26] Zigbee Alliance, (n.d.). http://www.zigbee.org (Accessed 31 December 2017).
[27] HART Communication Foundation Standard, WirelessHART Specification 75: TDMA Data-Link Layer, (2008).
[28] International Society of Automation, Wireless Systems for Industrial Automation: Process Control and Related Applications, (2011).
[29] WPAN IPv6 over Low power WPAN (6LoWPAN), (n.d.). https://datatracker.ietf.org/wg/6lowpan/documents/(Accessed 31 December 2017).
[30] IPv6 IPv6 Over the TSCH mode of IEEE802.15.4e (6TiSCH), (n.d.). http://tools.ietf.org/wg/6tisch/charters (Accessed 31 December 2017).
[31] Q. Wang, J. Jiang, Comparative examination on architecture and protocol of industrial wireless sensor network standards. IEEE Commun. Surv. Tutorials 18 (2016) 2197–2219, https://doi.org/10.1109/COMST.2016.2548360.
[32] M. Zheng, W. Liang, H. Yu, Y. Xiao, Performance analysis of the industrial wireless networks standard: WIA-PA. Mob. Netw. Appl. 22 (2017) 139–150, https://doi.org/10.1007/s11036-015-0647-7.
[33] M.R. Palattella, T. Watteyne, Q. Wang, K. Muraoka, N. Accettura, D. Dujovne, L.A. Grieco, T. Engel, On-the-Fly bandwidth reservation for 6TiSCH wireless industrial networks. IEEE Sensors J. 16 (2016) 550–560, https://doi.org/10.1109/JSEN.2015.2480886.
[34] T. Watteyne, P. Tuset-Peiro, X. Vilajosana, S. Pollin, B. Krishnamachari, Teaching communication technologies and standards for the industrial IoT? Use 6TiSCH!. IEEE Commun. Mag. 55 (2017) 132–137, https://doi.org/10.1109/MCOM.2017.1700013.
[35] A. Usman, S.H. Shami, Evolution of communication technologies for smart grid applications. Renew. Sust. Energ. Rev. 19 (2013) 191–199, https://doi.org/10.1016/j.rser.2012.11.002.
[36] Wi-Fi Alliance, Wi-Fi Simple Configuration Technical Specification, http://www.wi-fi.org, 2011.
[37] Wi-Fi Alliance, Wi-Fi Display Technical Specification, http://www.wi-fi.org, 2012.
[38] Wi-Fi Alliance, Wi-Fi Multimedia Technical Specification, http://www.wi-fi.org, 2012.
[39] E. Ferro, F. Potorti, Bluetooth and Wi-Fi wireless protocols: a survey and a comparison. IEEE Wirel. Commun. 12 (2005) 12–26, https://doi.org/10.1109/MWC.2005.1404569.
[40] J. Yick, B. Mukherjee, D. Ghosal, Wireless sensor network survey. Comput. Netw. 52 (2008) 2292–2330, https://doi.org/10.1016/j.comnet.2008.04.002.
[41] I.F. Akyildiz, X. Wang, A survey on wireless mesh networks. IEEE Commun. Mag. 43 (2005) S23–S30, https://doi.org/10.1109/MCOM.2005.1509968.
[42] T. Mackiewicz, Overview of IEC 61850 and benefits. in: 2005/2006 IEEE/PES Transmission and Distribution Conference and Exhibition, 2006, pp. 376–383, https://doi.org/10.1109/TDC.2006.1668522.
[43] L. Andersson, K.-P. Brand, The benefits of the coming standard IEC 61850 for communication in substations, in: Proceedings of Southern African Conference Power System Protection, 2000, pp. 8–9.
[44] P.P. Parikh, M.G. Kanabar, T.S. Sidhu, Opportunities and challenges of wireless communication technologies for smart grid applications. in: IEEE PES General Meeting, 2010, pp. 1–7, https://doi.org/10.1109/PES.2010.5589988.

[45] A. Patel, J. Aparicio, N. Tas, M. Loiacono, J. Rosca, Assessing communications technology options for smart grid applications. in: 2011 IEEE International Conference on Smart Grid Communications (SmartGridComm), 2011, pp. 126–131, https://doi.org/10.1109/SmartGridComm.2011.6102303.

[46] V.C. Gungor, D. Sahin, T. Kocak, S. Ergut, C. Buccella, C. Cecati, G.P. Hancke, Smart grid technologies: communication technologies and standards. IEEE Trans. Ind. Inf. 7 (2011) 529–539, https://doi.org/10.1109/TII.2011.2166794.

[47] A. Mahmood, N. Javaid, S. Razzaq, A review of wireless communications for smart grid. Renew. Sust. Energ. Rev. 41 (2015) 248–260, https://doi.org/10.1016/j.rser.2014.08.036.

[48] R. Riggio, T. Rasheed, S. Testi, F. Granelli, I. Chlamtac, Interference and traffic aware channel assignment in WiFi-based wireless mesh networks. Ad Hoc Netw. 9 (2011) 864–875, https://doi.org/10.1016/j.adhoc.2010.09.012.

[49] Commission of the European Communities, Exploiting the Employment Potential of ICTs, Staff Working Document, Commission of the European Communities, Strasbourg, 2012.

[50] C.X. Wang, F. Haider, X. Gao, X.H. You, Y. Yang, D. Yuan, H.M. Aggoune, H. Haas, S. Fletcher, E. Hepsaydir, Cellular architecture and key technologies for 5G wireless communication networks. IEEE Commun. Mag. 52 (2014) 122–130, https://doi.org/10.1109/MCOM.2014.6736752.

[51] D. Warren, C. Dewar, Understanding 5G: Perspectives on Future Technological Advancements in Mobile, GSMA Intelligence, 2014.

[52] A. Gupta, R.K. Jha, A survey of 5G network: architecture and emerging technologies. IEEE Access. 3 (2015) 1206–1232, https://doi.org/10.1109/ACCESS.2015.2461602.

[53] J. Rodriguez (Ed.), Fundamentals of 5G Mobile networks, First Published, Wiley, Chichester, West Sussex, 2015.

[54] R. Vannithamby, S. Talwar (Eds.), Towards 5G: Applications, Requirements & Candidate Technologies, John Wiley & Sons Inc, Chichester, West Sussex, 2017.

[55] Z. Pi, F. Khan, An introduction to millimeter-wave mobile broadband systems. IEEE Commun. Mag. 49 (2011) 101–107, https://doi.org/10.1109/MCOM.2011.5783993.

[56] T.S. Rappaport, J.N. Murdock, F. Gutierrez, State of the art in 60-GHz integrated circuits and systems for wireless communications. Proc. IEEE 99 (2011) 1390–1436, https://doi.org/10.1109/JPROC.2011.2143650.

[57] T.S. Rappaport, S. Sun, R. Mayzus, H. Zhao, Y. Azar, K. Wang, G.N. Wong, J.K. Schulz, M. Samimi, F. Gutierrez, Millimeter wave mobile communications for 5G cellular: it will work!. IEEE Access 1 (2013) 335–349, https://doi.org/10.1109/ACCESS.2013.2260813.

[58] C. Han, T. Harrold, S. Armour, I. Krikidis, S. Videv, P.M. Grant, H. Haas, J.S. Thompson, I. Ku, C.X. Wang, T.A. Le, M.R. Nakhai, J. Zhang, L. Hanzo, Green radio: radio techniques to enable energy-efficient wireless networks. IEEE Commun. Mag. 49 (2011) 46–54, https://doi.org/10.1109/MCOM.2011.5783984.

[59] ITU-R, IMT Vision—Framework and Overall Objectives of the Future Development of IMT For 2020 and Beyond, Geneva, Switzerlandhttps://goo.gl/UZpNpJ, 2015. Accessed 8 October 2017.

[60] NTT Docomo, 5G Radio Access: Requirements, Concept and Technologies, 2014.

[61] M. Torchia, U. Sindhu, Cellular and the Smart Grid: A Brand-New Day, IDC Energy Insights. EI230124 (n.d.) 1–13.

[62] P. Rengaraju, C.-H. Lung, A. Srinivasan, Communication requirements and analysis of distribution networks using WiMAX technology for smart grids, in: Wireless Communications and Mobile Computing Conference (IWCMC), 2012 8th International, IEEE, 2012, pp. 666–670.

[63] M. Paolini, Empowering the Smart Grid With WiMAX, Senza Fili Consulting, LLC., 2010. https://goo.gl/1c7qRN. Accessed 2 January 2019

Chapter 6

Cognitive radio based smart grid communications

Chapter outline

6.1 Introduction

One of the fast-developing areas of telecommunication industry is wireless communication technologies. In the last two decades, an exponential increase in spectrum demands has been appeared because of emerging wireless technologies and services that has led to an important problem in the allocation of wireless spectrum resources [1–3]. It is supposed that this problem was essentially

From Smart Grid to Internet of Energy. https://doi.org/10.1016/B978-0-12-819710-3.00006-5

based on increasing demand for wireless Internet access over mobile devices such as notebooks, tablets, e-book readers and smartphones, and this demand will constantly grow in the future. It is clear that wireless communication applications not only are being used for personal communication services but also are being utilized in metering, monitoring and control systems such as advanced metering infrastructure (AMI), health-monitoring systems, surveillance systems, traffic controlling systems and so forth. Nonetheless, the spreading of wireless systems is directly related to allocated spectrum that regulatory agencies have generally organized available spectrum by considering kinds of applications and services. Especially, the majority of available spectrum has been authorized to service providers in order to make long-term use possible in a large geographical area [4, 5]. This allocation task is performed by the regulatory authority of each country (i.e., Federal Communications Commission (FCC) in the United States, Information and Communication Technologies Authority in the Turkey). Regardless of the situation of data traffic, only allocated spectrum should be exploited by communication systems. On the other hand, the short-term licensing possibility of spectrum that provides an opportunity to satisfy transient traffic demand is restricted due to the allocation of the radio spectrum permanently. Furthermore, a dedicated spectrum may not be effectively used if there exist very few active users. The Spectrum Task Force (SPTF) of the FCC [5] has reported that most of the licensed spectrum bands are either partially or extremely unused at a certain time and in a particular area. It is also specified in [6] that the use of licensed spectrum may change from 15% to 85% depending on the time and place.

With respect to the present spectrum allocation procedures, new spectrum allocation for every wireless services should be performed by considering unemployed frequencies of the spectrum. A large part of the spectrum below 3 GHz is being assigned for special utilization because of these procedures [7]. In addition, several reports have clearly indicated that there is an unequal spectrum utilization [8–10] where both cellular bands and unlicensed bands are very crowded while the other parts of the spectrum is unemployed. Especially, when the unlicensed Industrial, Scientific and Medical (ISM) bands are considered, it is evident that there is an important bandwidth problem for license-free wireless systems. Therefore, bandwidth wastage, underutilization, and spectrum allocation are considered as the main challenges of wireless systems. Since the allocation scarcity problem is mainly based on the ineffective usage of spectrum, this issue can be solved through new approaches, regulations and technologies [11]. These challenges have encouraged researchers to develop new methods and systems for wireless communications that are able to present more reliable, efficient and effective communication platforms.

Cognitive radio (CR) technology intends to overcome these mentioned problems by employing limited spectrum more efficient without interfering to other users. CR is an emerging wireless communication technology where a radio is able to sense and analyze the spectrum, and it can reconfigure its system parameters to exploit best available wireless channels around it [12]. This approach was first introduced by Joseph Mitola III in 1998 and was reported in 1999 [13]. According to this approach, the CR was described as a smart wireless communication system that is able to perceive radio medium and employ artificial intelligence techniques to figure out detection results. Later, CR system can dynamically arrange its configuration parameters (i.e., waveform, operating frequency, protocol and networking) to enable communication over different frequencies at different times. Spectrum holes are opportunistically exploited by this technology for data transmission. The concept and practice background of this technology have evolved over the past few years. In 2003, the FCC has formally declared the definition of this technology as follows [14]:

> *A radio or system that senses its operational electromagnetic environment and can dynamically and autonomously adjust its radio operating parameters to modify system operation, such as maximize throughput, mitigate interference, facilitate interoperability, access secondary markets.*

While licensed spectrum access priority is always assigned to Primary Users (PUs), Secondary Users (SUs) that are authorized by the CR can also employ the licensed spectrum for data transmission, and no interference to PUs should be guaranteed by the SUs. Thus, effective usage of spectrum bands can be ensured by the CR technology [15, 16]. Up until now, numerous advances in CR technology have been attained thanks to the performed researches and promotion of regulators. In addition, various standards have been suggested to advance design and applications of CR technology. In 2004, the Institute of Electrical and Electronics Engineers (IEEE) has created a working group entitled IEEE 802.22 to improve international standards for this technology, and the working group developed IEEE 802.22 standard that focuses on wireless regional area network (WRAN) by considering white spaces in the television (TV) frequency spectrum. The main goals of this standard is to provide efficient usage of unused spectrum that are assigned to TV broadcast services, and the standard was published in 2011. The following sections will present technical features of the CR and CR network (CRN) applications in smart grids in a detail.

6.2 Cognitive radio technology

The CR technology has appeared as an emerging technology aiming efficient utilization of restricted spectrum in addition to supporting more wireless applications and services [17]. Therefore, this technology is generally considered as a crucial enabling technology for modern communication and network systems. In this technology, the use of spectrum can be also advanced through an opportunistic spectrum access method that exploits unoccupied part of spectrum allocated to the PUs for tentative usage of SUs [18, 19]. A CR transceiver can dynamically configure its system parameters to utilize unemployed parts of the spectrum and wireless networks [20]. The CR user is merely authorized to exploit unoccupied spectrum in order to prevent causing interference or collision with the PUs. In addition, several spectrum measurement reports indicate that it is not appropriate for existing wireless communication systems to consider only one spectrum policy [8–10]. Since the transceivers of CR systems can actively update their system configurations such as modulation type, bandwidth, and transmission power, they remarkably differ from conventional communication systems [21]. Before CR devices update their system configurations according to changes of the radio environments, they should obtain several information from the network. Therefore, CRs need to have two important features called "*cognitive capability*" and "*reconfigurability*" to accomplish opportunistic spectrum access (OSA) [2, 11]. The cognitive capability can be described as the ability to perceive radio environment, consider the perceived information and make a decision according to the considered information. The perceived information are related to network protocol, spectrum, transmitted waveform, communication network type, geographical information, locally available resources and services, user needs, security policy, and so forth. On the other hand, after these required information are collected from the surrounding radio environment, the CR devices reconfigure their system parameters in accordance with perceived data that is called as reconfigurability feature of the CR systems.

Inactive spectrum bands of the PUs are called as "spectrum holes" or "white spaces" that are illustrated in Fig. 6.1. A spectrum hole (or white space) is characterized in [2] as "*a band of frequencies assigned to a primary user (licensed user), but, at a particular time and specific geographic location, the band is not being utilized by that user*." In the event of the PU reclaims the spectrum, CR device(s) should move another white space or continues to employ this band by

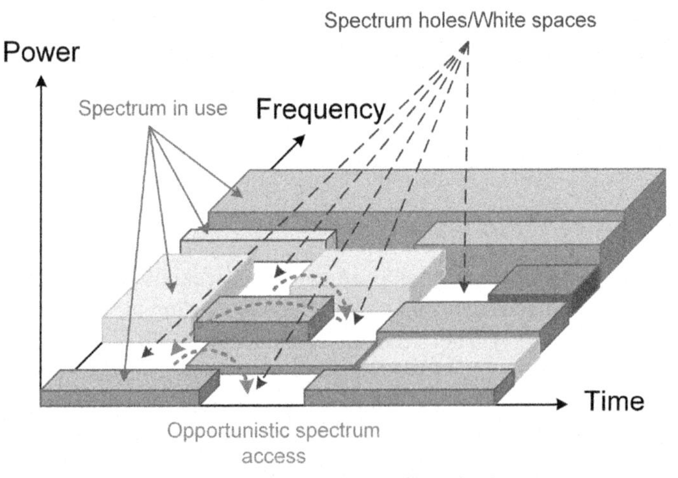

FIG. 6.1 Spectrum hole or white space concept in the CR technology.

changing its system configurations such as waveform, bandwidth, transmission power level to prevent inference with PUs. At the same time, the OSA of spectrum holes and altering of the operating frequency bands by a CR device can be shown from the Fig. 6.1.

The CRN improves wireless network services by adjusting system parameters, autonomously assigns spectrum bands to the users, and authorizes CR devices for adapting to wireless communication demands. Two important targets correlated the CR devices cover [22]:

- High accuracy, originality and communication feasibility on space-time effects
- Sufficient use of existing radio spectrum

Generally, spectrum bands can be classified into two categories as official and unofficial bands. Users employing the official bands with licenses have the highest channel access priority while users exploiting the unofficial bands are assumed to benefit from the band without affecting official users. While the number of available channels in conventional wireless network is constant, the number of available channels for the CR systems changes depending on the time and spectrum holes. The active number of PUs directly determines the number of accessible channels for the CR devices. The varying spectrum case of accessible channels may cause several challenges for the CR spectrum management systems. These challenges can be summarized as quality-of-service (QoS) awareness, reliable interference management and ensuring reliable communication for CRs independent of primary appearance. In order to overcome these issues, the CRNs should detect which parts of the frequency bands are accessible for CR devices, manage access to prevent collision and interferences, choose the best channel to reach the required QoS, and clear out the channel if the PUs return. A typical cognitive cycle of CR devices that contains detection of spectrum holes, choosing the best channel, managing spectrum access and clearing out the exploited frequency band when a PU returns as depicted in Fig. 6.2. This cycle is composed of the following stages: spectrum sensing, spectrum decision, spectrum sharing and spectrum mobility.

Spectrum Sensing: Spectrum sensing (SS) is a transaction realized in the CR devices to acquire radio or frequency information at a certain time and in a specific place. Sensing and analyzing processes are performed to detect signal strength over a particular spectrum, and CR devices decide whether there are available spectrum holes/white spaces. The acquired data thanks to the SS process contain information regarding channel status of PUs, interference rate affecting PUs, data traffic of PUs and channel gains among CR devices. In addition, the CRNs make spectrum access and spectrum sharing decision depending on these information. Depending on the SS time of transmission frame, the SS process can be performed by two methods called proactive method and reactive method [23]. While the reactive sensing is only accomplished on demand, the proactive sensing is realized periodically. On the other hand, CR devices

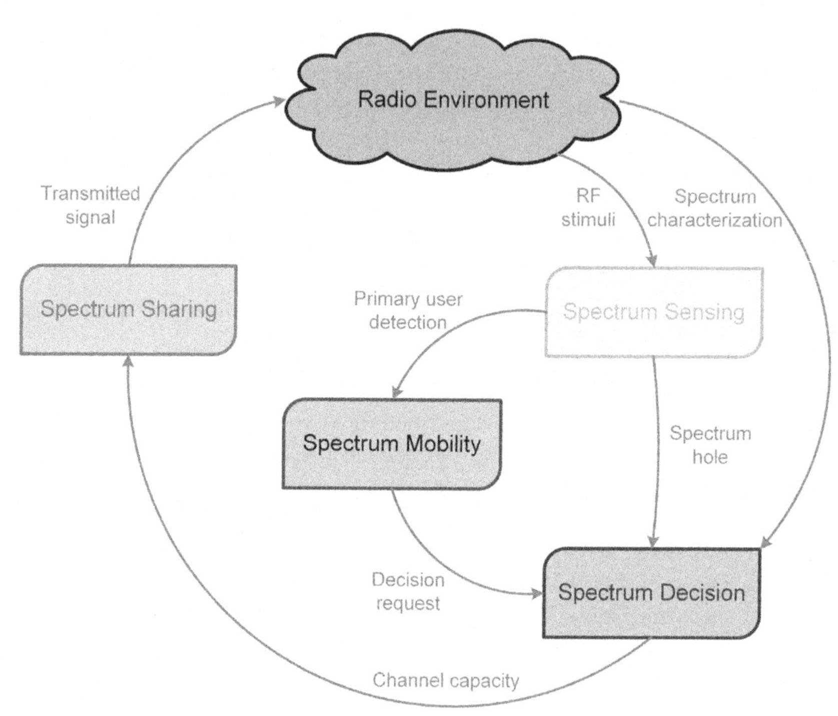

FIG. 6.2 Principles of cognitive cycle.

generally support two kinds of detection methods that are called feature detection and blind detection. In the first method, the CR devices exploit a priori knowledge of PU waveforms for idle channel detection. In the second method, detection process is blindly performed by using energy detection or autocorrelation detection where there is no requirement for a prior knowledge [24, 25]. In addition, the CRNs may exploit cooperative SS methods in order to enhance performance [26–29]. Besides, the CR devices should support well-timed SS process because of the fact that the CR devices should clear out the channel if a PU returns to its licensed spectrum band.

Spectrum Decision: This process is essential for choosing best idle channel (s) of the PUs to meet QoS needs of CR devices. After the SS results are analyzed, CR devices can additionally decide which idle channel can be exploited according to aims and requirements of particular applications. Therefore, the process firstly analyzes every idle channel depending on both the SS data and statistical data of each licensed user [30]. Later, spectrum management and handoff functions of CR devices authorize them for selecting the most appropriate channel(s) and channel switching opportunities in order to fulfill the QoS targets. After channel detection process is completed, each CR device reconfigures its system parameters for supporting communication over new

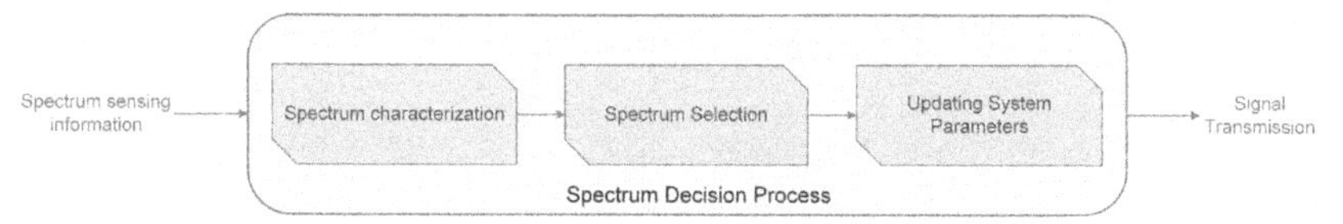

FIG. 6.3 Block diagram of the spectrum decision process.

determined channels. Process steps of the spectrum decision transaction are depicted in Fig. 6.3.

Also, it is important to note that each CR device should present high flexibility for system parameter update operation where several parameters such as transmit power, carrier frequency, mapping method, bandwidth, and channel coding schemes should be reconfigured to reach desired system performance in terms of the QoS [31, 32]. Furthermore, CR devices may also alter their access technology to further enhance the system performance [33].

Spectrum Sharing: This process is realized in cognitive media access control (MAC) protocols. The spectrum allocation method, spectrum access mechanism, and CRN architectures are crucial for spectrum sharing operation that authorizes the SUs to fairly use the licensed spectrum bands. In addition, transmit power level of the CR devices should be always limited to keep interference under a particular threshold value. Spectrum sharing process can be performed via two different ways. While one of them is called as centralized spectrum sharing, the other one is distributed spectrum sharing. Whereas a central controller manages the spectrum access and spectrum allocation processes in centralized spectrum sharing method [32, 34, 35], CR devices realize these processes themselves in distributed spectrum sharing method [19, 34]. Even though the centralized method presents higher performance than that of the distributed method, its disadvantages are increasing cost and higher system complexity.

On the other hand, spectrum allocation process can be divided into two groups as cooperative and non-cooperative technique [36, 37]. The former method can provide better performance by ensuring cooperation between entire active CR devices and primary networks, when it is compared with the latter method. The CR devices serving with non-cooperative method have no collaboration with other CR devices and primary networks, and they strive to increase their system performance as much as possible. Moreover, the cooperative method also presents several advantages in terms of accuracy, spectral efficiency and throughput. However, these advantages of cooperative method cause increment in both cost and system complexity. Conversely, the main advantage provided by the non-cooperative method is less energy consumption because of lower complexity.

Spectrum Mobility: One of the most important processes for CR devices is spectrum mobility that aims to autonomously change operating frequency of CR devices to ensure uninterrupted communication. In other words, spectrum

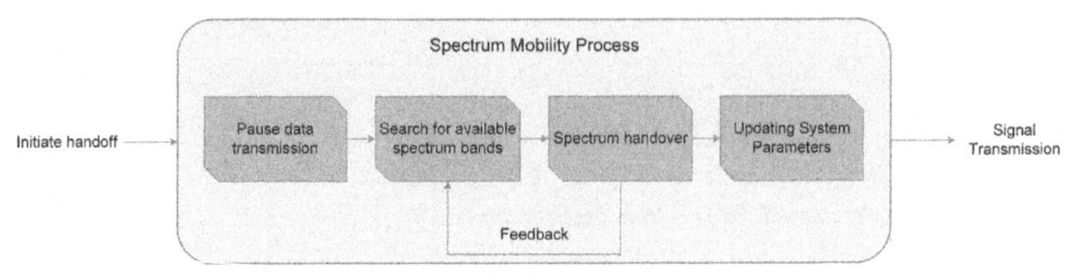

FIG. 6.4 Block diagram of the spectrum mobility process.

mobility is a process to sustain wireless communication by smoothly switching communication channels [38]. In order to accomplish this process, there are two required transactions called as spectrum handoff and connection management. Steps of the spectrum mobility process are illustrated in Fig. 6.4.

The connection management operation is related to adjusting or updating connection parameters while the spectrum handoff operation is related to changing an active transmission from a channel to a different idle channel. In addition, protocol stacks of CR device have to ensure easy access to various layers in order to decrease latency in spectrum mobility process. Regarding the spectrum handoff, the CR devices may use classic methods employed in conventional wireless networks because of decrease in channel quality.

6.2.1 Importance of CR for smart grid systems

The CR is considered as an encouraging technology to meet communication, standardization and security requirements of smart grid (SG) systems [39–42]. In addition, supporting several traffic types is very crucial for the SG systems [43]. Joint spatial and tentative spectrum sharing can be utilized in the SG applications, particularly in demand response management (DRM) applications [44]. Machine-to-machine (M2M) communication is a new type of communication aiming to enable connectivity among devices without human intervention. Nevertheless, a great number of connected devices may seriously restrict the spectrum of current communication networks. Therefore, the use of CR technology in the SG systems may be an appropriate alternative to reduce spectrum scarcity originated from the M2M communications [45]. Furthermore, the CRNs may be an excellent option for communication requirements of SG systems since they can support a wide variety of communication applications [41, 46–48]. The superiorities of CR technology utilization in the SG systems can be summarized as follows.

- Several wireless communication systems (i.e., Bluetooth, ZigBee, and Wi-Fi etc.) operate at 2.4 GHz license-free frequency band. Household appliances may also generate powerful electromagnetic waves. Further, wireless communication channels suffer from electromagnetic interferences generated electrical devices. The SG meters (smart meters (SMs)) that are home area network (HAN) components of SG systems also operate at 2.4 GHz license-free frequency band. The interference issue of license-free frequency band will adversely affect stable communication demand of the SG systems. Nevertheless, the CR devices can overcome this interference problem by reconfiguring their operating parameters to enable SG communication [49, 50].

- The gathering and transmission of big data regarding electrical energy systems may cause major problems for current communication networks. In addition to improving spectrum usage, the CR technology can handle for gathering and transmission of big data.
- The communication architecture of SG systems is composed of home areas, neighborhood areas, and wide areas. Hence, thanks to reconfigurability feature of CR devices, these devices can manage communication among all of these areas.
- Data transmission over spectrum holes in the CR technology can also support low-latency connections for time-sensitive applications of SG systems [40, 41, 51].
- Since the CR devices detect the radio environment and can dynamically configure the transmission parameters (i.e., transmission power level), this technology can decrease both power consumption and operating cost.
- The unoccupied TV spectrum (referred as TV White Space (TVWS)) is a sparse structure than that of the 2.4 GHz license-free frequency band. Hence, this band presents better signal propagation advantage. In addition, lower frequencies of the TVWS provide lower power consumption [52, 53]. Exploiting the TVWS spectrum in CRNs can yield high bandwidth opportunities for AMIs, Neighborhood Area Networks (NANs) and Field Area Networks (FANs).
- The IEEE 802.22 standard based CRNs may also provide installation cost advantage in SG networks since the long-range coverage feature of this standard (typically up to 100 km) decreases the number of required base stations [54, 55]. In addition, the IEEE 802.22 standard based CRNs can present advantages in collecting measurement data from wide area networks (WANs) [56].
- Metering data should be transmitted to control center in order to ensure an effective DRM system that is very crucial for the SG systems. Therefore, the DRM process can be efficiently maintained by exploiting the CRNs [47].
- The SGs contain several distributed generation systems such as solar cells and wind turbines that are generally located in far away. The effective utilization of these power systems directly affects electrical power quality, and collaboration between SG elements is necessary to achieve the desired electrical power quality. The CR can be considered as a promising technology in order to ensure this collaboration [47, 57].
- Low power quality and power outages may cause serious financial problems. An important reason of these issues is inadequate monitoring of electrical power systems. Since the electrical power systems may spread over

the wide areas, communication systems that provide the desired performance should be as cost-effective as possible. Combining monitoring systems and sensors with the CRNs can also improve the monitoring features of the SG systems. Therefore, the power outages can be significantly reduced through the CRNs [47, 57].

6.3 CR network architectures

Diversity in spectrum policies and communication technologies is one of the most important characteristics of the conventional wireless networks [58]. Several sections of spectrum are allocated for licensed communication systems whereas there still exist unlicensed spectrum bands. Hence, a good characterization for CRN architectures is required to improve communication protocols for these networks. To ensure reliable communication through CRNs, the CR devices should always track the network status and radio environment by providing coordination among the CR users. The mentioned coordination may cause some problems since the spectrum is time-dependent, and the CR users are generally mobile. The CRNs contain several types of radio communication systems operating together in the same geographic areas. Therefore, the CRNs can be taken into account as a kind of heterogeneous network containing numerous wireless communication systems. This diversity may be originated from connections among the base stations (BSs), access points (APs), mobile terminals (MTs), networking protocols, and CR terminals. Terminals and service providers can use the diversity of wireless access technologies. The CRNs authorize the users to continuously utilize spectrum bands. The main design target of the CRN architectures is to improve the use of communication networks. From the point of view of the users, the network usage indicates to accomplish place and time independent wireless communication through the CRNs. In terms of operators, the service providers should not only be obliged to provide high quality service to mobile users, but they should also use network resources efficiently and effectively [15]. A typical architecture of CRNs that is illustrated in Fig. 6.5 contains several elements regarding both the PU networks and the SU networks.

Typically, a primary network can be composed of PUs and primary BSs whereas a secondary network contains ensemble of the SUs with or without a secondary BS, as can be seen from the figure. A secondary BS that serves as the center of the secondary network generally arranges the OSA for the SUs. When various secondary networks employ the same spectrum band, a centric network equipment called *spectrum broker* manage the use of this spectrum band. This equipment gathers operation information from secondary networks and assigns the spectrum effectively and fairly. The network architecture schemes of the CRNs can be classified into three categories as infrastructure-based (centralized), ad hoc (distributed), and mesh architectures. The following subsections introduce these architectures in detail.

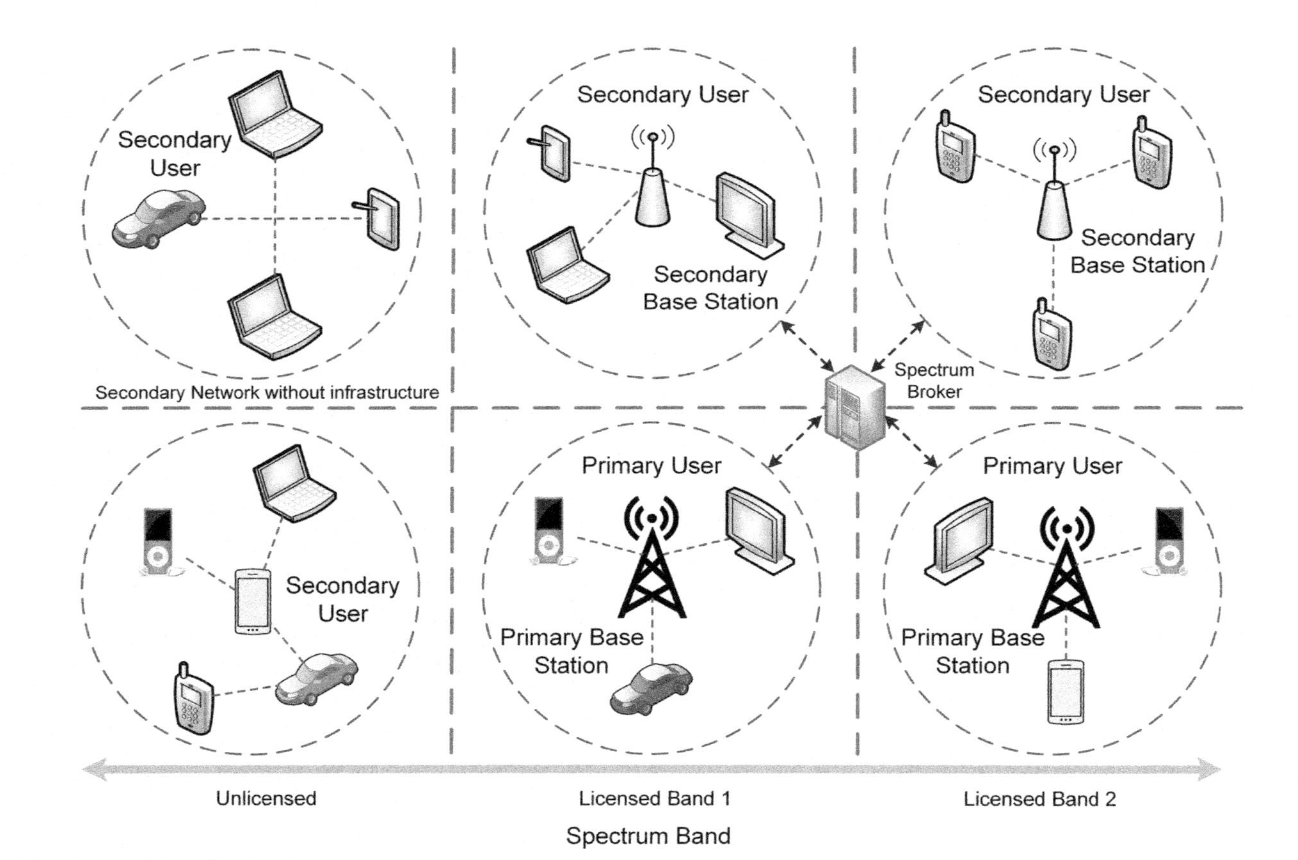

FIG. 6.5 Opportunistic utilization of spectrum bands.

6.3.1 Infrastructure-based (centralized) architecture

The structure of infrastructure-based (or centralized) CRN architecture that is illustrated in Fig. 6.6 consists of the primary network and the secondary network (the CRN). As stated earlier, the primary networks have priority for access the licensed spectrum bands while the secondary networks (the CRNs) can only access to the spectrum opportunistically [16]. A CRN including a cognitive BS (CBS) is considered as infrastructure-based CRN (or centralized CRN). The CBS behaves as a center gathering spectrum analysis information from all CR devices, and decides how to prevent interference with the primary networks. In accordance with this decision, each CR device updates its system parameters to continue data transmission. In addition, it is important to note that a central AP may also undertake this coordinator task [59]. In this network architecture, the CR devices can access both the CBSs and APs in a direct or multi-hop way. The CR devices in the same network can connect each other over the CBS/AP of the network. Communication among different networks are realized over a backbone or core networks. In order to satisfy requests of CR devices, the CBSs/APs carry out several transactions at the protocol layers. Furthermore, several applications and networking systems can be established by employing infrastructure-based network architecture. Moreover, the CBSs and/or APs can execute different communication protocols and standards to provide better services for CR devices. Conversely, the CR devices can connect to different type communication systems with permission of their CBS/AP.

6.3.2 Ad hoc (distributed) architecture

The distributed CRNs can be taken into account as ad hoc or point-to-point (P2P) communication systems in which the SUs can transmit their data over licensed or unlicensed frequency bands in an opportunistic manner. A typical

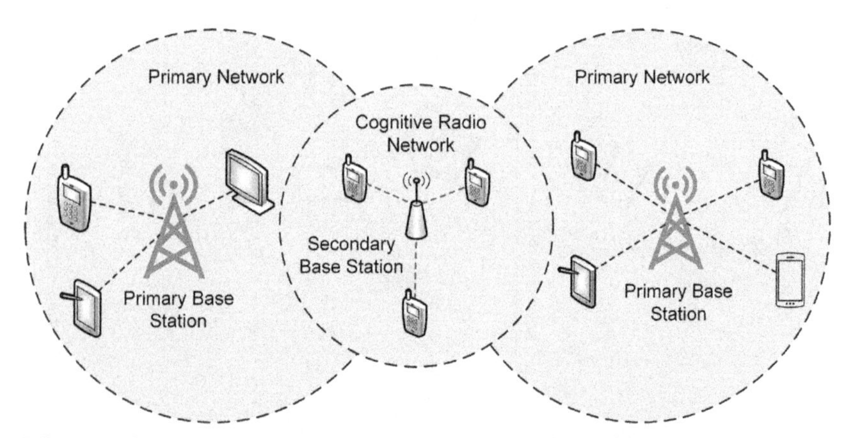

FIG. 6.6 Infrastructure-based (centralized) network example for the CR systems.

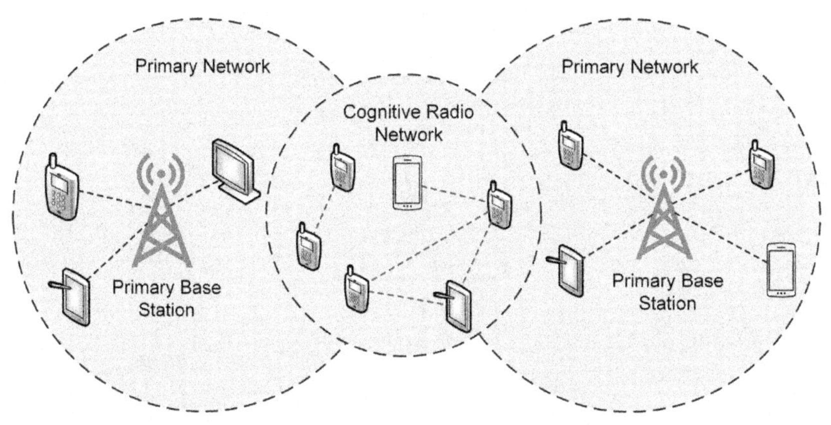

FIG. 6.7 Ad hoc (distributed) network example for the CR systems.

scheme of distributed network architecture is depicted in Fig. 6.7. This network architecture is essentially developed for supporting special-purpose applications. In addition, the spectrum access decisions that are created depending on the local information may not be optimum in this type of network architecture. The CR devices can expand their network information by sharing locally gathered data among themselves, and the devices realize their transactions depending on these sensed global information. The communication among the CR devices can be also organized by employing various network protocols and standards (i.e., Bluetooth, GSM, Wi-Fi, etc.). Different distributed network architectures may be available and each of them requires specific protocols for the CR devices. The SUs can continue their communication by using cooperative communication methods over both licensed and unlicensed frequencies thanks to only one transceiver. It is worth noting that a CR node belongs whole CR capabilities and has to decide its subsequent actions considering observed local information in this network architecture.

6.3.3 Mesh architecture

A combination of infrastructure-based and distributed architectures constitutes the mesh network architecture. A typical mesh network is illustrated in Fig. 6.8. This architecture authorizes users for both transmitting and receiving data between CR devices, APs and CBSs.

Each CR device not only behaves as a transmitter but also behaves as a router to control transmission of data packets. Thus, the CR devices may reach CBSs and/or APs directly or by employing cooperative communication methods in multi-hop manner. In addition, CBSs and APs generally operate as gateways to link the wired and wireless backbone networks together. Hence, this scheme is more secure and cost-effective than that of the other network

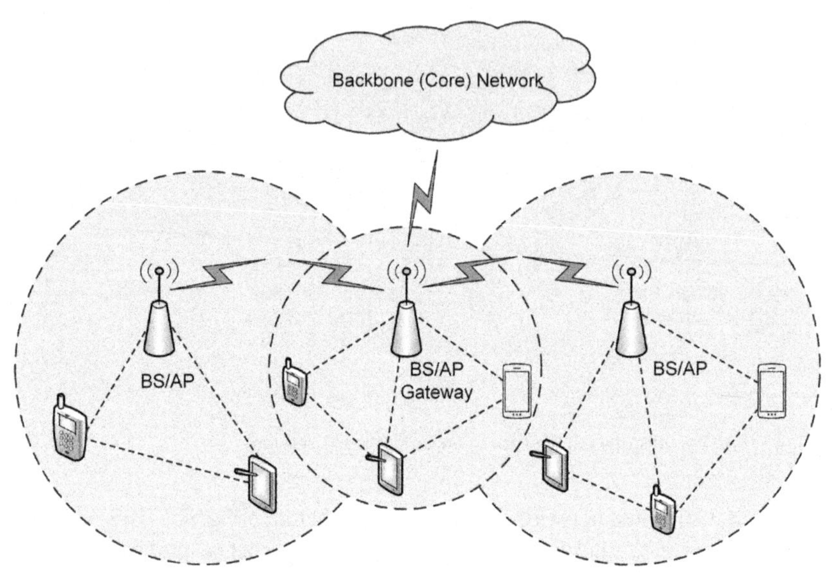

FIG. 6.8 Mesh network example for the CR systems.

schemes. The CBSs and APs sense and utilize white spaces to communicate with each other, and they have cognitive potentials in order that white spaces can be employed to achieve communication demands of CR devices.

An example of wireless mesh network is shown in Fig. 6.8 where the network contains mesh routers constituting the backbone (core) of the network. Each mesh router can be taken into account as an AP that serves several mobile users or mesh clients. A mesh client forwards its data traffic to its corresponding mesh router in order to reach the gateway connected to the Internet. When the present case of spectrum bands are considered, there may be a great number of white spaces. Therefore, the capacity of wireless connections among the CR devices and CBSs/APs may be enormous and the network may support more data traffic. This network schemes may be efficiently utilized in monitoring and data collecting applications.

6.4 Spectrum sensing (SS) techniques in CR networks

The SS process is required to determine unoccupied frequencies that can be opportunistically utilized to serve SUs. As this process consumes power, it is very important to use effective sensing methods. In addition, the used SS method should be optimized for battery-powered devices [60]. The SS operation contains the process of collecting data regarding the unoccupied frequency bands and active PUs in addition to determination of spectrum holes. Simplified hardware and/or SS operations with reduced sensing durations are generally utilized in order to provide cost-effective SS methods [49]. In a widespread application of CR based SGs, the SMs behave as SUs and look for the available frequency bands to convey measurement data to the data collector units. The CR systems authorize the SG components to perceive the unoccupied frequency bands and utilize these frequency bands without causing interference to PUs [46]. The SS transaction can be explained depending on the signal detection theory that can be simply expressed as follows [61].

$$y(k)=\begin{cases}\eta(k) & :\mathrm{H}_0\\ s(k)+\eta(k) & :\mathrm{H}_1\end{cases} \tag{6.1}$$

where $y(k)$ shows the sample to be investigated at each instant k, $\eta(k)$ stands for additive noise and $s(k)$ denotes the transmitted signal. There are two possible hypothesis which are noise-only (H_0) and signal-plus-noise (H_1). The Eq. (6.1) can be generalized by taking into account fading and shadowing effects of wireless channels as follows.

$$y(k)=\begin{cases}\eta(k) & :\mathrm{H}_0\\ h\cdot s(k)+\eta(k) & :\mathrm{H}_1\end{cases} \tag{6.2}$$

where h denotes a complex random variable. In addition, all possible results of the signal detection process can be listed below and can be illustrated as in Fig. 6.9.

$$\begin{array}{l}\text{Decide in favor of } H_0 \text{ when } H_0 \text{ is true } (\mathrm{H}_0|\,\mathrm{H}_0)\\ \text{Decide in favor of } H_1 \text{ when } H_1 \text{ is true } (\mathrm{H}_1|\,\mathrm{H}_1)\\ \text{Decide in favor of } H_1 \text{ when } H_0 \text{ is true } (\mathrm{H}_1|\,\mathrm{H}_0)\\ \text{Decide in favor of } H_0 \text{ when } H_1 \text{ is true } (\mathrm{H}_0|\,\mathrm{H}_1)\end{array} \tag{6.3}$$

While the ($\mathrm{H}_1|\mathrm{H}_0$) case is referred as *false alarm*, the ($\mathrm{H}_0|\mathrm{H}_1$) case is called as *missed detection* and the other cases are *correct detection*. The goal of a signal detector is to keep missed detection and false alarm detection rates as minimal as possible while trying to maximize correct detections rates.

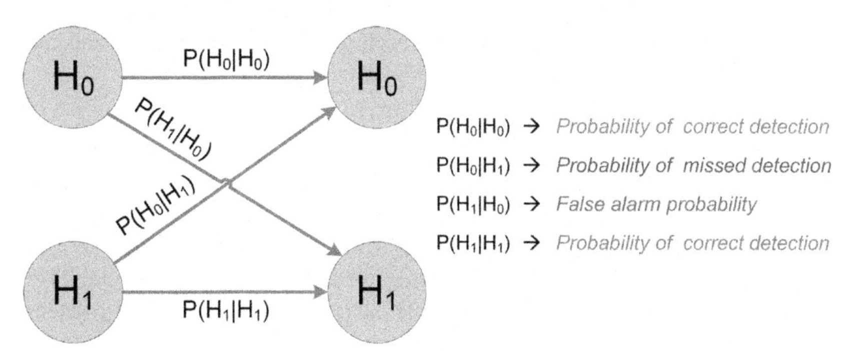

FIG. 6.9 Hypothesis tests and possible results for spectrum sensing process.

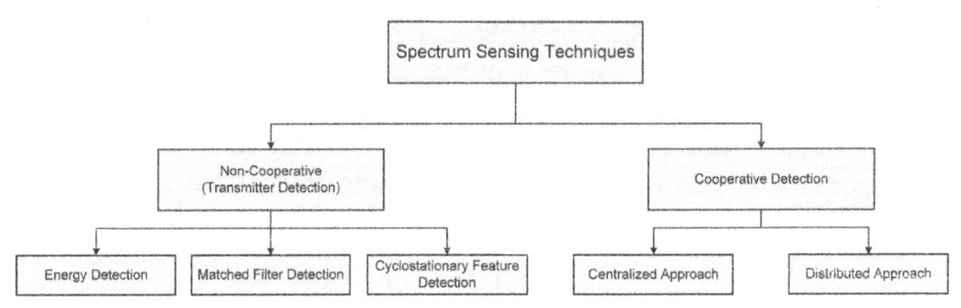

FIG. 6.10 Classifying of spectrum sensing methods.

The SS methods can be classified into two groups as non-cooperative (or transmitter detection) and cooperative detection as can be shown in Fig. 6.10. The non-cooperative methods are depending on the detection of transmitted signal from a PU via local observations of SUs. In this detection approach, it is assumed that the CR device does not know the location of the PU. Hence, the SUs should only depend on the detection of poor PU signals and exploit merely local observations to accomplish the SS process. A CR device does not possess full information regarding occupied spectrums in the coverage area. Therefore, there is no possibility to full prevent interference with the PUs. There are three popular detection approaches for non-cooperative detection method, which are referred as energy detection, matched filter detection, and cyclostationary feature detection.

An excellent line of sight (LOS) may be available between PU receiver and SU. However, the SU may not detect the PU due to shadowing problem that is a challenging problem for both urban and indoor environments. This issue is called as hidden terminal problem and cooperative detection methods aims to handle this problem. Cooperative detection can be divided into two categories as centralized and distributed approach. In former approach, a central unit gathers information from the CR devices; it determines the possible spectrum bands and shares this information with other CR devices. In latter approach, central node usage is not available, and CR devices similarly share sensing data with other CR devices. While the distributed approach is easily implementable and there is no backbone infrastructure requirements, the other approach provides more accuracy and can efficiently diminish shadowing and multi-path fading effects.

6.4.1 Energy detection method

Energy detection (ED) method that is a basic sensing technique do not need any prior knowledge about the PU signal. Therefore, this detection technique provides several advantages in terms of application and computation complexities. The received energy is a measure of a particular part of the spectrum. The detector compares the measured energy with a threshold value to decide whether the channel is available. This method needs longer sensing time in order to enhance signal to noise ratio (SNR), which leads to high power consumption, and

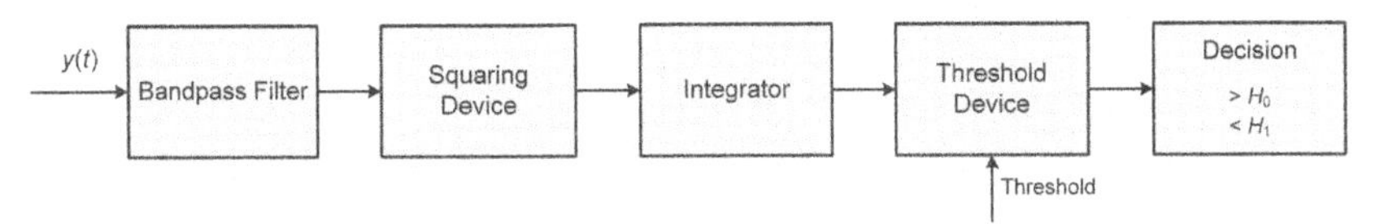

FIG. 6.11 Block diagram of energy detector system.

detector performance is greatly affected by the change of noise levels [61–63]. The decision metric ξ_{ED} of the ED can be expressed as follows.

$$\xi_{ED} = \frac{1}{N}\sum_{k=0}^{N-1} |y(k)|^2 \tag{6.4}$$

Alternatively, it can be stated through signal energy as follows.

$$\xi_{ED} = \int |y(t)|^2 dt \tag{6.5}$$

where N shows number of symbols and $y(k)$ shows the sampled signal, as expressed in Eq. (6.2). Therefore, the computed energy of signal can be compared with a threshold γ_{ED}. Later, the hypothesis is selected as H_0 if $\xi_{ED} < \gamma_{ED}$ or H_1 when $\xi_{ED} \geq \gamma_{ED}$.

A typical block diagram of ED system is shown in Fig. 6.11 where the received signal $y(t)$ is firstly fed to a bandpass filter (BPF) in order to limit received signal into the interest band. Later, the signal is applied a squaring device where all components of signal are squared to calculate the energy through the following integrator block. Afterward, a threshold device is utilized to decide whether channel is available.

The ED method has two important superiorities. The first of them is related to its low computational complexity and low application cost advantage while the second is related to no requirement for a prior knowledge regarding the PU signal. On the other hand, the performance of this method at low SNRs is poor.

6.4.2 Matched filter detection method

The matched filter (MF) technique is an optimum method to sense availability of spectrum since the MF can maximize the SNR even if there is additive white Gaussian noise (AWGN). This feature is accomplished via correlation process inherent of the method. Nonetheless, this method requires prior knowledge regarding PU signal. Depending on this prior knowledge requirement, CR devices should equipped with timing and carrier synchronization devices. Unfortunately, these requirements of the MF systems cause increasing in implementation complexity [62, 63]. It is important to note that this method may consume less power while sensing times is shorter. The block diagram of the MF method is shown in Fig. 6.12 where the received signal is correlated with a known signal at receiver device of the SU, and channel response is obtained depending on a particular threshold value.

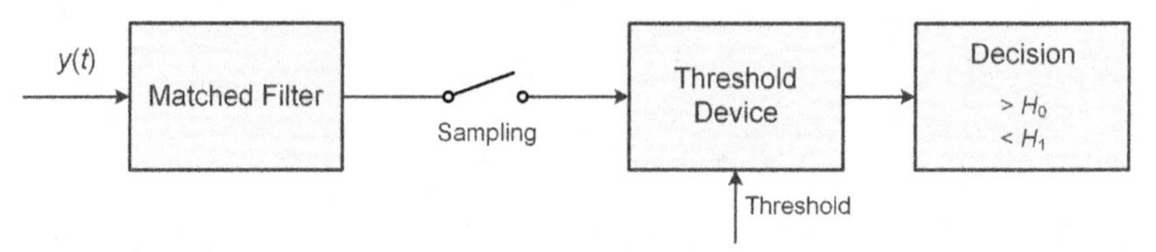

FIG. 6.12 Block diagram of matched filtering based SS system.

The correlation between received signal $y(t)$ and a copy of known signal $x(t)$ are examined to acquire the statistical results and the signal to be tested can be expressed in continuous time form as follows [61].

$$\hat{s}(t) = \int_0^T x^*(t-\tau)y(\tau)d\tau \tag{6.6}$$

Alternatively, it can be stated in discrete time as.

$$\hat{s}[n] = \sum_{k=1}^{N} x^*[n-k]y[k] \tag{6.7}$$

where $N = \frac{T}{T_s}$ with sampling period T_s, T is total sensing time. Moreover, the probabilities of detection and false alarm can be expressed as follows.

$$P_{false}^{MF} = P\{\hat{s} > \gamma_{MF} | H_0\} = Q\left(\frac{\gamma_{MF}}{\sqrt{\varepsilon\sigma_\eta^2}}\right) \tag{6.8}$$

$$P_{detection}^{MF} = P\{\hat{s} > \gamma_{MF} | H_1\} = Q\left(\frac{\gamma_{MF} - \varepsilon}{\sqrt{\varepsilon\sigma_\eta^2}}\right) \tag{6.9}$$

where $\varepsilon = \sum_{k=1}^{N} x^2[k]$ and σ_η^2 is the variance of the AWGN.

The MF detection method provides two important advantages. One of them is low detection time requirement while the other is high detection gain. On the other hand, this detector should have excellent prior knowledge regarding the PU signal, and a receiver should be assigned for each PU signal.

6.4.3 Cyclostationary feature detection method

Cyclostationary Feature Detection (CFD) method employs cyclostationary properties of the PU signals. Therefore, this method requires a priori knowledge regarding repetitive characteristics of PU signals. In other words, the CFD method handles the periodicity of the signals. The spectral correlation functions are employed by the technique to determine periodicity of the PU signal. Pulse trains, sinusoidal carriers, and spreading codes may be typically exploited to detect periodicity of signals [16, 62, 63]. A process can be considered as wide-sense stationary when autocorrelation and mean functions of this process are time-independent. In addition, if a process periodically indicates wide-sense stationary characteristic, it is also cyclic stationary. The periodicity of a PU signal and wide-sense stationary of an uncorrelated noise make it possible for the detector to simply distinguish noise from the PU signal. Nonetheless, computational complexity of this method is very high since the detector should compute two functions that are also depend on both frequency and cyclic frequency.

The cyclic spectral density (CSD) or cyclic periodogram of a received signal $y(t)$ can be calculated by using following equation [16].

$$S(f,\alpha)=\sum_{\tau=-\infty}^{\infty} R_y^\alpha(\tau)e^{-j2\pi f\tau} \tag{6.10}$$

where α shows the cyclic frequency (or frequency separation), $R_y^\alpha(\tau)$ denotes the cyclic autocorrelation function of the received signal. The $R_y^\alpha(\tau)$ can be also given as follows.

$$R_y^\alpha(\tau)=E\left[y(t+\tau)y^*(t-\tau)e^{j2\pi\alpha t}\right] \tag{6.11}$$

where $E[\cdot]$ shows the expectation operation while $(\cdot)^*$ is the complex conjugation. If there exist active PUs in the channel, it is shown that the CSD function expressed in Eq. (6.10) generates peaks in the event of cyclic frequency is same with the main frequency. Therefore, the availability of the channel may be determined by investigating peaks at the main frequency. A typical block diagram of the CFD detector is depicted in Fig. 6.13.

This detection method provides excellent detection results at low SNR values and it has a noise insensitive characteristic. On the other hand, the disadvantages of the CFD method contain the long observation period requirement and high computational complexity. A comparison table for non-cooperative detection methods are listed in Table 6.1.

6.4.4 Cooperative spectrum sensing based spectrum detection

The non-cooperative detection methods suffer from wireless communication problems such as shadowing, fading, multi-path propagation effects and hidden terminal problem. Unlike the use of non-cooperative methods, the CR devices can cooperate to provide better spectrum detection reliability and to handle hidden terminal problem that occurs in the presence of multipath fading and

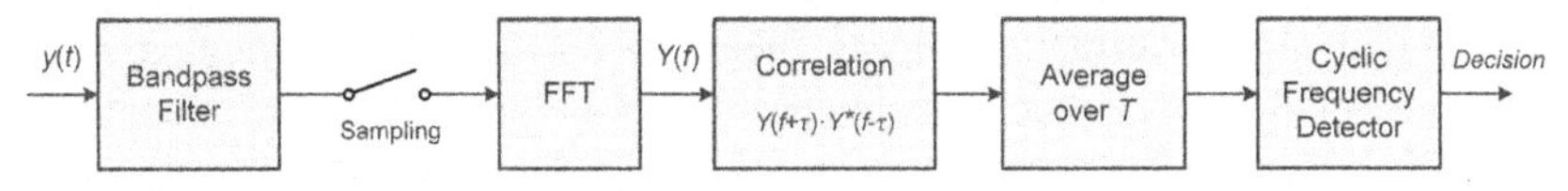

FIG. 6.13 Block diagram of cyclostationary system.

TABLE 6.1 Comparison of non-cooperative SS methods

Detection method	Advantages	Disadvantages
Energy detection	✓ No prior knowledge required ✓ Low computational cost	✖ Poor performance at low SNR values
Matched filter	✓ Optimum performance ✓ Low computational cost	✖ Needs prior knowledge regarding the PUs
Cyclostationary feature	✓ Good performance at low SNR values ✓ Noise insensitive characteristic	✖ Needs prior knowledge regarding the PUs ✖ High computational cost

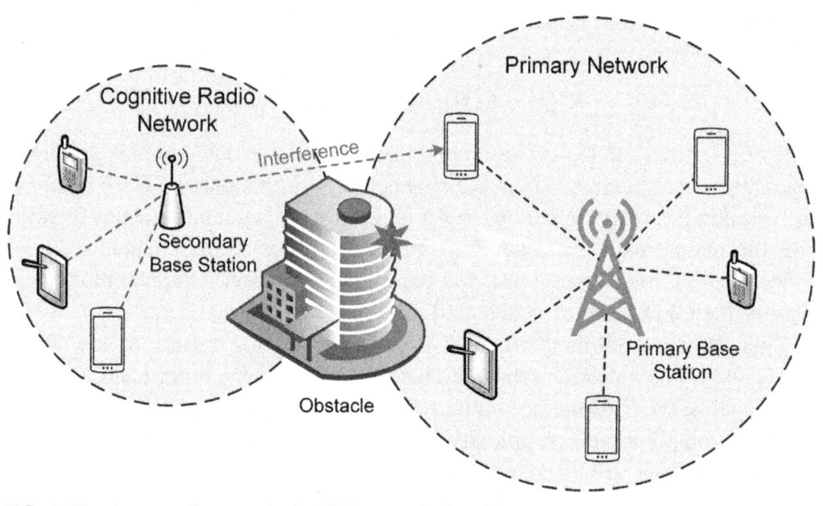

FIG. 6.14 An example scenario for hidden terminal problem.

shadowing as stated before [16, 64]. An example of hidden terminal problem based on shadowing is shown in Fig. 6.14.

The cooperative spectrum sensing (CSS) based detection process can be divided into two categories as centralized and distributed CSS. By using the diversity gains supplied by various CR devices, the CSS methods can generally present higher accuracy for the SS process. Even though this type of spectrum detection ensure performance improvement, the results of this advantage may cause several drawbacks such as more power consumption, higher system complexity and increased computational complexity [65]. A common control channel (CCC) is responsible for changing necessary information on the CSS. The signal/traffic load level in the CCC may change depending on the type of transferred data. For instance, one-bit decision from CR devices may be enough for some CSS algorithms while multiple decision bits may be required for other algorithms. In addition, insufficient clustering of CR devices for cooperation may not ensure the desired performance in the CSS. For instance, the sensing knowledge acquired from the CR devices concentrated in a narrow region may indicate high similarity because those devices may suffer from the same destructive effects. In centralized CSS, collaborated CR devices sense the spectrum bands and the sensed data are send to the central data center where the received data are analyzed to decide whether the spectrum is idle, or not. In distributed CSS, the CR devices change their perceived data between each other over the CCC, and each CR device creates its own sensing decision by combining the collected data.

6.5 CR based communication systems in smart grids

The communication infrastructure of SGs can be constituted by using several communication protocols that may be based on wired and/or wireless communication technologies. The power line communication (PLC) is a promising wired communication technology for the SG systems. This technology intends to efficiently exploit current power lines as a communication environment as well as their power delivery task. For instance, the PLC systems can ensure high data transmission rates up to the 200 Mbps over the single-phase networks. Although this technology is able to provide the advantage of canceling installation cost because of employing existing power lines, the power lines may lead to important performance degradation problems due to serving for many years. In addition, the changing channel impedances and dissimilar noise types of power lines also cause considerable problems [66–68]. Besides wired technologies, there exist several popular wireless communication technologies depending on wireless personal area network (WPAN) called IEEE 802.15.4 standard and the WRAN called IEEE 802.22 standard [69–75]. Furthermore, the insufficiencies of the PLC systems that are especially shown at high frequencies may be eliminated by exploiting wireless communication technologies [76].

The IEEE 2030-2011 standard, which is accepted as a main guideline for the SG systems, describes communication system requirements of the SG systems. The SG communication architecture characterized by the standard is illustrated in Fig. 6.15 where the communication infrastructure is formed according to a three-section model. The first segment of this definition that covers HAN, building area network (BAN) and industrial area network (IAN) is established on the consumer side through special networks. The second part of the definition that is based on the WAN containing NAN and FAN is constituted on the distribution side of the SGs. Both FANs and NANs perform monitoring and controlling processes of AMI, phasor measurement unit (PMU) and remote terminal units. The third one is referred as core network that is located at

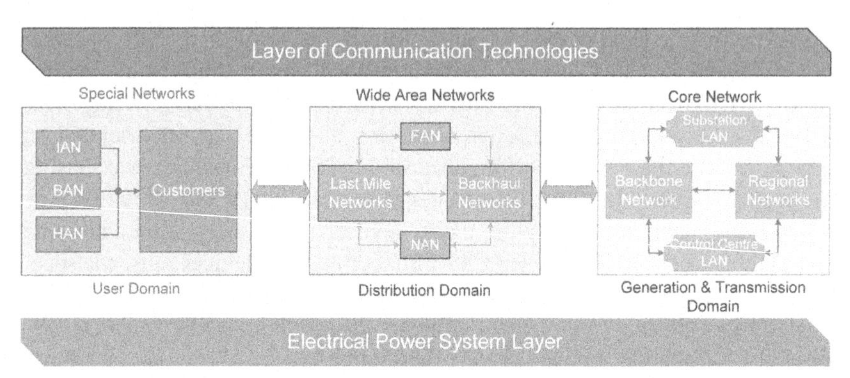

FIG. 6.15 Communication architectures of SG systems described in the IEEE 2030 standard.

generation and transmission stage of SG systems. Local area network (LAN), voice over internet protocol (VoIP), geographical information system (GIS), and virtual private network (VPN) are the main broadband communication infrastructures of core networks [68, 76].

In HAN/BAN/IAN architectures, sensor data obtained from various smart devices in home, buildings and industrial areas are gathered and transferred to data management and control center of SGs. Therefore, the HAN/BAN/IAN architectures are taken into account as premises networks. The HAN architecture can yield various significant properties such as managing household appliances based on their power consumption rates, reporting power consumption information to users, and assisting in prepaid client cards. The SMs are constituted in both user domain and industrial plant domain to present a gateway that aims to transfer information among the HANs and NANs. Since whole applications in these networks are realized inside of home and buildings, there is no need to exploit high-frequency communication systems. Hence, the HAN/BAN/IAN architectures ensure significant benefits such as low power consumption, simplicity, low cost and secure communication when they are efficiently utilized in realistic applications. The most preferred technologies in these networks are Bluetooth, ZigBee, Wi-Fi, PLC systems, and Ethernet. In addition, the coverage areas of HANs may be approximately 200 m^2 whereas data rates may change from 10 to 100 Kbps for each device. It is also important to note that these networks are generally considered as insensitive to latency [68, 76].

The NANs are responsible for ensuring the connection between users/consumers and data concentrator/substation where intelligent electronic devices (IEDs) are typically employed to collect information from the nearest data points. The NANs not only present to employ advanced communication technologies in both data concentrators and the SMs, but they can also transmit power consumption and control data depending on demands. The final component of a NAN may be a SM or a Data Aggregation Point (DAP) that collects information from the SMs and transmits obtained data to metering data

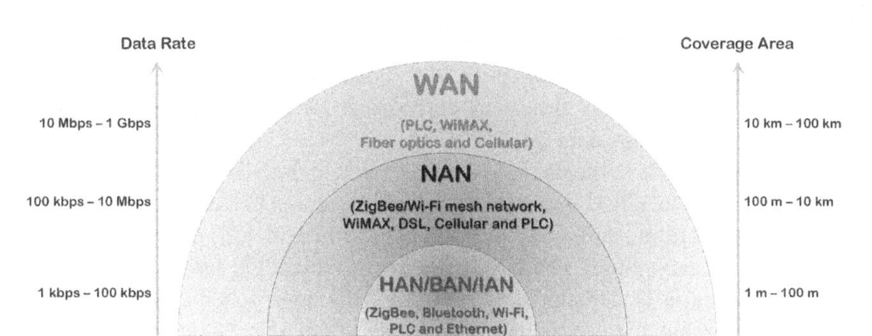

FIG. 6.16 Comparison of SG communication networks in terms of data rate and coverage.

management system (MDMS) over a backbone network. The SMs are able to execute numerous SG applications such as distribution automation, power outage management, and power quality monitoring. In addition, they can report power consumption rates in real time. The areas in the range of square kilometers may be covered by the NANs. The NANs are being utilized in the AMI systems and can improve the application ranges of the SGs. The most common technologies utilized in the NANs are ZigBee and Wi-Fi over mesh networks, PLC, DSL, cellular systems and WiMAX [68, 76]. Comparison of coverage and data rates for SG communications are illustrated in Fig. 6.16.

The FANs carry out the information exchange among grid control center, distribution substations, and feeders in order to empower monitoring, controlling and protection applications. The distribution substations transform high-voltage electricity into the low-voltage electricity to feed homes, offices and businesses. When a SG system is taken into account, it covers various metering, monitoring and control systems to realize substation control transactions. Otherwise, distribution feeders consist of power lines, towers and cable poles to deliver electricity to the user plants. Furthermore, the feeders are responsible for providing a common coupling point for microgrids. In order to enable metering and monitoring applications in the SG systems, a great number of sensors and actuators may be located in the distribution feeders [68].

The communication infrastructure among utilities and substations are based on the WANs that are typically composed of power generation systems, distributions stations and transformer systems. The WANs are utilized to enable real-time measurement and monitoring processes over wide areas. Hence, the WANs require maintaining secure backbone communication networks that should ensure high bandwidth features in order to deal with long-distance information transmission. In other words, the WANs establish a connection bridge among control centers and data concentrator of every NAN to enable data transmission through high-speed communication [68]. On the other hand, optical communication systems are commonly exploited between distribution substations and control center since they present the advantages of high capacity and decreased latency. As well as optical communication systems, cellular and WiMAX based communication systems are also utilized to widen coverage areas of the WANs. Typically, the WANs may almost cover thousands of square kilometers and can present about 10 Mbps data rates [76].

In order to improve the use of spectrum bands in the SG communications, it is believed that the CR is a promising technology for efficiently accessing and employing spectrum bands [68]. In 2010, Ghassemi et al. [39] proposed the utilization of CR technology in the SG systems. A typical CR based SG communication structure is depicted in Fig. 6.17. This approach includes a three-layered hierarchical structure that is composed of HANs, NANs, and WANs. This approach can also promote both energy-efficient designs and spectrum efficient designs. Moreover, the CR technology encourages the use of high bandwidths in the SG system applications that are required for conveying massive information including metering, monitoring and control information [68, 76]. The following sections present CR enabled networks in the SG applications.

6.5.1 CR enabled home area networks

The block diagram of a typical CR enabled HAN architecture is illustrated in Fig. 6.18. A HAN architecture carries out two important functions called commissioning and control. The former is responsible for detecting new appliances once they connect or disconnect from a HAN, and controls connecting or establishing of a self-organizing network. The latter is responsible for ensuring interoperability and manages the communication connections among devices in the SG networks. The mentioned communication connections generally utilize the unlicensed ISM bands. The popular technologies used in these connections are IEEE 802.11 WLANs, Wi-Fi, Bluetooth, and ZigBee that is widely used in the HANs due to the advantages of low cost and low power consumption. Therefore, the ZigBee may lead to interferences to other devices functioning in the license-free bands. The cognitive HANs that combine the CR with the ZigBee (or with IEEE 802.15.4 standard) to operate in a dynamic spectrum access manner may be a satisfactory solution in order to overcome this issue. Moreover, the CR devices in the HANs can also employ licensed spectrums if these bands are not in use by PUs. Consequently, the SG systems can efficiently present enhanced services in terms of metering, monitoring and controlling thanks to the CR technology. Sreesha et al. [77] reported to exploit wireless sensor networks (WSNs) with several changes in routing protocols in order to provide spectrum and energy efficient designs in the cognitive HANs. Moreover, Aijaz et al. [78] developed a MAC protocol for the CR systems that provides energy efficiency in addition to provided high reliability.

The HANs can be constituted either mesh topology or star topology where wired (i.e., PLC) and wireless communication technologies (i.e., ZigBee, Bluetooth, Wi-Fi, and CR) can be efficiently utilized [68, 76]. Therefore, an

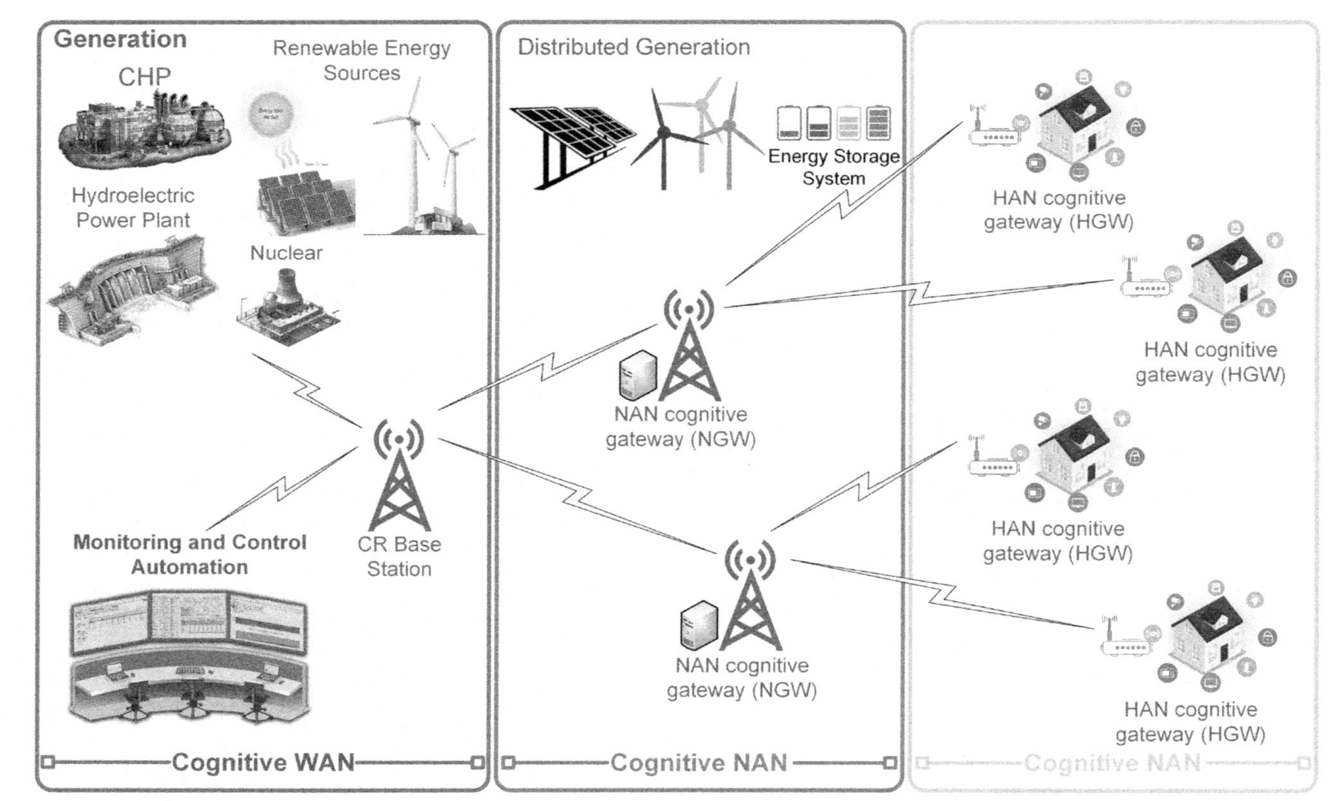

FIG. 6.17 CR enabled smart grid communication networks.

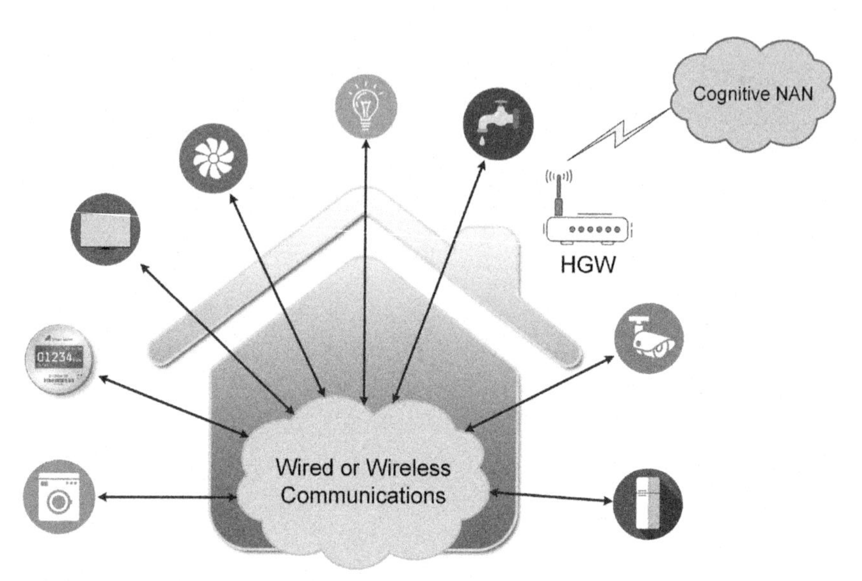

FIG. 6.18 Cognitive HAN architecture in the SG communications.

adjustable service gateway is highly required to control communication in the HAN and between other HANs. In the HANs, there may exist several components containing smart devices such as sensors, actuators, load control devices, SMs, cognitive home gateway (HGW) and plug-in electric vehicles [41, 79]. In order to systematically gather measurement data, a HGW can be combined with the SMs to enable communication between several devices and/or terminals in the HANs [80]. It is worth noting that smart devices can also connect to the HGW by employing either wired or wireless communication technologies. In order to combine with CR technology, the HGW should be improved via professional cognitive abilities that will authorize cooperation with the surroundings, transmission properties and transmission parameters depending on existing circumstances. The cognitive HGW examines the possible spectrum bands for detecting spectrum holes/white spaces that are unoccupied frequencies by the PUs, and it exploits these available frequencies to enhance communication performance by considering interference limits. In addition, the HGW regularly assembles power measurement data from several terminals located in the HANs to ensure two-way communication in the HANs [41]. Later, the information are transmitted to targets placed out of the NANs. In the reverse direction, the HGW acts as a center node in the NANs to acquire information (i.e., pricing, demand responses etc.) [41, 79]. After the information are obtained, they are delivered to the SMs or several terminals. Moreover, it can manage the communication among smart devices in the HANs. The HGW may exploit the unlicensed frequencies while functioning in the HANs [41, 80]. It has a significant task to simplify networking of sensors and devices existing in the HANs as well as controlling spectrum sharing. Lastly, the cognitive HGWs should effectively divide the possible spectrum bands between several smart devices existing in the HAN and should administer the uninterrupted access of new devices to the network by providing channels and IP addresses to each joined device.

6.5.2 CR enabled neighborhood area networks

A NAN architecture can be considered as the second communication layer platform for CR enabled SG systems as can be seen from the Fig. 6.17. This architecture may typically cover the areas in the range of several kilometers. The main goal of the NANs is the enabling a communication bridge among customer premises and electrical power systems in order to gather and control information from the near terminals. The NANs can communicate with HGWs of the HANs thanks to cognitive NAN gateway (NGW) structures that behave as cognitive APs to assemble data from various HANs. Typically, a NGW may be composed of a power substation, a utility pole-mounted device or a CR BS system that can access to the HGWs over only one-hop connection [41, 42]. This access opportunity can significantly enhance spectrum utilization efficiency in a NAN architecture. A NAN system gathers power-related measurement data from each customer connected to the network and delivers these data to the control center over the WANs in the SG systems.

Furthermore, in order to remove the leasing cost of licensed frequency bands, the HGWs convey measurement data over licensed spectrum bands by employing CR technology. Hybrid dynamic spectrum access method has been also developed to efficiently transmit measurement data from the HGWs to the NGWs [41, 79–81]. This access method employs licensed frequency bands assigned by the NGWs in an opportunistic way for connecting to the HGWs. Thus, the CR technology can improve the use of possible spectrum bands to enhance the performance of the NAN networks. On the other hand, the NANs are spread over very large areas including rural, urban and suburban areas, and cooperate with the AMI systems that may employ different communication technologies based on the SG applications [41]. Since the NANs should serve in large areas, various wired and wireless communication technologies are adopted by these network schemes. In addition, the CR enabled NANs should ensure a secure communication among the DAPs and the SMs [53].

6.5.3 CR enabled wide area networks

A WAN scheme that is the top layer of the SG communication architecture is typically composed of various NANs as can be seen from Fig. 6.19. It enables the broadband communication among distributed power generation systems, substations, NANs, and control center of utility. This network utilizes a bidirectional communication network with very high bandwidth that can accomplish

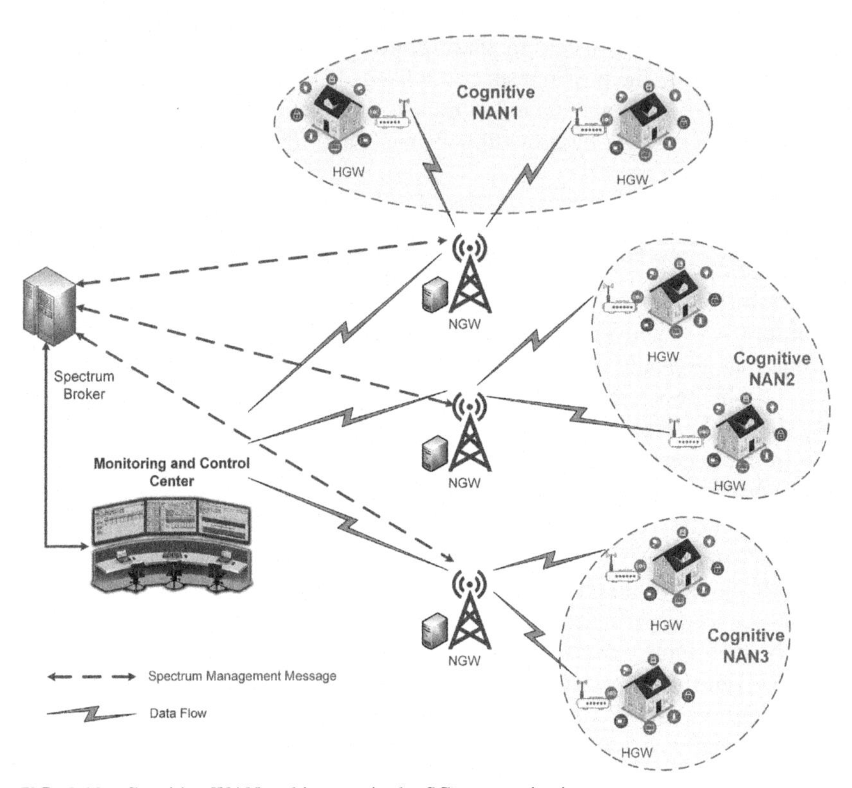

FIG. 6.19 Cognitive WAN architecture in the SG communications.

data transmission over long distances for monitoring and control applications of the SG systems. A WAN architecture comprises the combination of core network and backhaul network [68, 76]. The former is responsible for communicating with control center where the communication infrastructure is carried out by employing either fiber optics or cellular systems to provide low latency and high data rates. The latter aims to provide broadband communication links for both the NANs and monitoring devices. In addition, employing the CR technology in this network not only reduces installation costs, but also provides several improvements in terms of coverage, system capacity and flexibility [80]. Furthermore, the NGWs can effectively communicate with the CBSs since the each of them behaves as a SU in lieu of an AP in the CR-enabled backhaul networks.

In the cognitive WANs, spectrum brokers are intentionally utilized to control licensed spectrum bands allocated to the NANs as can be seen from the Fig. 6.19. When the dynamic changes of data in the network are taken into account, a spectrum broker should not only react to the demand promptly, but it should also efficiently deploy the available licensed frequency bands to satisfy data transfer needs. When a WAN serves in a wide area, some NANs may employ the same frequency bands devoid of leading to interference to each

TABLE 6.2 The features and techniques of CR enabled networks in the SG systems.

Cognitive area networks	HAN	NAN	WAN
Spectrum band	Unlicensed and licensed frequency band	Unlicensed and licensed frequency band	Unlicensed and licensed frequency band
Network topology	Centralized/ decentralized	Centralized	Centralized
Network users	Sensors, SMs, HGWs	HGWs, NGWs	NGWs, spectrum brokers
Featured strategy	Cross-layer spectrum sharing	Hybrid dynamic spectrum access	Optimal spectrum leasing
Key techniques	Access control, power coordination	Guard channel, spectrum handoff	Join spectrum management

other. For instance, a scenario is illustrated in the Fig. 6.19 where there exist three NANs and 12 different licensed spectrums that are leased from a telecommunication operator. A spectrum broker may allocate five bands to NAN1 and seven bands to NAN2 by considering data traffic demands. As long as there is no interference, the spectrum broker may also deploy five bands to NAN3 that is the same as assigned to the NAN1 since these networks are far from each other. Table 6.2 summarizes the features and techniques of the CR enabled networks in the three subareas.

6.6 CR enabled smart grid applications

The stability and security of bidirectional communication among control centers and SMs are critical significance for the total performance of the SG systems. The utilization of WSNs in the SG systems has caused the development of novel wireless communication technologies that are being widely employed in communication networks of the SG systems, and these new wireless communication systems have caused to occur highly heterogeneous networks in the SG systems. Simultaneously, the number of applications has greatly risen, which should handle widespread challenges faced in the SG systems. These challenges may be related to metering, monitoring, communications and automation systems characteristics of the SG systems [82–85]. Furthermore, the SG applications may create numerous types of data traffic containing heterogeneous QoS demands such as appropriate reliability level, maximum allowable delay, and minimum needed throughput.

On the other hand, the use of CR technology in the SG systems can provide remarkable improvements to reduce interference effects which are generated by the presence of a variety of wired and/or wireless communication systems that are functioning at the same frequencies in the HANs. In order to accomplish this, the fundamental characteristics of the CR devices to revise their capacities depending on the particular criterions will be employed to present more efficient organizing of various data transfers. Furthermore, the CR technology can be exploited in the NANs in which this technology can provide connections between the NANs with the opportunistic use of unoccupied frequency bands [41, 51]. By the following subsections, the application examples of SG systems where CR technology can be efficiently utilized to improve overall system performance will be introduced in the context of HAN, NAN, and WAN as previously depicted in the Fig. 6.17.

6.6.1 Home energy management systems

One of the most important SG applications in the HANs is home energy management systems (HEMSs). The HEMSs carry out home automation and control processes thanks to home appliances that can transmit various data to SMs, indoor monitors and other smart devices. They can decrease power consumption rates by employing adaptive control techniques [73]. In addition, commercial and industrial customers can utilize the HEMS for building automation, heating, ventilation, and air conditioning (HVAC) control and other energy management practices [86]. Typically, a SM or an Internet gateway is exploited for connecting a HAN to SG services or service providers. Therefore, real-time data transfer should be performed to ensure that HEMS transactions are performed effectively and securely. The CRN architectures may provide several advantages for the real-time data sharing between HEMSs [73, 75]. It is also important to note that the access technologies should guarantee the QoS with a considerable bandwidth for the HEMS applications.

6.6.2 Real-time pricing and demand response management applications

In meter reading applications, several quantities are measured by using the SMs and the obtained data are transmit to a central database for analysis and billing purposes [68]. The meter reading process can be classified into three categories as on demand, scheduled and bulk transfer. While the meter reading task is performed based on the requests in the first one, the second one is achieved for predetermined time durations. In third one, the measurements are gathered from the whole SMs in a predetermined service area. A CRN can ensure the information transmission services for all of these meter reading methods [68, 73,

87]. Since the CR devices may be spread over wide areas in electrical power systems of the SGs, the users/consumers and utilities can also perceive and observe power outages [88]. This provides a utility to notice a power outage once the power blackout happens over the SMs and outage detection units. It is worth noting that the crucial measurement data (i.e., metering data, monitoring data, SCADA information etc.) will be conveyed over a committed communication network utilizing either licensed frequency bands or wired communication systems due to the real-time needs. The dynamic spectrum allocation in the CR technology is aimed to be utilized for comparatively less important data (i.e., obtained data from the SMs). Therefore, more bandwidth on the committed networks can be provided for applications with higher priority. Pricing applications aim to transmit billing information to the SMs and smart devices, and pricing information can be divided into three groups as real-time pricing (RTP), time-of-use (TOU) pricing, and critical peak pricing (CPP). In addition, connection or disconnection of a SM to the SG services is carried out based on these pricing groups. Furthermore, service providers can compute billing information for costumers and start required transactions such as sending notification messages to the customers [68, 73, 87].

The DRM (or sometimes referred as Demand Side Management (DSM)) is realized on the customer sides, and it collaborates with markets, operational regions and service providers. Thanks to the DRM systems, the service providers can connect to smart appliances situated at customer premises to decrease the load on the distribution systems in the time of peak demand periods. Demand response applications are developed to change the power consumption habits of the customers in reaction to prices and other types of encouragements to efficiently use system capacity in order to avoid capacity expansion requirements [89, 90]. In addition, two-way communication among customers and service providers is required for ensuring effectiveness on the DRM systems [91]. The SMs established at customer plants enable bidirectional communication among the customers and the service providers. Furthermore, this communication infrastructure allows to the service providers to form load profiles of users autonomously and efficiently [92]. When the DRMs are considered from the point of view of the users, they can enable customers to turn desired devices on or off by conveying command signals to a load controller situated at customer plants [73, 75]. Smart control systems based on sensors (i.e., smart lighting systems [93]) may also help to the DRM systems. Utilization of secure and effective communication technologies in demand response applications is highly needed to provide robust DRM systems. In order to enable two-way communication between the customers and the SMs, the IEEE 802.15.4 based, IEEE 802.11 based or PLC based technologies are preferred for DRM applications [68, 76]. Alternatively, cellular communications systems may be also used for the applications that are able to provide low-latency advantage. The DRM systems also assist the service providers to improve transaction capabilities and the usage of distributed energy resources (DER) in power systems. The

CR technology is considered as a promising technology to improve communication infrastructure of the DRM system since it is one of the most important components of the SG systems.

6.6.3 Distribution automation applications

Distribution Automation (DA) is responsible for monitoring, controlling and managing the power distribution grids in the SG systems. It can present real-time operational notifications regarding components of the distribution grid such as voltage regulators, capacitor bank controllers, fault detectors, and switches. These notifications are also shared with the other intelligent field devices (i.e., the IEDs), and the DA combines them with customer transactions and transmission systems. The communication infrastructure needs of the DA systems may usually change based on the service provider demands. When the characteristic features of the DA systems are taken into account, the CRNs have seamless capabilities to meet requirements of the DA systems.

6.6.4 Distributed generation system applications

Electrical power can be produced by using distributed renewable energy sources (RESs) (i.e., solar, wind, fuel cell etc.) constructed close to the customer plants to avoid overload problems and to adjust power demand and supply [49]. The integration of the RESs into the SG systems has gained a great importance depending on the development of clean energy concept. On the other hand, it is important to note that the distributed power generation is merely possible as long as the voltage, frequency and power quality parameters of the grids remain stable. Because of the widespread use of the distributed RESs, the utilization of advanced control methods are necessary to sustain the robustness and safety of entire power grid [94]. The agile DR, effective data transmission and intelligent energy storage can be considered as examples of these control methods. All of them require the using of secure and efficient communication systems in the SG systems. Moreover, these energy sources also need the using transmission systems since they are widely located away from distribution stations. In addition, remote monitoring systems of the energy sources can be faced several challenges owing to congestion of the unlicensed frequency bands. Since conventional wired communication systems require high installation costs, easy deployment and low cost superiorities of wireless communication technologies make them more feasible for this type of SG applications. Nevertheless, wireless communications have to overcome various difficulties such as interference, multi-path propagation, noise and shadowing. Fortunately, CR enabled SG systems can deal with these issues and it can opportunistically utilize the spectrum bands to enhance the performance of remote monitoring systems and the SG systems [95].

6.6.5 Wide area monitoring applications

Wide area monitoring system (WAMS) covers the real-time data collection from several components of SG systems such as IEDs and PMUs. The PMU that is also referred as synchrophasor is an essential component of a Wide Area Monitoring Protection and Control (WAMPAC) system. Typically, a PMU obtains voltage and current waveforms of the power system by employing transformers and produces their phasor by sampling measured data. The PMU is one of the measurement methods, which is usually established and is swiftly being utilized in increasing numbers. This system is developed for observing load sharing, power flows, voltage stability, phasor synchronizations, restoring the power systems, and estimating the effective algorithms to recover lost power. It extracts the phasor of the line voltage and line current in power systems in which entire components of phasor are compared to the period of signal. Furthermore, a PMU can estimate the frequency of the transmission line and handles the correct detection of frequency security [68].

On the other hand, wide area control systems can present several advantages such as self-repairing, faster response and local control opportunity when compared to the classical control performed by a control center. Wide area protection is able to preserve the power systems against the power outages and unforeseen situations completely and autonomously [73, 75]. The WAMS cannot only observe the all stages of power grids (i.e., power generation, transmission and distribution), but it can also collect information regarding system devices such as protection devices, capacitor banks, and transformers. Various communication technologies operating over licensed or license-free spectrum bands may be employed for the monitoring applications in the SG systems. Nonetheless, it is obvious that the utilization of licensed frequency bands leads to increased costs. The CR technology can be efficiently exploited in the wide area monitoring applications based on its advantages explained previously.

6.6.6 Advanced metering infrastructure applications

One of the most important applications of the SG systems is the AMI system that is utilized to measure, obtain and analyze power-related data regarding energy consumption and power quality of each user. The AMI system can be also supposed as an advanced version of conventional automated meter reading (AMR) and automatic meter management (AMM) systems because of the fact that it includes various improved devices and networks types such as the SMs, HANs, WANs, and NANs [68]. Since the gathered information are very crucial, a communication backbone that will provide reliability, security, scalability and cost effectiveness should be established between customers and services providers. A SM supporting the AMI infrastructure has the ability to communicate with measurement devices on demand. The AMI characteristics of the SMs allow controlling of the two-way communication that makes the SMs accomplished to execute commands transmitted by the service providers. Furthermore, this two-way communication among service providers and customers can be utilized to enhance demand management, maintenance, and planning capability of service providers.

It is also possible in any SG systems that there will be exist a large number of the SMs as well as many APs. Therefore, mesh network structures are widely taken into account in the AMI communication infrastructures. Since the unlicensed ISM bands are utilized by numerous communication systems in urban areas, communication infrastructure of the AMI systems highly suffers from the interferences. Therefore, performances of the AMI systems operating in the unlicensed frequency bands decrease and this situation causes to high latency and data losses as well as the security problems. The utilization of licensed spectrum for enabling communication among the APs and service providers generally requires high costs. The CR technology can be efficiently used in backhaul networks of the AMI systems since its advanced spectrum access characteristics will ensure to improve system performance [49]. In addition, self-configuration and easy deployment properties of the CR technologies make them potential candidates for employing in communication infrastructures of the AMI systems [51, 95].

6.6.7 Wide-area situational awareness

The last considered SG application is wide-area situational awareness (WASA) that realizes remote monitoring, control and protection of the SG systems by employing the PMUs and IEDs. In addition, it has self-repairing feature through local control and quick response characteristic compared to the conventional control managed by a control center. Furthermore, the WASA can completely maintain the power systems against the power outages and unforeseen situations. The communication networks of the SG systems should face up to several important problems such as interference, latency, and insufficient capacity that will lead to significant issues on security, reliability and efficiency of the SG systems. Moreover, deployed and large-scale characteristics of wide areas also cause various challenges for the WSNs such as maintenance, the necessity of using different spectrum, fading, interference, shadowing, and multi-path propagation. These challenges require utilizing the improved communication technologies to transmit predicted state of the all system. Therefore, the CR technology can also enhance the system performance of WASA and it not only improves the spectrum efficiency, but also enhances the system capacity through advanced spectrum management methods.

References

[1] B. Wang, K.J.R. Liu, Advances in cognitive radio networks: a survey. IEEE J. Sel. Top. Sign. Proces. 5 (2011) 5–23, https://doi.org/10.1109/JSTSP.2010.2093210.

[2] S. Haykin, Cognitive radio: brain-empowered wireless communications. IEEE J. Sel. Areas Commun. 23 (2005) 201–220, https://doi.org/10.1109/JSAC.2004.839380.

[3] I.F. Akyildiz, W.-Y. Lee, M.C. Vuran, S. Mohanty, Next generation/dynamic spectrum access/cognitive radio wireless networks: a survey. Comput. Netw. 50 (2006) 2127–2159, https://doi.org/10.1016/j.comnet.2006.05.001.

[4] E. FCC, Docket no. 02-135 Spectrum Policy Task Force Report, Federal Communications Commission (FCC), 2002.

[5] F.E.D. No, No 02–155 Spectrum Policy Task Force Report, Federal Communications Commission (FCC), 2002.

[6] C. Powell, ET Docket no 03–222 Notice of Proposed Rule Making and Order, Federal Communications Commission, Washington, 2003.

[7] H. Kim, Efficient Identification and Utilization of Spectrum Opportunities in Cognitive Radio Networks, 2010.

[8] M. McHenry, Spectrum white space measurements, in: New America Foundation Broadband Forum, 2003June.

[9] S. Yin, D. Chen, Q. Zhang, M. Liu, S. Li, Mining spectrum usage data: a large-scale spectrum measurement study, IEEE Trans. Mob. Comput. 11 (2012) 1033–1046.

[10] X. Ying, J. Zhang, L. Yan, G. Zhang, M. Chen, R. Chandra, Exploring indoor white spaces in metropolises, in: Proceedings of the 19th Annual International Conference on Mobile Computing & Networking, ACM, 2013, pp. 255–266.

[11] Q. Zhang, J. Jia, J. Zhang, Cooperative relay to improve diversity in cognitive radio networks, IEEE Commun. Mag. 47 (2009) 111–117.

[12] E.Z. Tragos, S. Zeadally, A.G. Fragkiadakis, V.A. Siris, Spectrum assignment in cognitive radio networks: a comprehensive survey, IEEE Commun. Surv. Tutorials 15 (2013) 1108–1135.

[13] J. Mitola, G.Q. Maguire, Cognitive radio: making software radios more personal. IEEE Pers. Commun. 6 (1999) 13–18, https://doi.org/10.1109/98.788210.

[14] F.C. Commission, Notice of Proposed Rule Making and Order: Facilitating Opportunities for Flexible, Efficient, and Reliable Spectrum Use Employing Cognitive Radio Technologies, ET Docket, 2005, p. 73.

[15] J. Ren, N. Zhang, X. Shen, Introduction. in: Energy-Efficient Spectrum Management for Cognitive Radio Sensor Networks, Springer International Publishing, Cham, 2018, pp. 1–13, https://doi.org/10.1007/978-3-319-60318-6_1.

[16] E. Hossain, K.G.M. Thilina, Cognitive radio networks and spectrum sharing. in: Academic Press Library in Mobile and Wireless Communications, Elsevier, 2016, pp. 467–522, https://doi.org/10.1016/B978-0-12-398281-0.00013-2.

[17] X. Hong, C. x Wang, H. h Chen, Y. Zhang, Secondary spectrum access networks. IEEE Veh. Technol. Mag. 4 (2009) 36–43, https://doi.org/10.1109/MVT.2009.932543.

[18] I.F. Akyildiz, W.-Y. Lee, K.R. Chowdhury, CRAHNs: cognitive radio ad hoc networks, Ad Hoc Netw. 7 (2009) 810–836.

[19] Q. Zhao, L. Tong, A. Swami, Y. Chen, Decentralized cognitive MAC for opportunistic spectrum access in ad hoc networks: a POMDP framework. IEEE J. Sel. Areas Commun. 25 (2007) 589–600, https://doi.org/10.1109/JSAC.2007.070409.

[20] P. Ren, Y. Wang, Q. Du, CAD-MAC: a channel-aggregation diversity based MAC protocol for spectrum and energy efficient cognitive Ad Hoc networks. IEEE J. Sel. Areas Commun. 32 (2014) 237–250, https://doi.org/10.1109/JSAC.2014.141205.

[21] U.F.C. Commission, others, Spectrum Policy Task Force Report, ET Docket No. 02–135, Nov. (2002).

[22] G. Tomar, A. Bagwari, J. Kanti, Introduction to Cognitive Radio Networks and Applications, Taylor & Francis, a CRC Title, Part of the Taylor & Francis Imprint, a Member of the Taylor & Francis Group. The Academic Division of T&F Informa, plc, Boca Raton, FL, 2017.

[23] C. Cordeiro, K. Challapali, M. Ghosh, Cognitive PHY and MAC layers for dynamic spectrum access and sharing of TV bands, in: Proceedings of the First International Workshop on Technology and Policy for Accessing Spectrum, ACM, 2006, pp. 1–11.

[24] H. Kim, K.G. Shin, In-band spectrum sensing in cognitive radio networks: energy detection or feature detection? in: Proceedings of the 14th ACM International Conference on Mobile Computing and Networking, ACM, 2008, pp. 14–25.

[25] M. Oner, F. Jondral, On the extraction of the channel allocation information in spectrum pooling systems. IEEE J. Sel. Areas Commun. 25 (2007) 558–565, https://doi.org/10.1109/JSAC.2007.070406.

[26] W. Han, J. Li, Z. Li, J. Si, Y. Zhang, Efficient soft decision fusion rule in cooperative spectrum sensing. IEEE Trans. Signal Process. 61 (2013) 1931–1943, https://doi.org/10.1109/TSP.2013.2245659.

[27] J. Ma, Y.G. Li, soft combination and detection for cooperative spectrum sensing in cognitive radio networks. in: IEEE GLOBECOM 2007—IEEE Global Telecommunications Conference, 2007, pp. 3139–3143, https://doi.org/10.1109/GLOCOM.2007.594.

[28] K.M. Thilina, K.W. Choi, N. Saquib, E. Hossain, Machine learning techniques for cooperative spectrum sensing in cognitive radio networks. IEEE J. Sel. Areas Commun. 31 (2013) 2209–2221, https://doi.org/10.1109/JSAC.2013.131120.

[29] K.M. Thilina, K.W. Choi, N. Saquib, E. Hossain, Pattern classification techniques for cooperative spectrum sensing in cognitive radio networks: SVM and W-KNN approaches. in: 2012 IEEE Global Communications Conference (GLOBECOM), 2012, pp. 1260–1265, https://doi.org/10.1109/GLOCOM.2012.6503286.

[30] B. Canberk, I.F. Akyildiz, S. Oktug, Primary user activity modeling using first-difference filter clustering and correlation in cognitive radio networks, IEEE/ACM Trans. Netw. 19 (2011) 170–183.

[31] A.T. Hoang, Y.-C. Liang, A two-phase channel and power allocation scheme for cognitive radio networks, in: Personal, Indoor and Mobile Radio Communications, 2006 IEEE 17th International Symposium on, IEEE, 2006, pp. 1–5.

[32] P. Kaur, M. Uddin, A. Khosla, Adaptive bandwidth allocation scheme for cognitive radios, Int. J. Adv. Comput. Technol. 2 (2010) 35–41.

[33] P. Demestichas, G. Dimitrakopoulos, K. Tsagkaris, K. Demestichas, J. Adamopoulou, Reconfigurations selection in cognitive, beyond 3G, radio infrastructures, in: Cognitive Radio Oriented Wireless Networks and Communications, 2006. 1st International Conference On, IEEE, 2006, pp. 1–5.

[34] Q. Liang, S. Han, F. Yang, G. Sun, X. Wang, A distributed-centralized scheme for short- and long-term spectrum sharing with a random leader in cognitive radio networks. IEEE J. Sel. Areas Commun. 30 (2012) 2274–2284, https://doi.org/10.1109/JSAC.2012.121219.

[35] K.B.S. Manosha, N. Rajatheva, M. Latva-aho, Overlay/underlay spectrum sharing for multi-operator environment in cognitive radio networks. in: IEEE 73rd Vehicular Technology Conference (VTC Spring), 2011, pp. 1–5, https://doi.org/10.1109/VETECS.2011.5956425.

[36] H.A.B. Salameh, M. Krunz, O. Younis, Cooperative adaptive spectrum sharing in cognitive radio networks, IEEE/ACM Trans. Netw. 18 (2010) 1181–1194.

[37] H. Zheng, L. Cao, Device-centric spectrum management. in: First IEEE International Symposium on New Frontiers in Dynamic Spectrum Access Networks. DySPAN 2005, 2005, pp. 56–65, https://doi.org/10.1109/DYSPAN.2005.1542617.

[38] I. Christian, S. Moh, I. Chung, J. Lee, Spectrum mobility in cognitive radio networks. IEEE Commun. Mag. 50 (2012) 114–121, https://doi.org/10.1109/MCOM.2012.6211495.

[39] A. Ghassemi, S. Bavarian, L. Lampe, Cognitive radio for smart grid communications. in: 2010 First IEEE International Conference on Smart Grid Communications, 2010, pp. 297–302, https://doi.org/10.1109/SMARTGRID.2010.5622097.

[40] R.C. Qiu, Z. Hu, Z. Chen, N. Guo, R. Ranganathan, S. Hou, G. Zheng, Cognitive radio network for the smart grid: experimental system architecture, control algorithms, security, and microgrid testbed. IEEE Trans. Smart Grid 2 (2011) 724–740, https://doi.org/10.1109/TSG.2011.2160101.

[41] R. Yu, Y. Zhang, S. Gjessing, C. Yuen, S. Xie, M. Guizani, Cognitive radio based hierarchical communications infrastructure for smart grid. IEEE Netw. 25 (2011) 6–14, https://doi.org/10.1109/MNET.2011.6033030.

[42] T. Nghia Le, W.-L. Chin, H.-H. Chen, Standardization and security for smart grid communications based on cognitive radio technologies—a comprehensive survey. IEEE Commun. Surv. Tutorials 19 (2017) 423–445, https://doi.org/10.1109/COMST.2016.2613892.

[43] K. Tan, K. Kim, Y. Xin, S. Rangarajan, P. Mohapatra, RECOG: a sensing-based cognitive radio system with real-time application support, IEEE J. Sel. Areas Commun. 31 (2013) 2504–2516.

[44] Q. Li, Z. Feng, W. Li, T.A. Gulliver, P. Zhang, Joint spatial and temporal spectrum sharing for demand response management in cognitive radio enabled smart grid. IEEE Trans. Smart Grid 5 (2014) 1993–2001, https://doi.org/10.1109/TSG.2013.2292528.

[45] A. Aijaz, A.-H. Aghvami, PRMA-based cognitive machine-to-machine communications in smart grid networks, IEEE Trans. Veh. Technol. 64 (2015) 3608–3623.

[46] V. Dehalwar, M. Kolhe, S. Kolhe, Cognitive radio application for smart grid, Int. J. Smart Grid Clean Energy 1 (2012) 79–84.

[47] N. Ghasemi, S.M. Hosseini, Comparison of smart grid with cognitive radio: solutions to spectrum scarcity, in: 2010 The 12th International Conference on Advanced Communication Technology (ICACT), 2010, pp. 898–903.

[48] R. Amin, J. Martin, X. Zhou, Smart grid communication using next generation heterogeneous wireless networks, in: Smart Grid Communications (SmartGridComm), 2012 IEEE Third International Conference on, IEEE, 2012, pp. 229–234.

[49] V. Gungor, D. Sahin, Cognitive radio networks for smart grid applications: a promising technology to overcome spectrum inefficiency. IEEE Veh. Technol. Mag. 7 (2012) 41–46, https://doi.org/10.1109/MVT.2012.2190183.

[50] T. Li, J. Ren, X. Tang, Secure wireless monitoring and control systems for smart grid and smart home. IEEE Wirel. Commun. 19 (2012) 66–73, https://doi.org/10.1109/MWC.2012.6231161.

[51] Y. Zhang, R. Yu, M. Nekovee, Y. Liu, S. Xie, S. Gjessing, Cognitive machine-to-machine communications: visions and potentials for the smart grid. IEEE Netw. 26 (2012) 6–13, https://doi.org/10.1109/MNET.2012.6201210.

[52] P. Ferrari, E. Sisinni, A. Flammini, A. Depari, Adding accurate timestamping capability to wireless networks for smart grids, Comput. Netw. 67 (2014) 1–13.

[53] Z. Zhu, S. Lambotharan, W.H. Chin, Z. Fan, Overview of demand management in smart grid and enabling wireless communication technologies. IEEE Wirel. Commun. 19 (2012) 48–56, https://doi.org/10.1109/MWC.2012.6231159.

[54] Ş. Temel, V.Ç. Gungor, T. Koçak, Routing protocol design guidelines for smart grid environments, Comput. Netw. 60 (2014) 160–170.

[55] R. Ma, H.-H. Chen, Y.-R. Huang, W. Meng, Smart grid communication: its challenges and opportunities. IEEE Transactions on Smart Grid 4 (2013) 36–46, https://doi.org/10.1109/TSG.2012.2225851.

[56] M. Erol-Kantarci, H.T. Mouftah, Energy-efficient information and communication infrastructures in the smart grid: a survey on interactions and open issues. IEEE Commun. Surv. Tutorials 17 (2015) 179–197, https://doi.org/10.1109/COMST.2014.2341600.

[57] J. Huang, H. Wang, Y. Qian, C. Wang, Priority-based traffic scheduling and utility optimization for cognitive radio communication infrastructure-based smart grid. IEEE Trans. Smart Grid 4 (2013) 78–86, https://doi.org/10.1109/TSG.2012.2227282.

[58] I. Akyildiz, Y. Altunbasak, F. Fekri, R. Sivakumar, AdaptNet: an adaptive protocol suite for the next-generation wireless Internet. IEEE Commun. Mag. 42 (2004) 128–136, https://doi.org/10.1109/MCOM.2004.1273784.

[59] I.F. Akyildiz, W.Y. Lee, M.C. Vuran, S. Mohanty, A survey on spectrum management in cognitive radio networks. IEEE Commun. Mag. 46 (2008) 40–48, https://doi.org/10.1109/MCOM.2008.4481339.

[60] P. Rawat, K.D. Singh, J.M. Bonnin, Cognitive radio for M2M and internet of things: a survey. Comput. Commun. 94 (2016) 1–29, https://doi.org/10.1016/j.comcom.2016.07.012.

[61] L. Claudino, T. Abrão, Spectrum sensing methods for cognitive radio networks: a review. Wirel. Pers. Commun. 95 (2017) 5003–5037, https://doi.org/10.1007/s11277-017-4143-1.

[62] P. Dibal, E. Onwuka, J. Agajo, C. Alenoghena, Application of wavelet transform in spectrum sensing for cognitive radio: a survey, Phys. Commun. 28 (2018) 45–57.

[63] H. Sun, A. Nallanathan, C.-X. Wang, Y. Chen, Wideband spectrum sensing for cognitive radio networks: a survey, IEEE Wirel. Commun. 20 (2013) 74–81.

[64] S. Pandit, G. Singh, Spectrum sensing in cognitive radio networks: potential challenges and future perspective, in: Spectrum Sharing in Cognitive Radio Networks, Springer International Publishing, Cham, 2017, pp. 35–75, https://doi.org/10.1007/978-3-319-53147-2_2.

[65] S. Atapattu, C. Tellambura, H. Jiang, Energy detection based cooperative spectrum sensing in cognitive radio networks. IEEE Trans. Wirel. Commun. 10 (2011) 1232–1241, https://doi.org/10.1109/TWC.2011.012411.100611.

[66] E. Kabalci, Y. Kabalci, I. Develi, Modelling and analysis of a power line communication system with QPSK modem for renewable smart grids. Int. J. Electr. Power Energy Syst. 34 (2012) 19–28, https://doi.org/10.1016/j.ijepes.2011.08.021.

[67] E. Kabalci, Y. Kabalci, A measurement and power line communication system design for renewable smart grids. Meas. Sci. Rev. 13 (2013)https://doi.org/10.2478/msr-2013-0037.

[68] Y. Kabalci, A survey on smart metering and smart grid communication. Renew. Sust. Energ. Rev. 57 (2016) 302–318, https://doi.org/10.1016/j.rser.2015.12.114.

[69] Y. Yan, Y. Qian, H. Sharif, D. Tipper, A survey on smart grid communication infrastructures: motivations, requirements and challenges. IEEE Commun. Surv. Tutorials 15 (2013) 5–20, https://doi.org/10.1109/SURV.2012.021312.00034.

[70] W. Wang, Y. Xu, M. Khanna, A survey on the communication architectures in smart grid. Comput. Netw. 55 (2011) 3604–3629, https://doi.org/10.1016/j.comnet.2011.07.010.

[71] A. Usman, S.H. Shami, Evolution of communication technologies for smart grid applications. Renew. Sust. Energ. Rev. 19 (2013) 191–199, https://doi.org/10.1016/j.rser.2012.11.002.

[72] X. Lu, W. Wang, J. Ma, An empirical study of communication infrastructures towards the smart grid: design, implementation, and evaluation. IEEE Trans. Smart Grid 4 (2013) 170–183, https://doi.org/10.1109/TSG.2012.2225453.

[73] M. Kuzlu, M. Pipattanasomporn, S. Rahman, Communication network requirements for major smart grid applications in HAN, NAN and WAN. Comput. Netw. 67 (2014) 74–88, https://doi.org/10.1016/j.comnet.2014.03.029.

[74] R.H. Khan, J.Y. Khan, A comprehensive review of the application characteristics and traffic requirements of a smart grid communications network. Comput. Netw. 57 (2013) 825–845, https://doi.org/10.1016/j.comnet.2012.11.002.

[75] E. Ancillotti, R. Bruno, M. Conti, The role of communication systems in smart grids: architectures, technical solutions and research challenges. Comput. Commun. 36 (2013) 1665–1697, https://doi.org/10.1016/j.comcom.2013.09.004.

[76] J.A. Cortes, J.M. Idiago, Smart metering systems based on power line communications. in: E. Kabalci, Y. Kabalci (Eds.), Smart Grids and Their Communication Systems, Springer Nature, Singapore, 2019, pp. 1–43, https://doi.org/10.1007/978-981-13-1768-2_4.

[77] A.A. Sreesha, S. Somal, I.-T. Lu, Cognitive radio based wireless sensor network architecture for smart grid utility, in: Systems, Applications and Technology Conference (LISAT), 2011 IEEE Long Island, IEEE, 2011, pp. 1–7.

[78] A. Aijaz, S. Ping, M.R. Akhavan, A.-H. Aghvami, CRB-MAC: a receiver-based mac protocol for cognitive radio equipped smart grid sensor networks. IEEE Sensors J. 14 (2014) 4325–4333, https://doi.org/10.1109/JSEN.2014.2346430.

[79] J.K. Thathagar, et al., Cognitive radio communication architecture in smart grid reconfigurability, in: Emerging Technology Trends in Electronics, Communication and Networking (ET2ECN), 2012 1st International Conference On, IEEE, 2012, pp. 1–6.

[80] F. Liu, J. Wang, Y. Han, P. Han, Cognitive radio networks for smart grid communications, in: Control Conference (ASCC), 2013 9th Asian, IEEE, 2013, pp. 1–5.

[81] A.A. Khan, M.H. Rehmani, M. Reisslein, Cognitive radio for smart grids: survey of architectures, spectrum sensing mechanisms, and networking protocols. IEEE Commun. Surv. Tutorials 18 (2016) 860–898, https://doi.org/10.1109/COMST.2015.2481722.

[82] V.C. Gungor, B. Lu, G.P. Hancke, Opportunities and challenges of wireless sensor networks in smart grid. IEEE Trans. Ind. Electron. 57 (2010) 3557–3564, https://doi.org/10.1109/TIE.2009.2039455.

[83] Y. Yang, F. Lambert, D. Divan, A survey on technologies for implementing sensor networks for power delivery systems. in: 2007 IEEE Power Engineering Society General Meeting, 2007, pp. 1–8, https://doi.org/10.1109/PES.2007.386289.

[84] S. Ullo, A. Vaccaro, G. Velotto, The role of pervasive and cooperative sensor networks in smart grids communication. in: Melecon 2010—2010 15th IEEE Mediterranean Electrotechnical Conference, 2010, pp. 443–447, https://doi.org/10.1109/MELCON.2010.5476236.

[85] R.A. Len, V. Vittal, G. Manimaran, Application of sensor network for secure electric energy infrastructure, IEEE Trans. Power Delivery 22 (2007) 1021–1028.

[86] A. Ahmad, S. Ahmad, M.H. Rehmani, N.U. Hassan, A survey on radio resource allocation in cognitive radio sensor networks. IEEE Commun. Surv. Tutorials 17 (2015) 888–917, https://doi.org/10.1109/COMST.2015.2401597.

[87] K. Sharma, L. Mohan Saini, Performance analysis of smart metering for smart grid: an overview. Renew. Sust. Energ. Rev. 49 (2015) 720–735, https://doi.org/10.1016/j.rser.2015.04.170.

[88] A.D. Dominguez-Garcia, C.N. Hadjicostis, N.H. Vaidya, Resilient networked control of distributed energy resources, IEEE J. Sel. Areas Commun. 30 (2012) 1137–1148.

[89] G. Mauri, D. Moneta, C. Bettoni, Energy conservation and smartgrids: new challenge for multimetering infrastructures, in: PowerTech, 2009 IEEE Bucharest, IEEE, 2009, pp. 1–7.

[90] P. Dongbaare, S.D. Chowdhury, T. Olwal, A. Abu-Mahfouz, Smart energy management system based on an automated distributed load limiting mechanism and multi-power switching technique, in: Proceedings of the 51st International Universities' Power Engineering Conference, 2016, pp. 06–09.

[91] G. Salami, O. Durowoju, A. Attar, O. Holland, R. Tafazolli, H. Aghvami, A comparison between the centralized and distributed approaches for spectrum management, IEEE Commun. Surv. Tutorials 13 (2011) 274–290.

[92] J.M. Guerrero, P.C. Loh, T.-L. Lee, M. Chandorkar, Advanced control architectures for intelligent microgrids—part II: power quality, energy storage, and AC/DC microgrids, IEEE Trans. Ind. Electron. 60 (2013) 1263–1270.

[93] C. Basu, J.J. Caubel, K. Kim, E. Cheng, A. Dhinakaran, A.M. Agogino, R.A. Martin, Sensor-based predictive modeling for smart lighting in grid-integrated buildings, IEEE Sensors J. 14 (2014) 4216–4229.

[94] E. Kabalci, Design and analysis of a hybrid renewable energy plant with solar and wind power, Energy Convers. Manag. 72 (2013) 51–59.

[95] A.O. Bicen, O.B. Akan, V.C. Gungor, Spectrum-aware and cognitive sensor networks for smart grid applications. IEEE Commun. Mag. 50 (2012) 158–165, https://doi.org/10.1109/MCOM.2012.6194397.

Chapter 7

Internet of things for smart grid applications

Chapter outline

7.1 Introduction

The smart grid which is transformed from the conventional utility involves bidirectional communication infrastructure and allows using information and communication technologies (ICT) at any stage of generation, transmission, distribution, and even consumption levels. This improved grid concept has been

From Smart Grid to Internet of Energy. https://doi.org/10.1016/B978-0-12-819710-3.00007-7

described as intelligent grid or inter-grid, but the smart grid term has been a standard for ICT supported grid structure [1]. The sudden or planned changes at generation, transmission and distribution stations, and consumption substations are handled by improved control and ICT technologies that are reacting with smart grid infrastructure. The reaction capability is frequently provided by the sensor assisted power electronics such as intelligent transformers, converters, intelligent electronic devices (IEDs), and remote terminal units (RTUs) that are based on wired or wireless communication systems [1–3].

Smart grid involves several devices to perform measurement, metering, smart appliance control, distributed generation (DG) management, and renewable energy source (RES) integration tasks. Although numerous surveys and researches have been presented in the current literature, their focus can be categorized into threefold as *smart infrastructure systems*, *smart management systems*, and *smart protection systems* as in [1, 2]. ICT and power networks comprise smart infrastructure system for generation, transmission, distribution, and consumption levels. Therefore, the metering, monitoring, and enhanced communication systems are included in the context of smart infrastructure system. The power quality and reliability of entire smart infrastructure is sustained by *smart management system* that is based on numerous control and management subsystems. The objectives of a smart management system can be listed as energy quality and demand side management (DSM), power control, sustainability and cost management, and DG control. These objectives are handled by several management infrastructures and control services to maintain the reliability and security of smart grid. The management services and applications have been listed in terms of optimization algorithms and soft computing methods. The last category of smart grid surveys includes *smart protection system* based on services and applications such as fault detection, self-healing processes, fault protection issues in the context of power systems [2, 3]. On the other hand, security and privacy protection services are provided by smart protection systems in terms of ICT protection.

Regardless of any smart system type that is used, the communication infrastructure plays a vital role in smart grid applications and services since all subsystems are coordinated and controlled on the communication, power systems and ICT bases. IEEE Std. 2030-2011 provides a concurrence among various frameworks and architectures that have been independently used in the existing grid environment. The standard defines the smart grid as a "system of systems" and provides guidance to numerous applications.

The suggested interoperability architecture along applications and reference models is illustrated in Fig. 7.1 where it provides a wide area for any smart grid application and services such as advanced metering infrastructure (AMI), plug-in electric vehicles (PEVs), and similar smart grid applications. The interoperability is the ability of any networks, applications, services, devices, and systems to communicate by using the ICT in a secure and reliable way. Thus, smart grid interoperability infrastructure is incorporated with hardware and

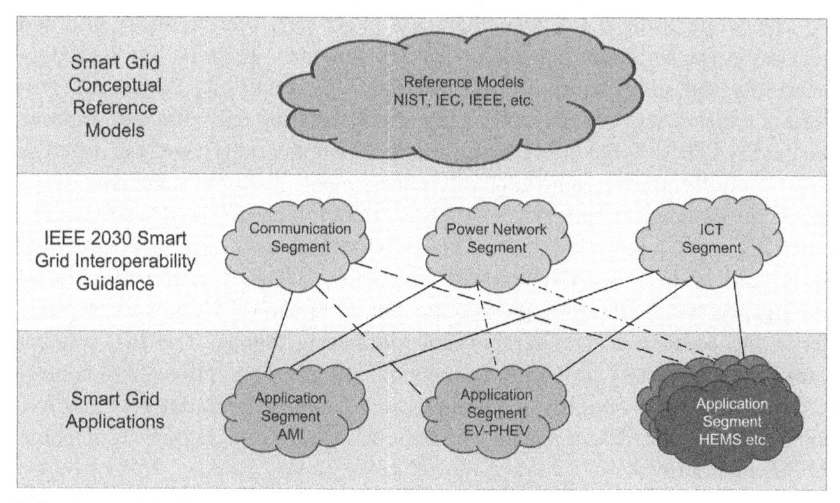

FIG. 7.1 IEEE Standard 2030-2011 smart grid interoperability.

software systems, several data transmission systems, and data exchange networks. The ICT interaction of interoperability is improved regarding to Open Systems Interconnect (OSI) reference model where functions are placed into seven layers, and layers are connected with service interfaces as done in the internet infrastructure [4]. The communication architecture providing the connection between applications and reference models of smart grid is comprised by core network, wide area networks, and private networks from generation level to consumption level, respectively.

On the other hand, a most recent term that is related with internet-based communication along smart grid infrastructure is known as Energy Internet (EI). The EI concept proposes an improved vision of smart grid applications into internet-based communication system facilitating and enhancing data transmission and control operations. The smart grid is transformed to a widespread cyber-physical system (CPS) with the integration of IoT technology. Besides contributions of IoT to smart grid in the context of AMI, DG, RTU, IED and other emerging applications, it can lead to unpredicted collapses such as cyber-security or denial of service (DoS) attacks. Therefore, IoT and smart grid integration requires greatly established communication and control architecture to ensure data security against threats. The IoT based smart grid infrastructure is composed of three-layer framework that includes sensing layer, gateway layer, and control layer. In this generalized structure, the sensing layer includes several measurement nodes to inherit the required data from related areas, and to transmit to certain application or receiving nodes. The gateway layer provides several routing protocols and router devices to manage provided data from sensing layer. The control layer, which is comprised by control nodes manage the gateway nodes and ensures efficient use of the resources.

The cooperation of IoT and smart grid is not only seen in utility grid and generation-transmission-distribution cycles but also in daily life including smart city and smart building infrastructures. The smart city framework provides a wide variety of applications that are facilitating daily life and improve the quality of life. Some of IoT based smart city applications have been listed as smart mobility, traffic control, health management, waste management, street and environmental monitoring, and smart building monitoring systems. The urban IoT applications require strict precautions in terms of security and privacy as well as in IoT based smart grid architecture. The operation of such an intelligence-based infrastructure is quite similar to natural framework of internet in the context of data servers and monitoring clients. The IoT concept implies for a network providing connections, monitoring options, and control features. The basic features of internet can be adopted to smart grid and IoT applications by providing machine-to-machine (M2M) and human-to-machine (H2M) interactions.

The IoT is one of the most recent communication paradigms that perform required objectives by using microprocessors, various communication mediums for digital data transmission, and several protocol and layer structures to fulfill M2M and H2M operations as being a part of internet. It is hard to generate a reference framework and layer description since an industry-wide acceptance has not been defined yet. Some of the IoT frameworks proposed by institutes can be listed as Arrowhead Framework, ETSI architecture for M2M, Industrial Internet Reference Architecture (IIRA), IoT-A, ISO/IEC WD 30141 IoT reference architecture (IoT RA), Reference Architecture Model Industrie 4.0 (RAMI 4.0), and the IEEE Standard for an Architectural Framework for the IoT [5]. On the other hand, researchers have proposed some other IoT architectures, i.e., a seven-layer OSI-like IoT framework [6], a five-layer reference model including physical, data link, internetwork, transport, and application layers [7], and middleware based framework [8]. It is assumed that OSI reference framework can be associated with IoT layers along Transmission Control Protocol/Internet Protocol (TCP/IP) layers as shown in Fig. 7.2. TCP/IP provides end-to-end connection and network layer of IoT provides the internetwork routing for data transmission.

The improvements of several protocols such as IP from IPv4 to IPv6 developed the capabilities of automation systems. The IPv6 addressing protocol provide a dedicated infrastructure for small link-layer frames in IoT framework for novel devices. Moreover, the constrained application protocol (CoAP) is a downscaled version of Hyper-Text Transfer Protocol (HTTP) on the User Datagram Protocol (UDP) for constrained resource applications in IoT environment.

There several studies of IoT have been proposed for smart cities, medical purposes, wireless sensor networks (WSNs), metering processes, and environmental monitoring. The interoperability of smart grid is an important topic for communication infrastructures. The IoT services and applications require a web-enabled smart grid architecture that is compatible with emerging web

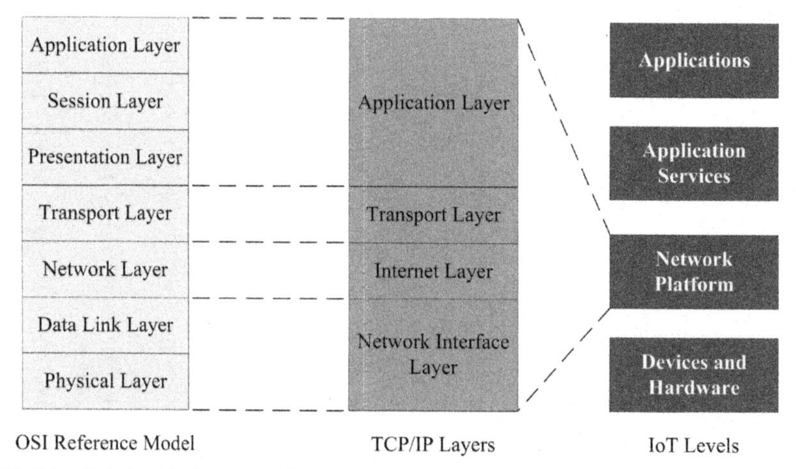

FIG. 7.2 Relationship between OSI reference model, TCP/IP, and IoT layers.

services. In recent years, the integration of IoT and smart grid has attracted more attention for a novel network type named Internet of Energy (IoE). The integration of smart grid and IoT application is described as web-enabled use of smart grid since it is improved with IP-based features of internet. The internet-based smart grid provides a number of benefits to generation and distribution operators, consumers, and stakeholders owing to intelligence-based management opportunities. Another significant contribution of IoT to smart grid is the monitoring, management, control, and information applications that almost all of them can be operated any mobile devices and various operating systems. Thus, consumers inherit increased control on smart grid components. The most recent IoT and smart environment researches have been focused on IoE networks focusing on WSN in [9] by Rana. The most significant contributions of WSNs to smart grid and IoT integration include cognitive radio-assisted sensor networks (CRSNs), end-user applications, data analysis and security and privacy issues.

The potential benefits of CRSNs to WSN applications as dynamic spectrum access, opportunistic channel usage, adaption to reduce the power consumption, and flexible communication band use including licensed and unlicensed infrastructures. It is obvious that smart grid applications produce huge amount of data and communication traffic where IoT and big data analytics are convenient to tackle this issue. In Table 7.1, a summary of introduced applications of smart grid and IoT integration have been expressed. In addition to summarized topics, several studies have been done for DSM, AMI, and smart monitoring applications in the context of smart grid and IoT infrastructure. The layer frameworks, protocol stacks, applications, and services of IoT are presented in the following sections in detail.

TABLE 7.1 Selected studies relating to IoT and smart grid

Grid level/ application area	Index terms/conducted content	Communication technology
Consumption	Residential demand response, EV, real-time pricing, DR, AMR, energy management system	6LoWPAN, CoAP, HTTP, UDP
Consumption	Smart energy meters, condition monitoring, last-meter, home energy management systems	IoT, M2M, H2M, IoE, WSN, WiFi
Daily life, mobile application	Health information system, mobile applications, industrial IoT	Low power wide area networks (LPWAN), IoT, smart grid, ZigBee, Bluetooth low energy
Smart cities, consumer applications	IoT for smart cities, urban IoT, cloud computing, smart home, energy harvesting, distributed generation, renewable energy sources, intelligent control system, PHEV	Radio frequency, power line communication, mobile communication, ZigBee, WSN, TCP/IP, 6LoWPAN
Transmission	Smart metering, smart transmission system, substation monitoring, remote control, ICT	IoT, PMU, ZigBee, wide area measurement system (WAMS), supervisory control and data acquisition (SCADA)
Generation, transmission, distribution	Smart grid management, smart metering, networking applications, data management, centralized and decentralized control	Cognitive radio sensor network (CRSN), IoT, WSN, LPWAN, WAMS
All levels	Big data, IoT, smart cities, smart buildings, EVs, IEDs, renewable energy sources, cloud/fog/edge computing, WSN, sensor and actuators, monitoring, metering, data storage, energy big data	Bluetooth, WiFi, ZigBee, mobile, WAMS, RFID, light detection and recognition (LIDAR), UDP, M2M, TCP
Generation, consumption	Estimation algorithms on DG sources potential, load demand forecasting, microgrid state estimation, IoT based smart grid control	WSN, TCP, UDP, HTTP, wireless mesh network (WMN), ZigBee, WiFi
Communication, data transmission	Security and privacy preserving applications for PHEV, generation, remote monitoring and metering, physical layer security, intrusion detection, critical system protection, SCADA security, M2M communication security, IoT and smart grid infrastructure security	WSN, M2M, TCP, UDP, Mobile, SCADA, ZigBee, 6LoWPAN, CoAP, IPsec, WAN, NAN, HAN

The communication infrastructure should be resistant against interference, latency, noises and fading sources for the presented applications in Table 7.1. The component of a communication system along smart grid may be compliant with wireline or wireless transceivers regarding to operation area and requirements. The prominent wireline communication system used in smart metering applications is known as power line communication (PLC) while wireless communications are based on radio frequency (RF), microwave or mesh transceivers at various frequency and data rates. The smart metering and monitoring applications perform required control services due to AMI technology. In the literature, very few papers comprehensively surveying IoT applications and services in smart grid has been presented so far.

However, it is possible to find featured studies presenting IoT based smart grid applications considering recent communication opportunities. The recent papers on IoT and smart grid topics have been focused on specified applications as generation, transmission or consumer side applications. Comparing to other recent surveys and researches, the primary aim is to present IoT applications for all aspects of smart grid infrastructure and their contribution to smart grid systems. In this chapter, smart metering concepts and systems are presented in a deeper description of smart meters, AMI technologies, and smart monitoring systems. It has been summarized the featured and emerging applications, services, architectural perspectives and challenging technologies of whole smart grid infrastructure in the following sections. Afterwards, the driving factors of IoT for smart grid and IoT applications in smart grid are presented in this chapter.

7.2 Driving factors of IoT for smart grid

The remote monitoring and control ability of smart grid infrastructure increases the capability of power plants to achieve more robust DSM and DG while decreasing the losses. Hence, the power generation cycle requires some control structures such as demand forecasting (DF) and automatic generation control (AGC) in smart grid. The load and source variety owing to wide usage of EVs, RESs, energy storage systems, and smart consumers with their own distributed energy resources (DERs) adversely affect the common demand management approaches. The gradual and increasing integration of these source-load duos to the electricity grid requires robust DF to manage the generation cycle. Another control method to tackle the problems caused by penetration of intermittent RESs and network requirement is AGC which is known as secondary frequency control method. In fact, the main idea behind the AGC was to decouple generation and load balance for several years. The preliminary applications were known as automatic voltage regulator (AVR) that is named as the load frequency control (LFC). These control methods were effective on slow and limited changed in the load profile. Therefore, they become inadequate against the recent integration of RESs and distributed generation sources drawing nonlinear dynamics. The AGC manages the power demand by tracking and compensating the frequency of entire systems since a fluctuation on the load causes equivalent changes on the system frequency, and the load demand is compensated by stabilizing the system frequency again [10].

Transmission networks, substations and management centers of utility network comprise core components of smart grid infrastructure where the grid term stands for all the parts of transmission system as a backbone. The increased use of ICT and advanced metering-monitoring applications on the smarter grid involved automation systems at substations and feeders, data acquisition and management solutions, demand management operations and marketing challenges. Therefore, some solutions are improved based on classical methods such as hierarchical grid control, active and reactive power quality (PQ) control, DSM, DER penetration control, virtual power plant (VPP) control, cyber-secure communication, and soft computing methods. The smart grid infrastructure can be analyzed in three technical perspectives that are infrastructure, management, and protection.

In addition to the legacy devices, the distribution network is integrated with several novel devices regarding to the improvement of smart grid applications. One of the most significant contributions of smart grid to the distribution network is to increase the system flexibility and reducing the losses. Although the demand response is complicated, the development of IEDs and ICT enable smarter distribution network. The EV integration to the existing grid also brings numerous improvements in terms of smart grid applications in distribution network level. The grid-to-vehicle (G2V) and vehicle-to-grid (V2G) applications are involved with charge and discharge cycles where the energy storage systems (ESSs) are gradually developed. The smart grid infrastructure involves the highest share of smart generation, transmission and distribution, metering, monitoring, management and communication sections.

Due to its ubiquitous computing infrastructure, IoT provides sophisticated contributions to smart grid at any levels. The WSN, RFID, SCADA, and M2M communication comprise four main integration instrument of IoT to smart grid [8]. The middleware that is a generic description for software and services providing interaction between two layers accommodate heterogenous devices and applications to operate in a harmony. IoT middleware are also required to integrate communication systems in the context of smart grid for generation, transmission, and distribution levels. The cloud based IoT applications that are being extensively improved provides stable and flexible CPS for SCADA and WSN networks.

7.2.1 Smart grid applications in generation level

The critical applications, measurement and control topics, ICT and devices used in smart grid are summarized in Table 7.2. The table is organized regarding to the generation, transmission and distribution stages and additional consumption

TABLE 7.2 A summary of critical applications and ICT in smart grid

Power system level		Critical applications	Measurement and controls	Communication technology and applications	Devices used
Utility	Generation	Real-time monitoring	• Energy efficiency • Demand profile • Voltage, current, frequency, phase, and power control • Droop control • Active/reactive power • Demand side management	• Metropolitan area network (MAN) • Wide area network (WAN) • Local area network (LAN) • Backbone and core networks • Supervisory control and data acquisition (SCADA)	• Automatic voltage restorer (AVR) • Phasor measurement units (PMU) • Intelligent electronic devices (IED) • Smart transformers
		Power plant control			
		Alternative energy sources			
		Distributed generation			
		Quality of services (QoS)			
	Transmission	Substation automation	• Power loss and leakage • Localization, fault prevention • Phasor measurement • Network topology analysis • On-line power flow • Security monitoring	• Wide area network (WAN) • Local area network (LAN) • Energy management systems (EMS) • Supervisory control and data acquisition (SCADA)	• Phasor measurement units (PMU) • Smart fault passage indicator (FPI) • Smart remote terminal units (RTUs) • Global positioning system (GPS)
		Line control and monitoring			
		Power monitoring			
		Quality of services (QoS)			
	Distribution	Substations automation	• Cost and pricing • Phasor measurement • Optimization periods • Machine learning and control services • Demand profile • Demand side management and control	• Extended area network (EAN) • Neighborhood area network (NAN) • Field area network (FAN) • Wireless regional area network (WRAN)	• Intelligent electronic devices (IED) • Phasor measurement units (PMU) • Wide area monitoring, protection and control devices (WAMPAC) • Smart MV/LV controller • Smart remote terminal units (RTUs)
		Smart transformer control			
		Direct load control (DLC)			
		Advanced metering infrastructure (AMI)			

Continued

TABLE 7.2 A summary of critical applications and ICT in smart grid—cont'd

Power system level		Critical applications	Measurement and controls	Communication technology and applications	Devices used
		Automatic meter reading (AMR)	• Distribution management system controller (DMSC)	• Wireless personal area network (WPAN) • Metering data management system (MDMS)	
Consumer	Consumption	Home (residential) energy management	• Consumption control • Microsource generation control • Monitoring and automation • Home automation • Security control • Advanced metering infrastructure (AMI)	• Home area network (HAN) • Industrial area network (IAN) • Building area network (BAN) • Digital subscriber lines (DSL)	• Smart appliances • Smart meters • HomePlug • Sensors (noise, temperature, humidity etc.) • Smart plugs • Ethernet
		Microgrid management			
		Prediction and forecast			
		Electric vehicle integration			

level to the Fig. 1.9. The critical applications and requirements on each stage are tabularized considering the most widely researched studies. In the following subsections, the specified topics are surveyed in order to present further reading and understanding about the progress of smart grid applications. One of the most important requirements of the generation stage is real time monitoring [1]. There are several parameters of the power generators involving real time monitoring to ensure the generation security. The integration of DERs and microgrid improvements has provided several alternative monitoring systems in addition to well-known SCADA system. Lu et al. presents a microgrid monitoring system by using 4G Long-Term Evolution (LTE) mobile communication platform as an example of IoT and smart grid interaction in [11].

The noted study presents a middleware integrating heterogenous device and communication protocols in smart grid and IoT architecture. Similarly, some novel monitoring studies based on Ethernet and microprocessor interaction have been presented in [12] that Garcia et al. introduces a local area network (LAN) based microgrid monitoring system in as a IoT-assisted smart grid application. Power plant control is also required to improve generation reliability in smart grid as well as in conventional grid applications. However, the bidirectional communication requirement of smart grid has induced new researches on power flow control. Firouzi et al. proposed a unified interphase power controller (UIPC) study in [13] which is required in wind farms. The conventional droop control that manages the power sharing does not include any communication method in conventional grid applications. However, there are some novel studies have been proposed to improve the conventional decentralized control method and the wired and wireless communication infrastructures. The DER and alternative energy sources integration to generation and transmission levels sometimes involve particular solutions regarding to source type.

The small sized DERs are mostly connected at the distribution level and the control of them is relatively easier comparing to the large penetration of renewable and alternative energy sources. The voltage and frequency stabilization of large DERs forces generation suppliers to improve rapid monitoring and reacting solutions. The computational analysis methods, algorithms, agent-based controllers and data centers are improved to provide more reliable and faster control on demand.

One of the most important contributions of IoT to smart grid is on improving ICT based requirements including semantic web, agent-based control, and enhanced connectivity along M2M interaction. The responsibility, behaviors, targets, and operation systematic of each agent are determined in order to provide connectivity and communication features with their operation area and each other agents. This software-based approach facilitates detecting the faults and system failures rapidly. It is noted in several papers that agent-based and multi-agent system (MAS) oriented programming leverages intelligent programming opportunities and facilitates system control in energy generation and management operations in IoT and smart grid interacted infrastructure [14]. Another significant requirement of smart grid applications is the quality of service (QoS) which should be provided by ICT technologies at any stage ranging from generation to consumption [1]. Regardless of communication medium or communication layer, QoS should ensure the integrity of the monitored data, security of the communication, response time and control command delivery as expected in any smart grid infrastructure. Therefore, several QoS researches and surveys have been conducted including physical and medium access control (MAC) layer communications, various area network communications, WAN placements, and interference security.

7.2.2 Smart grid applications in transmission and distribution levels

The transmission and distribution levels include substations to adjust the voltage levels in a power network. The substations are located between generation and consumption levels where interconnection is composed by transformers, circuit breakers, regulators, power factor controllers, phasor measurement units and numerous sensors are done. The conventional substation automation has been realized by SCADA systems as a standard for long years. After several control and data acquisition systems have been implemented and have been integrated to substations, a standard is required to ensure interoperability of different communication protocols and vendor-featured devices. The IEC 61850 standard which is improved for IEDs became a standard scheme for substation automation including remote measurement functions, authentication, cyber-security, PQ monitoring and measurements, DER integration, and energy management applications [15, 16]. It should be noted that all these applications and requirements are being converted to IoT-based smart grid applications thanks to intelligence-based middleware and MAS architectures.

A secure and reliable transmission system requires monitoring and fault detection infrastructure in addition to substation automation. Moreover, the fault detection systems are required to ensure the reliability of transmission lines. The fault detection systems are focused on power conversion systems and physical security of entire transmission system while the WAMS and WMNs have provided widespread usage in transmission and distribution systems. These power flow monitoring technologies have attracted self-healing capability of smart grid where the automatic restoration and reacting to unusual situations have been enabled. Mousavi-Seyedi et al. proposed a PMU with WAMS control in [17] while some other researchers have proposed power line communication (PLC) based power monitoring, PMU applications and fault detections that all are possible applications to be performed by IoT and smart grid interaction.

The featured smart grid applications include substation automation as in transmission level, monitoring of underground cables, smart transformer control, Direct Load Control (DLC), AMI and Automatic Meter Reading (AMR) in distribution level as listed in Table 7.2. Although the substation automation is inherited from transmission level systems and infrastructures, it has been improved regarding to the requirements of distribution level and DG sources usage. Therefore, it is called as distribution system automation (DSA) which controls the DSM. This automation system is based on smart meter control, communication control, and multi-agent controls. These control mechanisms provide fault detection, isolation of fault points, reconfiguration of distribution system and restoration that all enables to have a self-healing system.

Another novel research area in the smart grid control of distribution level is smart transformers, which are inherited from conventional line frequency transformers. This new transformer type called solid state transformers (SST) or bidirectional intelligent semiconductor transformer (BIST) is an emerging technology. The researches have shown that this technology improves the reliability and stability of the electric distribution grid even in high frequency (HF) conversions. Now, the smart transformers (ST) compete with SST and conventional transformers in reliability and efficiency issues. The ST can be easily interfaced with microgrid generators to adjust the line frequency. Besides, ST manages active power request by controlling the current fluctuations on the medium voltage (MV) side, and the reactive power can be handled independently by ST to enable unity power factor [18, 19].

The DLC is a featured control method used in smart grid for power resiliency at distribution level. The penetration of photovoltaics (PVs), wind turbines, combined heat and power (CHP), fuel cells and even EVs to the distribution system have introduced several challenges on system reliability and resiliency. This challenge is handled by using VPP design that allows assembling the capacity of these DERs into a single source profile. The reliability of distribution system is not only based on automation and control but also on the measurement and metering that provide the required monitoring data. A robust measured infrastructure minimizes the fault and outage probability on the transmission and distribution levels. The measurement infrastructure quality is related to a number of standards where some of them are IEC61000-4-30 for supply quality, IEEE Std. 1588 for precision time protocol in distributions systems, IEEE Std. 2030-2011 for the smart grid interoperability references, IEEE Std. 1547.4 for planned islands/micro-grids and IEEE Std. 1547.6 for interconnection to distribution secondary networks [1].

The improvement of AMI that is also listed in Table 7.2 as a featured application of smart grid is one of the most important technological challenge in energy monitoring. The AMI has brought bidirectional communication and metering capability to distribution service providers and consumers and thus, the mechanical electricity meters have been converted to smart meters (SMs). The communication medium used by AMI can be selected from numerous variety of wired or wireless communication methods such as PLC, IEEE 802.15.4 based wireless schemes, Worldwide Interoperability for Microwave Access (WiMAX), and even satellite communication.

Inga et al. presented wireless heterogeneous network architecture in [20] similar to shown in Fig. 7.3 where multi-hop capabilities for routing traffic from the SMs to the universal data aggregation points (UDAP) are considered. The

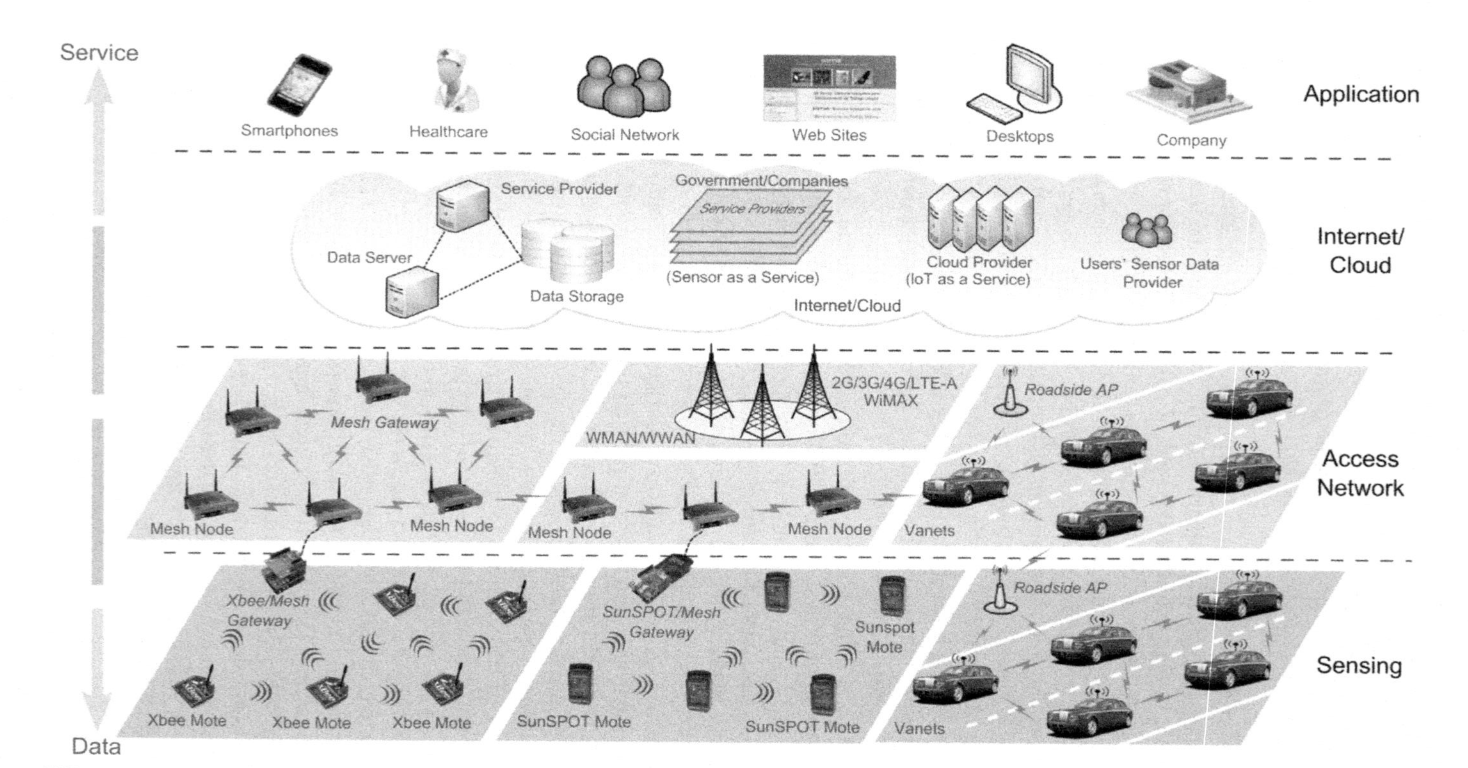

FIG. 7.3 Wireless heterogeneous network architecture for AMI [21].

proposed communication network eliminates the cellular or satellite communication by connecting the nearest UDAP for data transmission.

In addition to the Metropolitan Area Network (MAN) shown in Fig. 7.3, the AMI applications require building area network (BAN) and neighbor area network (NAN) that are designed by using WMN in urban places. The topological design and management methods implemented for AMI applications are presented and are concluded in the following related subsection. The measurement section of a SM involves pricing control based on time-of-use (ToU), the data management interface, and AMR infrastructure [1]. Although it is one of the components of SM, the AMR term is sometimes used to mention the whole system. However, AMR allows transmitting the collected metering data to other aggregated devices or data management centers by using several communication methods. The AMR provides a number of infrastructures to provide real time monitoring and pricing, power quality analysis, instant pricing, direct control of load management, DSM, remote on-off control and troubleshoot notifications. The consumer requirements and applications that are implemented in the context of smart grid are introduced in the following subsection where the consumption level issues are surveyed in brief.

7.2.3 Smart grid applications in consumption levels

The consumption level is assumed as the last part of smart grid cycle where generation, transmission, and distribution interactions has been presented in previous sections. The consumer side of a regular grid system includes a wide variety of loads. The DSM and DR issues have been carried to a new approach due to these type applications and opportunities. The integration of RESs may cause some power problems if there is not a well-planned management system that does not exist on the consumer side. On the other hand, the generation potential and intermittent structure of unbalanced sources require prediction and forecasting management under these circumstances. The increased use of EVs and their integration to the utility grid is another research area in the context of consumer side control issues. Therefore, the consumption level applications and requirement of smart grid have been presented with four main topics that are home or residential energy management, microgrid management, prediction and forecasting, and EV integration. The required infrastructures, driving factors, and integration technologies of the aforementioned application groups are presented in terms of IoT and smart grid interaction.

Regardless of consumers' behaviors and habits, any of these applications can be seen in any consumer site. It is noted in [22] that household appliances in United States have realized 42% of the overall electricity consumption. Hence, there are new regulations are made by local authorities to manage the DR and DSM issues for consumers. The home energy management system (HEMS) is one of the proposals to improve the PQ against load and source profile of consumers and a solution to DSM. Although the residential energy management systems cause to several concerns about consumer's privacy, decentralized managements systems have been proposed for privacy protection. There several DR control methods are improved where widely accepted three types are ToU, real time pricing, and critical pricing. A variety of researches and several home energy management systems have been proposed in the context of MAS, a multi-objective mixed integer nonlinear programming (MO-MINLP) model, and mixed-integer linear programming (MILP) by considering the DERs and load profiles of consumers. A significant control system at the consumption level is microgrid controller that manages the PQ of the generated energy by consumers. The controlled microgrid can be one of residential, commercial, institutional or industrial as illustrated in Fig. 7.4.

The microgrid controllers should comply with IEEE Standard P2030.7 (Standard for the Specification of Microgrid Controllers) for transition and dispatch requirements that enable DERs to operate autonomously or grid-connected modes. The communication methods are crucial parts of microgrid managements systems (MMS). Optimal Power Flow (OPF) control that detects the optimal operation state of DERs is one of the key components of MMS. The operation modes of DERs are islanded operation and grid-connected that is managed by selecting centralized or decentralized control methods in a microgrid. The centralized controller obtains all the measurement data and transmits the detected parameters to MMS. This control method is preferred due to its economic features in small-scale DER penetrations, but it may cause communication errors since all the DERs share same transmission channel.

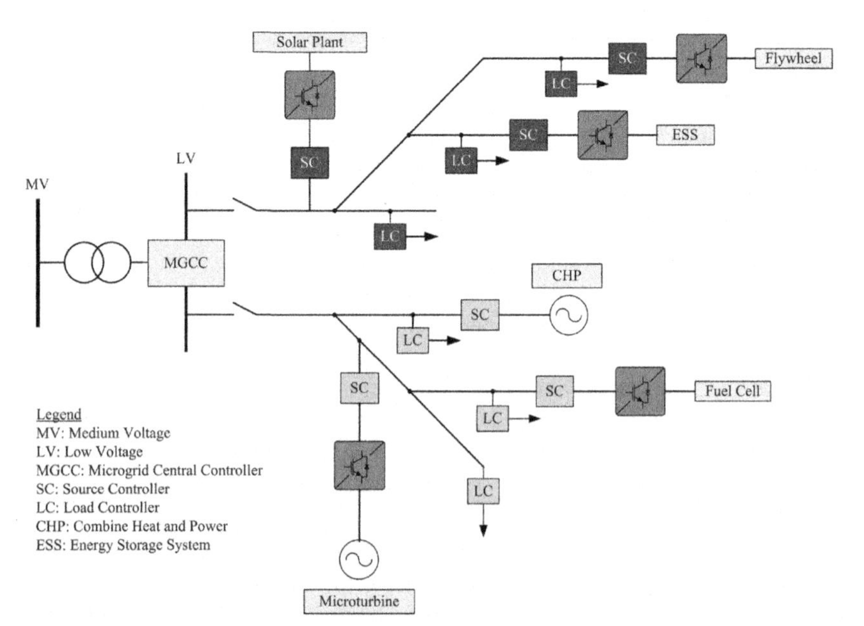

FIG. 7.4 Block diagram of a microgrid scheme and MGCC.

The integration of large RES plants improves the capabilities of utility grid. However, the power quality and control operations should also comply with interconnection standards such as IEEE-P1547-2003, which presents a benchmark model for integration. The integration of DERs to utility grid causes several challenges in terms of technical, economical, and management aspects. Therefore, several control and management procedures have been implemented for microgrid and DG integration to utility grid.

The microgrid central controllers (MGCC) that are improved for microgrid management are dedicated systems to accomplish DSM, DR control, and generation control duties. It regulates the voltage and frequency of the microgrid and sustains the system stability. MGCC, source controller (SC), and load controllers (LC) illustrate the block diagram of a MGCC in the microgrid scenario in Fig. 7.4 where LV section is controlled. The MGCC compares load demand and generation level of microgrid and decides to increase the generation level or leaving some non-critical loads from microgrid to supply critical loads. MGCC accomplish this operation by controlling SC and LC subsystems in the microgrid. It is noted that MGCC can save 21.56% of daily energy consumption by managing DR and generation control [23]. Furthermore, MGCC can detect the instant power quality of microgrid at point of common coupling (PCC) and can improve power factor by connecting and disconnecting to the utility grid. The synchronization operation during reconnecting is also performed by MGCC. Regardless of controller type used among MGCC, micro-source controller or decentralized controller; the microgrid controller improves resiliency and flexibility of power network to supply load without any curtailment or blackout. Besides, controllers decrease the operation costs, DER integration and usage rate, limits the carbon emission, and enhances reliability, sustainability, and security of sources. The microgrid controllers enable the power network to interface distributed management systems, DER aggregators, and distributed metering systems [23–26].

On the other hand, the decentralized control provides separate communication channels dedicated to each DERs. Thus, all the controllers transmit measurement data on their own channel, and the errors caused by the high penetration of widespread sources eliminated in this way. The DER controllers are no longer subjected to central controllers in decentralized control infrastructure while adaptive control methods are also proposed for ESS and energy conversion devices in the context of MMS. The intermittent structure of RES and charge-discharge cycles of energy storage devices used microgrids require prediction and forecast systems to ensure the resiliency and DSM reliability of residential systems to improve IoT applications. The computational methods that can be applied in predicting and forecasting studies are surveyed in [27] where selecting, sizing, optimization, power source variety analysis are researched. The coordination of central controllers also provides forecasting assistance in addition to operational cost analysis, load forecast in short and very short terms, and power demand detection. The optimization plays vital role in predicting and forecasting objectives of microgrid controller middleware in IoT-based smart grid applications. The most widely used optimization algorithms perform the objectives regarding to the predicted data where the process is based on probabilistic methods such as Monte Carlo, Support Vector Machine (SVM), Dynamic line rating (DLR), and Markov Decision Process. The soft computing methods can provide better forecasting techniques by analyzing the stored data from past to now, and the performance of entire system may be optimized.

The EVs are accepted as an environmental friendly technology to reduce the adverse effect of internal combustion engine (ICE) cars. The high use of EVs expedites their integration to utility grid as a source and load at discharge and charge cycles. The increased usage of EVs may play destructive role on utility grid if the uncontrolled power demand of charging EVs is integrated to the mains. Two types of EV integration to the utility network are defined as V2G and G2V. In the V2G type of integration, power flow occurs in bidirectional mode where one is performed from grid to vehicle as charging and the second is from vehicle to grid during discharging the batteries. On the other hand, the power flow is always unidirectional since grid charges the batteries in G2V operation. The bidirectional V2G capable EVs ease the DR control by scheduling the discharge cycles at peak hours and charging cycles at off-peak hours. As a consumption level application, the EV integration to the smart grid is an emerging research area in terms of DG, DR, microgrid, and communication topics. Once a huge number of EVs are connected to the grid for charging, it causes several PQ problems drawing highly loaded grid profile. Several voltage deviations, frequency fluctuations, degraded PQ, instability on the power grid, inadequate efficiency are some of the deficiencies occurred by uncoordinated integration of EVs to the grid. Therefore, several solutions are provided owing to smart communication and coordination features of smart grid. These coordination and control issues are mostly related with battery management systems (BMSs) that are responsible to track some parameters of batteries such as state of charge (SoC) and state of health (SoH). The communication methods are listed into three categories as dedicated lines, wired systems such as Ethernet, telephone lines etc., and wireless technologies based on Bluetooth or IEEE 812.15.4 type communication mediums. Regardless of communication method, the main purpose of EV communication is to implement a CPS that is based on bidirectional communication network across the AMI base of EV and utility grid. The CPS facilitates the coordination of EVs by using many type of sensors together, and by installing a transmission system for inherited data. Since the CPS corresponds with the digital communication medium and analog power transmission medium, it is vulnerable to intrusions and attacks caused by the natural structure of power system. Thus, a protection and authentication system is required to improve reliability and security of CPS. The confidentiality, authenticity, availability, and integrity are researched in the context of privacy aware CPS.

7.2.4 Driving factors of IoT for smart grid

It is clearly seen in Table 7.2 that transformation of conventional grid to smart grid has been accomplished by contributions of recent ICT systems. The smart grid infrastructure has brought an improved ecosystem with communication interfaces at each level of conventional grid starting from generation to consumption. Moreover, each node of this infrastructure has been equipped with increased intelligence-based technologies and autonomous communication and control applications. The widespread monitoring and decentralized control have provided to increase efficiency, resiliency, and flexibility of power network. These improvements have facilitated grid management for transmission system operator (TSO) and distribution system operators (DSOs) with the recent advances in low-cost and reliable sensor technologies, wired and wireless communication systems, data management and control operations and remote measurement devices. The increased DSM and DR control capabilities of TSO and DSOs have leveraged grid management operations. On the other hand, regulations provided by policymakers enables consumers' transformation to prosumer by integrating microgrid and DER integration to power network. The increased communication ability of smart grid has found application areas at the consumption level due to smart meters, HEMS, and remote control of DERs.

As discussed earlier, smart grid provides two-way data flow besides its two-way power flow substructure in its ecosystem. Such a system equipped with ICT technologies and generated colossal data should be operated in secure, scalable, and interoperable ways. The data and system security are one of the most important issue to prevent vulnerabilities to cyber-attacks in the context of smart grid. The IoT is a rapidly emerging technology by integration of wide variety of CPSs. The widely used internet applications provide increased connectivity between M2M and H2M communications. Therefore, several advantages of internet communication are adopted to smart grid infrastructure by improving IoT-based smart grid applications. Moreover, recent improvements in IoT technology increases mobility, energy efficiency, security, accessibility, and interoperability of M2M and H2M communication interfaces.

The widespread deployment of smart meters, sensors, and actuators generate large scale and colossal data on the monitoring and control centers. The generated data should be instantly received, stored and processed in a short while to detect situation of grid or to act against faults, failures, losses or leakages. These operations require high-capacity data processing capability, reliable machine learning and decision-making algorithms for management. Therefore, big data and cloud computing services facilitate management of smart grid. The smart grid cycle will become a more robust and reliable CPS with the aid of IoT technology that provides sophisticated communication technologies with high-bandwidth and data rates, security protocols, interoperability protocols and services, data management opportunities, and big data management applications. Therefore, an increasing interest is seen on IoT-based smart grid infrastructure to enhance smart grid applications including CPS security, smart home and smart city applications, asset management, data management and processing topics [28–30].

The driving technologies of IoT in the context of smart grid can be basically grouped into three categories as *(i) data acquisition technologies producing contextual information from "things", (ii) ICT, data processing and management technologies,* and *(iii) security and privacy aware technologies.* The things term denotes any measurement and monitoring device based on sensor technologies with communication capability. Thus, entire infrastructure can be equipped with monitoring and control devices that can be accessed rapidly and securely due to IoT technology. The generated measurement data are transmitted to monitoring and control center where big data and cloud-computing technologies can be easily adopted to smart grid system. On the other hand, these operations are sustained by providing security and privacy requirements with the aid of IoT. A summary of essential objectives required to create CPS environment in any IoT-based smart grid system are listed as follows;

7.2.4.1 Identification

Things in an IoT system should have unique ID. The identification is required to recognize the objects and to respond according to their service demand. Many methods can be used for identification i.e., RFID, bar codes, near field communication (NFC), electronic product codes (EPC) and ubiquitous codes (uCode). In addition to assigning an ID to things, addressing is another important point to identify the objects. The IPv6 taking the place of IPv4 is an intelligent solution for addressing with increased number of headers and compression mechanism. The extensive address space of IPv6 is comply with billions of devices for open RF mesh wireless (IEEE 802.15.4g, DECT Ultra Low Energy) and PLC infrastructures (IEEE 1901.2) using the IPv6 over low-power wireless personal-area network (6LoWPAN) [31, 32].

7.2.4.2 Data acquisition

The instrumentation and measurement play vital role in IoT-based smart grid systems. The sensing devices and other instrumentation components comprise the data acquisition interface of IoT in a smart grid application. The sensors, actuators, and smart components acquire the required data from related things and transmit to the database or cloud interfaces. Although wireline transmission has been a legacy for a long while, wireless networks and meshes are now widespread. The improving WSN and communication systems such as WCDMA, LTE, 4G, 5G facilitates to connect many sensors to transmit data together [32, 33]. Besides, the intelligent microcomputers i.e., Arduino, Raspberry PI, BeagleBone Black, etc., associated with sensors improves the functionality and integrated to IoT and TCP/IP layer applications in monitoring and control requirements of any smart grid application indicated in Table 7.2.

7.2.4.3 Communication

An IoT system is comprised by several heterogeneous systems and components. Many wireless technologies and protocols such as GSM (2G), Universal Mobile Telecommunications System (UMTS 3G), LTE (4G), LTE-Advanced (LTE-A), Wi-Fi (IEEE 802.11), WiMAX (IEEE 802.16) Bluetooth (IEEE 802.15.1), Low-Rate Wireless Personal Area Networks (LR-WPAN, IEEE 802.15.4), Z-wave, and LoRaWAN R1.0—LoRa are being used in IoT communication instead of lossy RFID and NFC frameworks. The recent CPSs including smart phones, handheld devices, and cloud integrated monitoring terminals require rapid and secure communication architectures such as 4G and 5G networks. The massive databases have been carried to cloud applications that are one of the most important driving factors of 5G infrastructure. Since the smart grid and IoT interaction requires M2M communication, cellular communication is state-of-the-art in enabling technologies of CPSs [31, 32].

7.2.4.4 Data processing and management

Another important segment of IoT framework required for smart grid applications is data processing stage. The ICT applications represent fundamental data processing section of IoT. This segment includes hardware and software interaction where the microprocessors and microcomputers are responsible to obtain the transmitted data signals, to analyze the context, to perform the required objectives, and to store the processed data by using their own operating system. The processor interfaces are implemented with high-level digital signal processors (DSPs), field programmable gate arrays (FPGAs) or system on chip (SoCh) processors such as ARM, Intel Galileo, Raspberry PI, Gadgeteer, BeagleBone, Cubieboard, Z1, WiSense, Mulle, and T-Mote Sky. Moreover, particular operating systems, namely Real-Time Operating Systems (RTOS), are featured for IoT applications to process and manage the inherited data. The Contiki-OS, TinyOS and ChibiOS are specified and widely known RTOS that are specialized for WSNs. However, the novel and lightweight operating systems such as LiteOS, MagnetOS, AmbientRT, MANTIS OS, and SOS have been presented for IoT-based smart grid applications due to limited resources and energy efficiency.

7.2.4.5 Services

The ubiquitous structure of IoT requires a services-oriented architecture (SOA) to meet the requirements of smart grid applications. A huge number of IoT applications have been improved for any aim ranging from residential usage to industrial areas based on WSNs. The applications need services to optimize the operation, integration, and interoperability along smart grid devices and services. The IoT services are listed into four categories as identity-related services, information aggregation services, collaborative-aware services and ubiquitous services [34, 35]. The identity-related services, either active or passive, is the most essential service that is used to identify RFID tags. The information aggregation services are required to acquire, process, and transmit the data as its name implies. Collaborative-aware services manage the inherited data to accomplish an action. Thus, this service retrieves the data and decides. Ubiquitous services are defined as omnipotence and omnipresence service to provide communication needs at any time and at anywhere.

7.2.4.6 Security

The IoT framework should satisfy security requirements as confidentiality, authentication, integrity, non-repudiation and anonymity at any stage of the smart grid infrastructure. The security issues include the environment starting from sensor networks to data servers and cloud servers. Confidentiality requires protection of transmitted data between sensors and servers, while authentication is required for transmitting and receiving sections. Non-repudiation is needed to prevent the denial of previous data transmission, and anonymity is for privacy-aware operation.

7.3 Communication infrastructures of IoT

Many communication technologies have been improved for different application planes and requirements. While some of these technologies are common for particular application planes such as Bluetooth in personal networks and ZigBee in home and device automations, the other communication technologies provide wide application areas such as WiFi, LPWAN, and cellular technologies. The headmost requirements of smart grid infrastructure are interoperability and accessibility. Therefore, advanced communication technologies enable smart grid to be ubiquitous as its services and applications. The communication and coverage planes of smart grid can be classified into three groups as HAN, NAN, and WAN. The HAN is essential for residential units with smart appliances, energy management systems, power control tools, ESSs, PV panels, small-scale wind turbines, electric vehicles and smart meters. NAN deals with distribution level of smart grid where a group of residential or industrial loads have been aggregated in a substation or transformer while WAN associates several NAN areas for management. Due to coverage area and monitored data sizes, each area network requires featured communication technology to meet their requirements. Accordingly, IEEE 802.11 or IEEE 802.15.4 based communication technologies can be adequate for HAN and NANs while WANs require fiber optic, cellular, UMTS, LTE, LTE-A type wide coverage communication technologies due to their wider operation areas. First communication method used for IoT can be dated back to RFID technology in the late 1990s while WSN technology provided an achievement for IoT communication technologies. However, the most significant improvements have been seen in the last decade with the aid of standardized sophisticated communication technologies such as Bluetooth Low Energy, IEEE 802.15.4 based advanced modems, IEEE 802.11 based area network technologies, and mobile communication generations such as UMTS, LTE, LTE-A and 5G.

Moreover, the advanced computer and internet technologies such as IPv6, security and interoperability standards, and improved communication protocols have leveraged IoT communication and its integration to smart grid infrastructure for several application requirements. The increased data transmission rates, decreased transmission latency, and improved coverage features of these advanced communication technologies meet the most demanded communication requirements of IoT applications [33, 36].

IoT provides several technologies that can deal with data transmission and processing requirements of smart grid applications. The architecture should be appropriate for interconnection of things on the internet cloud. Thus, some prominent abilities are required to be provided by communication infrastructure in IoT framework. First, the communication system is expected to have low power consumption since high majority of IoT devices are powered by batteries. On the other hand, communication system should cope with security and reliability issues. The protocols and communication standards incorporate several control and detection methods. Moreover, and most important, the communication system should be complying with internet network in terms of data transmission. In other words, the communication system is required to be associated with TCP/IP protocols [37].

The open standards reference model represents five-layer IoT architecture as shown in Fig. 7.5. The communication standards seen on physical layer (Layer 1) and data link layer (Layer 2) have significantly improved on IoT architecture owing to the contributions of IPv6. Although the layers are organized independent from each other, the reference model allows optimizing by cross-layer connection, and by using application-programming interfaces (APIs). The data link layer provides MAC enhancements, Logical Link Control (LLC) for 6LoWPAN, IPv6 over Ethernet, and IP or Ethernet convergence sublayer features. The network layer of reference model includes addressing, routing, QoS, and security architecture. The transport layer manages the security based on Datagram Transport Layer Security (DTLS) protocol that provides communications privacy for datagram transmission.

The protocol allows client/server applications to communicate in a way that is designed to prevent eavesdropping, tampering, or message forgery. The privacy and communication security of IoT reference model provide a tailored architecture for smart grid applications where reliability is crucial. The DTLS protocol is based on the Transport Layer Security (TLS) protocol and provides equivalent security guarantees [38, 39]. The application layer also provides security as network and transport layers do. The encryption methods of Layer 5 are ANSI C12.22, Device Language Message Specification (DLMS/COSEM) that is a standard for electricity metering data exchange under IEC 62056 set which plays vital role in smart metering and AMI applications of smart grid

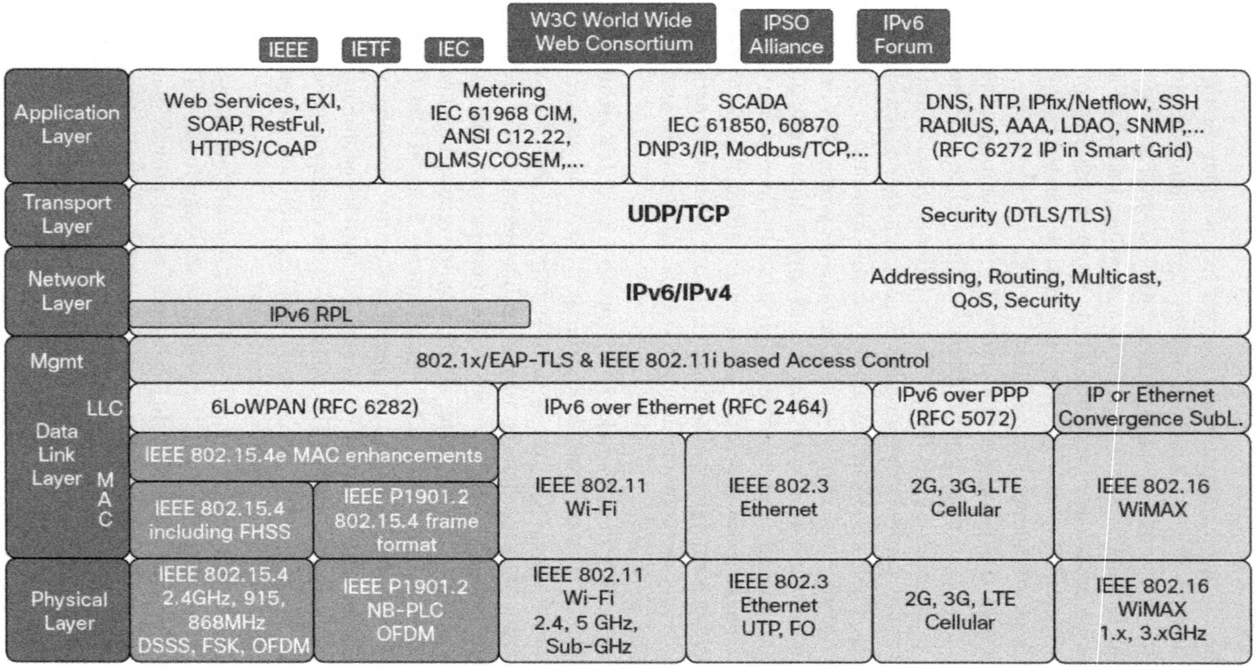

FIG. 7.5 Open standards reference model and standards.

infrastructure. Besides the encryption, application layer accommodates secure communication protocols (HTTPS), application protocols (CoAP), access protocols (Simple Object Access Protocol-SOAP, Lightweight Directory Access Protocol-LDAP), and secure network protocols (Distributed Network Protocol-DNP3, Secure Shell-SSH). The communication infrastructures of conventional grid are based on legacy methods and equipment. However, the featured communication technologies of IoT technologies are addressed according to the development of infrastructures, services and protocols, security, privacy, interoperability, energy efficiency, and sensor networks.

7.3.1 Software-defined networks (SDNs)

SDN provides several advances to enhance flexibility, reliability, scalability, and interoperability of IoT-based smart grid communication infrastructure as an encouraging model. It allows to dispatch control and data communication devices into different planes. Principally, SDN offers an open architecture model in three stages by separating control and data planes, enabling centralized logical control and incorporating network programming capability. Thus, SDN copes with communication problems occurred in conventional architecture combining protection control, billing, and monitoring data transmission. The open architecture of SDN prevents inefficiency of M2M communication, and provides facilitated design, deployment, management, and maintenance of communication networks. This recent networking architecture has been separated into three layers as application, control, and physical layers. The application layer performs system operations and managements as the highest layer, and control layer interfaces application layer with physical layer. The APIs performs communication between application and control layers. Thus, control layer becomes a network operating system due to APIs and manages physical layer regarding to instructions. SDN enables operators to improve network function virtualization (NFV) applications that provides aggregation of several DERs and microgrids that are using different communication technologies on a virtual network. The integration of SDN and smart grid provides several benefits for improving communication networking, real-time monitoring applications, increased latency management, and bandwidth control [40, 41].

In addition to benefits of SDN supplied to communication infrastructure, it also improves resiliency of power network against cyber-attacks. SDN enables communication system for intrusion detection, isolating selected devices upon detection and protection approach, decreasing malicious traffic and denial of service attacks, and remote control of sensors and smart meters.

7.3.2 IEEE 802.x based communication technologies

Almost all IoT communication methods that are presented in the following are based on RF wireless transmissions. The communication infrastructure is in the first two layers of Fig. 7.5 that are physical layer (PHY) and MAC layer. Both layers meet measurement requirements of smart grid and are primarily improved for sensor networks. The improvement of PHY and MAC layers provided energy efficiency and increased reliability in the transmission medium. PHY layer manages the sensor data transmission and receive operations while MAC layer operates the protocols concerning medium allocation among the sensor nodes. Due to variety of sensor networks, numerous technologies operating on PHY and MAC layers have been developed [42]. Some outstanding and widespread communication technologies are presented as follows:

IEEE 802.15.1 and Bluetooth: RFID and IEEE 802.15.1 technology that Bluetooth has also been included performs the data transmission at 2.45 GHz Industrial Scientific and Medical (ISM) frequency band. RFID tags and systems use shift keying methods as for amplitude shift keying (ASK), frequency shift keying (FSK), and phase shift keying (PSK). On the other hand, Bluetooth decreases power consumption since it transmits data between devices in a short coverage range which is proper for energy efficient smart grid applications. After the first implementation of Bluetooth technology, the most recent improvements have been performed by the development of Bluetooth 4.1, or Bluetooth Low Energy (BLE) in another name, that provides higher speed and IP support for IoT. It is noted that BLE has been benefited from power saving mode features of IEEE 802.11b and coverage area has been extended at least 10 times of its antecedent versions. The PHY data rate of BLE 5.0 extends up to 2 Mb/s while its variant BLE PSM provides up to 11 Mb/s PHY data transmission rate, and coverage area up to 100 m [32].

IEEE 802.15.4: IEEE 802.15.1 and its improved variant BLE provides high data rate for transmission, but they lack on supporting WSN and large network connections that is particularly required for smart metering and monitoring applications. IEEE 802.15.4 WPAN protocol has been developed to create a cross-layer between PHY and MAC for low-rate WPAN (LR-WPAN). IEEE 802.15.4 has been accepted as a suitable infrastructure for IoT, M2M, H2M, and WSN application due to key features as low power consumption, low cost, sufficient data transmission rate, and operability that are crucial for smart grid applications. The PHY layer of IEEE 802.15.4 presents success on short-range communication and obtained a widespread acceptance that results the adoption to four major standards (ZigBee, WirelessHART, ISA100.11a, and WIA-PA). However, it is not same for MAC layer in terms of transmission success and reliability in industrial application. Therefore, WirelessHART, ISA100.11a, and WIA-PA improved IEEE802.15.4 MAC layer by adding different technologies according to specified industrial applications. It can be said that the most prominent contribution to widespread use of IEEE 802.15.4 is provided by ZigBee technology, which deals with low rate data transmission and is capable to manage large number of sensor nodes for WSNs in a smart grid environment. ZigBee supports three frequency channels at 868 MHz for Europe, 915 MHz for America, and 2.4 GHz for worldwide utilization.

The data transmission rates at PHY layer for each frequency channel are 20, 40, and 250 Kbps, respectively. MAC layer of beacon-enabled mode uses Carrier-Sense Multiple Access with Collision Avoidance (CSMA/CA) access protocol. The lately improved IEEE 802.15.4e has brought two operation modes as Time Slotted Channel Hopping (TSCH) and Deterministic and Synchronous Multi-Channel Extension (DSME) for automation processes. Both of these operation modes enable the protocol to mitigate the fading problems and increase the reliability, security, and integrity [32, 43].

IEEE 802.11: It is a set of MAC and PHY layer designation to implement Wireless Local Area Network (WLAN) communication operating at several frequency bands as 900 MHz, 2.4 GHz, 3.6 GHz, 5 GHz, and 60 GHz. WLAN technologies play important role in IoT infrastructure. Wi-Fi is a featured technology that is implemented for WLAN devices based on IEEE 802.11. The legacy IEEE 802.11a/b/g/n WLAN technologies increased the channel bandwidth of 20/40 MHz to 80/160 MHz in IEEE 802.11ac standard that reaches up to 1 km transmission range in the outdoor. Orthogonal frequency-division multiplexing (OFDM) which is 10 times slower has also been adopted by PHY to the IEEE 802.11ac to extend the range. On the other hand, headers in MAC layer have been decreased and the energy efficiency has been increased by this improvement.

A specified version of IEEE 802.11b is known as IEEE 802.11 power save mode (PSM) that is designed to operate in power-saving mode to increase energy efficiency by idle and sleep mode support for devices. The IEEE 802.11b determines the data bandwidth at 11 Mbps, which is built by 8 bps segmented Complementary Code Keying (CCK) infrastructure. While the IEEE 802.11a/b/g/n/ac versions have a generic usage for smart and IT devices, recently improved IEEE 802.11af/ah/ac protocols are used to extend transmission range and for specialized applications of WSNs, AMI devices, IoT applications, and M2M communication purposes. Although IEEE 802.11ac is older than IEEE 802.11af/ah protocols, there a PSM mode optimization enabled it to be used PHY layer sub-GHz IoT applications. IEEE 802.11af/ah provides higher energy efficiency in sub-GHz applications. IEEE 802.11af uses many of the recent operational enhancement techniques adopted by the most recent IEEE 802.11 standards, such as multiple-input multiple-output (MIMO), OFDM, and channel bonding. By the improvement of IEEE 802.11af-2013 standard in 2014, the protocol has been adopted to cognitive radio (CR) applications in addition to legacy analog or digital television usages. The frequency of IEEE 802.11af ranges from 54 to 790 MHz where the maximum bandwidth is 35.6 Mbps for 8-MHz channels where 16 channels can be spared.

IEEE 802.11ah which is in sub-GHz band as IEEE 802.11af operates at 900 MHz band. It provides different channels at lower bandwidths as 1, 2, 4, and 8 MHz. Hence, IEEE 802.11ah provides longer range, better transmission sensitivity, and lower energy consumption than other WLAN standards. Moreover, IEEE 802.11ah utilizes Distributed Coordination Function (DCF) as in IEEE 802.11 PSM [44, 45].

IEEE 802.3: This IEEE standard has been presented in 1985 as Ethernet standard specifying CSMA/CD MAC protocol with 10 Mbps bandwidth. Several revisions have been published and Ethernet versions defined regarding to the bandwidth. The prominently improved versions are IEEE Std 802.3u fast Ethernet at 100 Mbps, IEEE Std 802.3z Ethernet over Fiber-Optic at 1 Gbps, IEEE Std 802.3ae Ethernet at 10 Gbps, IEEE Std 802.3af Power over Ethernet (PoE), and IEEE 802.3az Energy-Efficient Ethernet (EEE). The Ethernet technology is much more spread in industrial applications at any level of automation systems communication. IEEE 802.3az amendment that is introduced in 2010 has provided an optional power saving feature among other IEEE 802.3 standards. It provided to achieve power saving method called low power idle (LPI) mode during periods of low network traffic that enables power save for devices. The industrial applications of Ethernet enhanced the use and growth of this standard that real-time Ethernet (RTE) has been accepted as real-time communication method in automation and industry applications. The regular transmission rate of 100 Mbps has been increased to 1 Gbps and up to 10 Gbps with 1GBASE-T and 10GBASE-T [46].

Cellular Technologies (GSM, UMTS, LTE, LTE-A): The most recent middleware of IoT are based on smart phone applications in the context of M2M communication that refers to data transmission between machine type communication (MTC). The cellular communication systems were one of outstanding driving technology for IoT systems since they provide large coverage areas among any wireless communication system described up to now. The Third Generation Partnership Project (3GPP) associates the leader telecommunication groups to create a global communication system. The technologies that are improved by 3GPP are known with the names of Global System for Mobile Communications (GSM) as second generation (2G), UMTS as third generation (3G), LTE as fourth generation (4G), LTE-A as beyond 4G or pre-fifth generation (5G) that all are enhancements of cellular technologies. The improved cellular technologies are intended to lower device cost, improve the battery efficiency, improve the coverage and bandwidth, and decreasing the device complexity at each next generation communication systems. As of the improvement of 4G technologies and networks, the cellular technologies have dominated M2M communication that facilitates autonomous application improvement in IoT and smart grid infrastructure. The 4G suitably complies with IP protocol, and brought broadband communication and applications provided improvement of LTE-A and WiMAX usage in IoT. Since 4G technology is reaching its maturity, extensive studies are being performed for the next generation cellular technology that is arranged as 5G and the standards are expected to be introduced in 2020. It is anticipated that 5G will provide a thousand of times capacity comparing to 4G, which enables millions of devices to connect [47].

IEEE 802.16: Although it has been amended as wireless MAN, the commercial definition WiMAX is more familiar. IEEE 802.16 series include a number of wireless broadband standards. However, IEEE 802.16e that is announced in 2005 mobile broadband wireless access system and IEEE 802.16e assigned for multihop relay in 2009 are prominent technologies improving IoT applications. WiMAX systems employ several stations as base station, subscriber station, and relay station that interconnects the other stations. The wide coverage range of WiMAX triggers the researches of LPWAN technologies that are specified for IoT. Some of the recent low power WAN technologies such as SigFox, LoRa, Weightless, and Ingenu find wide usage chance in IoT applications. On the other hand, Internet Engineering Task Force (IETF) has designed and proposed the IPv6 stack for 6LoWPAN to perform IP-based connectivity with ultra-low power devices and applications. The IEEE 802.16 provides convenient solutions to smart grid applications in geographically spanned areas and energy efficiency [45, 48].

7.4 IoT protocols and services

A huge number of general and special purpose applications are available in IoT infrastructure that provides specific contributions to smart grid applications. These applications and middleware perform various operations and require featured services. As discussed earlier, the IoT services are listed into four categories as identity-related services, information aggregation services, collaborative-aware services and ubiquitous services. The identity-related services fulfill the identification process for any type of thing located in the IoT environment. These services can be active or passive where the passive services do not require any power source for operating, while the active services include a power source. RFID tags are essential example to these services where passive RFID tags are operated by using external electromagnetic field to identify the thing while active RFID tags have power sources to transmit the identification signal to external devices. Some of the identity-related services are passive payment systems, internet access and bankcard services, mobile wallet applications, ticket and charge services etc.

Once any application supported by identity-related services identifies the device or thing, information acquisition services act to get uniform resource identifier (URI) from name definition server. The IoT information server is operated by a series of information aggregation services that acquire and store the data and transmit the data to communication platform. The information aggregation services collaborate with many types of communication systems to deploy the obtained data from sensing devices. M2M nodes, communication medium, middleware applications, and OS comprise the information aggregation system. In a comprehensive communication medium, the wireless or mobile communication network includes gateway devices to manage access and applications. These gateways aggregate terminals, nodes, networks, and service interfaces for users [48, 49]. The detailed service modules and functions have been illustrated in Fig. 7.6 where the M2M platform is interfaced by

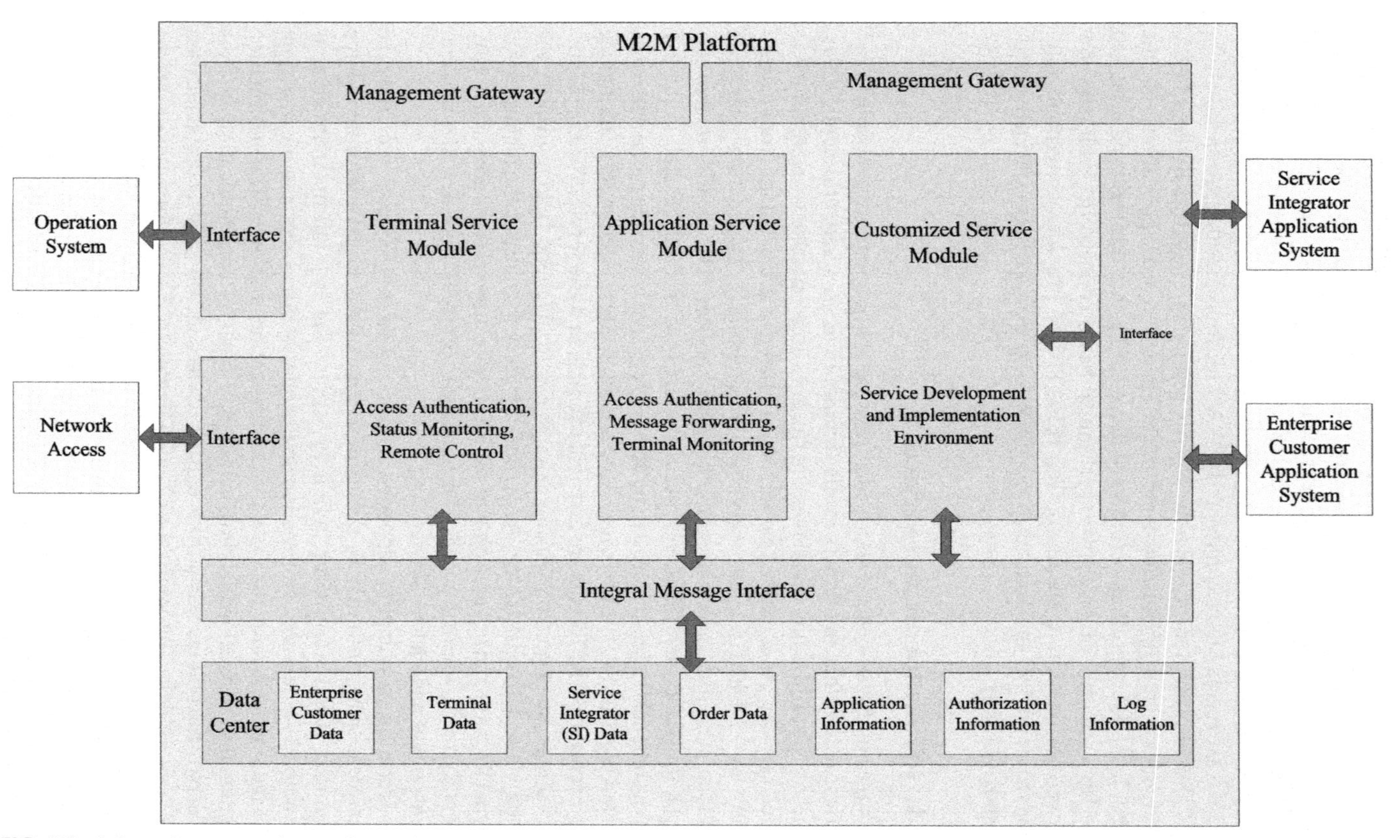

FIG. 7.6 Information aggregation services and interactions.

management and application gateways [34]. The terminal and application service modules interact with internal message interface that is fed by data center. On the other hand, the customized service module interacts with data center and it interfaces the enterprise customer and service integrator (SI) application systems. The third significant service type that is known as collaborative-aware services is dedicated to analyze, and to generate decisions by using the inherited data from information aggregation services.

This service requires node-to-node or node-to-user communication to perform required tasks which is appropriate for M2M and H2M communication in smart grid applications. The fourth component of services infrastructure in IoT are ubiquitous services that are omnipotence and omnipresence service to provide communication requirements at any time and at anywhere. Ubiquitous services utilize internet and cellular systems as communication medium. International Telecommunication Union (ITU) defines the objective of ubiquitous services to provide the seamless communication of anything in the context of IoT. The networking should encompass the connectivity to omnipotence and omnipresence capability, omnipresent reality to create a real communication infrastructure, and intelligence to provide adequate communication to meet the increased requirements. Therefore, the high-level capabilities expected from ubiquitous services are identified as open web-based services, context-awareness, multi-networking capabilities, and end-to-end connectivity in Y.NGN UbiNet recommendation. The communication devices operated in an IoT environment require lightweight protocols in order to increase the interoperability and to prevent the high resource requirements. The software developing kits such as C, Java, Message Queue Telemetry Transport (MQTT), Phyton or some script-based APIs are used to implement IoT-based smart grid applications [34, 50].

7.4.1 IoT protocols

There are two major IoT protocol classifications proposed in the literature where one is based on data exchange protocols as bus-based and broker-based while the other one lists the protocols into three sections as application protocols, service discovery protocols, and infrastructure protocols [32]. The bus-based protocol architecture enables clients to transmit a particular message to the assigned recipients of that message. The protocols used for this objective, which is also identified as service discovery protocol approach, include Data Distribution Service (DDS), Representational State Transfer (REST) and Extensible Messaging and Presence Protocol (XMPP). In the broker-based architecture, broker that saves, transmits, filters, and ranks the priority level controls the distribution of message. The prominent broker-based IoT protocols are Advanced Message Queuing Protocol (AMQP), CoAP, MQTT and Java Message Service API (JMS) protocols. Another classification method for these protocols is to decide that if they are message-centric or data-centric. The message-centric protocols are same with the broker-based classifications that are AMQP, MQTT, JMS and additionally REST included. On the other hand, the data-centric protocols are similar to bus-based classification where CoAP is included to DDS and XMPP that focus on successful data delivery. The CoAP is a web-based, and XMPP and AMQP are application-based protocols. The protocols and their functions required for IoT-based smart grid applications are presented as follows.

Constrained Applications Protocol (CoAP): CoAP is a web-based client and server model protocol that is based on REST architecture on HTTP and operates in the application layer (APP). The web transfer protocol based on REST represents a simpler way to exchange data between clients and servers over HTTP. The constrained devices such as sensors or sensor nodes are utilized as servers for IoT applications. Although many web-based applications inherit the congestion control by TCP, CoAp does not interact with TCP and operates regarding to UDP protocol and performs the congestion control by itself. The UDP is easier to be implemented by using microcontrollers, but it does not support security tools such as Secure Sockets Layer (SSL) and TLS.

Advanced Message Queuing Protocol (AMQP): AMQP is also an APP protocol as CoAP that is based on message-centric architecture. Despite of CoAP, AMQP uses TCP and it ensures the successful message delivery owing to its authentication and encryption methods based on SSL/TLS controls.

Message Queue Telemetry Transport (MQTT): MQTT is a broker-based open source protocol managing the message transmission from clients to a central broker. It uses TCP protocol and operates a routing mechanism for M2M connection. It connects up to a thousand of device nodes with enough performance and low power consumption.

Java Message Service API (JMS): JMS is a middleware API similar to open source Java Call Control (JCC) that is responsible of managing message transmission and control. JMS is a web-based API operating on TCP/IP protocol to provide secure, reliable and asynchronous communication. It separates the APP and transport layer objectives and supports the communication between different applications.

Data Distribution Service (DDS): The DDS is a recent publish/subscribe protocol used in M2M communication to enable real-time, high performance and interoperable data transmission. Object Management Group (OMG) has implemented it. DDS is featured on QoS policies such as urgency, priority, security, and reliability etc. DDS can inherit the location, redundancy, time, message flow, and operating platform of the system. The protocol includes Data-Centric Publish and Subscribe (DCPS) model, and a DDS Interoperability Wire Protocol (DDSI) that DCPS identifies the DDS architecture, and DDSI defines interoperability structure.

Representational State Transfer (REST): REST provides a simple way for clients and subscribers to communicate over HTTP protocol. It is a point-to-point protocol and operates on TCP/IP. REST enables HTTP methods such as get, put, post, and delete for clients and subscribers on a web-based architecture. The protocol supports up to a thousand of devices with the proper efficiency and performance.

Extensible Messaging and Presence Protocol (XMPP): XMPP is a message-oriented middleware to transmit voice and video data in decentralized client-subscriber architecture. XMPP enables users to communicate by instant messages on web-based platform regardless of any OS. There are some extensions have been included in the recent versions of XMPP to improve its usage in IoT applications [32, 49].

7.4.2 Services and security of IoT

The integration of IoT and smart grid enable to data acquisition in a large manner that almost all the data are vital for industrial, residential, and daily life applications. Besides the acquisition, obtained data are stored in several local or remote databases. Thus, the security and privacy issues play a crucial role for IoT and smart grid infrastructures. The security studies focus on typical basic features to increase the durability against attacks. These features can be listed as *confidentiality* that defines to protect data and private information secure against unauthorized accesses, *integrity* referring to prevention of modifications on stored data, *authentication* that means to enable someone to manage the data and to refuse the illegal access, and *availability* that refers to the ability of authorized users to access the data whenever required. Other related security requirements are privacy, anonymity, liability and non-repudiation.

The attacks are categorized into four groups as interception that is a passive attack method, and interruption, modification, and fabrication that are active attacks. Interception targets the confidentiality and identifies an antagonistic eavesdropping on a content of recipient. It performs traffic analysis and release of message. The effect of interception can be quite important since attacks are directed to basic infrastructure such as network base and M2M gateways. Interruption is an active attack threat that blocks the transmission of a message by jamming, man-in-the-middle attack (MITM) or DoS attacks. In DoS attack, adversary makes any service, machine or network unavailable for recipients. Modification defines the action of intercepting and transmitting the modified message to a specific service that is something similar with MITM attack. Fabrication attack refers to adjusting a message and inserting to the original message to transmit to recipient [51, 52].

The cyber-security threats are defined as passive and active attacks where the passive attacks target confidentiality of system. The attack types under this group are traffic analysis and release of message contents. The widely known interruption type attacks can be listed as channel jamming, routing attacks, DoS, and breaking the communication line. The attacks targeting authenticity are message forgery and spoofing of smart meter, power plant, and household devices. The attacks on availability are performed in several ways as selective forwarding attack, flooding, overloading and jamming attacks. These attacks overload the communication infrastructure and targets to increase energy consumption in sensor nodes to disrupt functionality.

The traffic analysis attacks can be active or passive according to its traffic injection act to network. The active traffic analysis attack can be seen in WirelessHART, ISA100 and ZigBee networks of smart grid and can be resolved by using authentication mechanisms. On the other hand, passive traffic analysis attack is seen in ZigBee networks. The industrial IoT standards such as WirelessHART, ISA100.11.a, and 6LoWPAN require particular attention on security as ZigBee Pro standard due to their application contents, wireless network infrastructure, and WSN security. Although M2M communication methods have not been faced to special threats, they have been exposed to existing threats since they are more vulnerable comparing to conventional network infrastructures and devices. The attacks targeting M2M communication systems can be analyzed in three groups as physical, logical, and content attacks. The physical attacks are listed as side channel attacks, software modification attacks, and destruction of the M2M device. The logical attacks are impersonation, DoS attacks, and relay attacks that adversary targets communication devices and network. The content attacks are based on privacy, modification, and interception intrusions. The industrial automation systems are classified into two groups as process automation (PA) and factory automation (FA). The PA applications use industrial wireless communication standards that are presented below in detail. The FA applications use wireless interface for sensors and actuators (WISA) and the wireless sensor actuator network for factory automation (WSAN-FA) standards that both are operated at PHY layer. The PA applications that are utilized for monitoring and control acquire the data and send to data concentrators at predefined intervals. In despite of PA application, the FA applications are quite sensitive to latency and delays. Therefore, PA applications have been paid much attention comparing to FA applications.

IoT services and protocols supported the improvement of a novel industrial standard 6TiSCH that is proposed by IEEE and IETF in addition to legacy industrial standards. ZigBee Alliance has introduced ZigBee Pro in 2007, which has become a widespread technology even in industrial applications as WSNs. The star, mesh, and tree topologies can be comprised in ZigBee networks that mesh provides more reliable and secure communication infrastructure comparing to others. Nevertheless, star topology is more suitable to meet controllability requirements of industrial networks and it is preferred to mesh topology in industrial applications. After that ZigBee Pro has been widely used, International Electrotechnical Commission (IEC) approved the WirelessHART as an international and industrial wireless communication standard with IEC 62591Ed.1.0. The main components of a WirelessHART are field devices, gateway, access points, the network manager and mobile devices. The field devices are installed at the industrial plant for data acquisition and routing processes that the data is transmitted to gateway through access points. The network manager configures the devices and topologies of network to operate the transmission.

On the other hand, Chinese Industrial Wireless Alliance has introduced another industrial standard entitled Wireless network for Industrial Automation-Process Automation (WIA-PA) almost at the same time in 2010. A regular WIA-PA network associates star and mesh topologies to compose a hierarchical network. The mesh topology includes router and gateways, while router and field devices compose star topology. The components of WIA-PA networks are same with WirelessHART where the field devices are integrated at industrial stages to obtain and to transmit the data. ISA100.11a has been presented by International Society of Automation (ISA) in 2009 and has been approved as IEC 62734 standard by IEC in 2014. Despite the previous wireless standards, ISA100.11a is aimed to be a new approach with wider coverage and broader connectivity features. The devices are spread into two main groups as field devices and infrastructure devices in ISA100.11a that data acquisition devices, routers, and mobile devices are defined as field devices and backbone routers, gateways, and security devices are accepted as infrastructure devices [42, 53].

The improvements of TSCH and DSME approaches brought by IEEE 802.15.4e promoted IPv6 utilization in industrial networks. 6TiSCH is a crucial standard performing IPv6 communication over TSCH by interconnecting MAC layer and network layer. IEEE 802.15.4e is not capable to fill the gap between scheduling and managing the networks traffic that is accomplished by 6TiSCH Operation Sublayer. Thus, the top stack of 6TiSCH enables IPv6 that allows converting long 6LoWPAN packets to short IEEE 802.15.4. Three major open source wireless devices; OpenWSN, Contiki, and RIOT, support 6TiSCH to implement WSN and wireless communication infrastructure. The features and differences of industrial IoT (IIoT) protocol architectures are illustrated in Fig. 7.7 regarding to the wireless industrial standards. The layer structure is comprised by considering OSI seven-layer structures where some standards do not include respecting protocols are left blank.

Security and privacy issues should be taken into consideration at the adoption stage of IoT devices. The security environment includes all sides of infrastructure as users, things, technological system, and processes. The PHY and MAC layer security is related to communication devices where IEEE 802.15.4 is prominent standard enabling security services operating at MAC layer with various encryption methods. The 32-, 64-, and 128-bit Advanced Encryption Standard (AES) keys are used to increase security in Cipher Block Chaining (CBC) mode that generates a Message Integrity Code (MIC) or Message Authentication Code (MAC) inserted to the transmitted data. The security mode AES keys are identified as AES-CBC-MAC-32, AES-CBC-MAC-64 and AES-CBC-MAC-128 at MAC layer. These keys prevent the vulnerabilities to integrity and authentication attacks. Cipher blocking and counter modes are associated to increase confidentiality of IEEE 802.15.4 standard at link layer transmission. The associated encryption modes are defined with CBC-MAC AES/CCM and AES keys are titled as AES-CCM-32, AES-CCM-64 and AES-CCM-128 with different heap sizes and encryption levels.

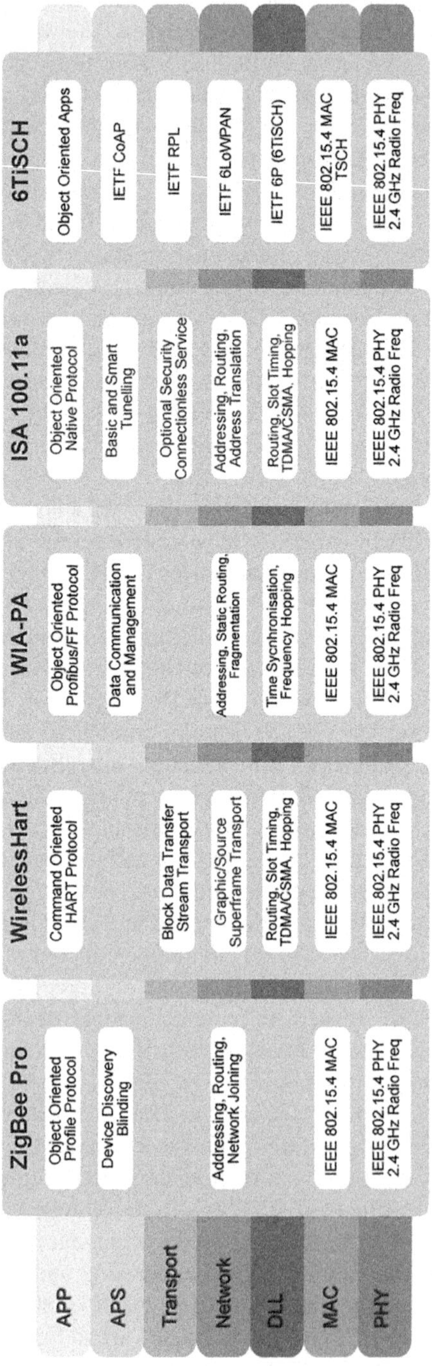

FIG. 7.7 Comparison of IIoT standards throughout the layers.

Smart objects and code segments can provide the IP security of IoT applications in the context of 6LoWPAN usage at link layer. The link layer security that is inherited from IEEE 802.15.4 supports data encryption and integrity check. On the other hand, security can be supported by using TLS and SSL that includes a key mechanism to control the authentication, confidentiality, and integration at the transport layer. The UDP based TLS namely DTLS is used in 6LoWPAN enabled networks to increase the resiliency against attacks at network layer [52].

Several sensors comprise sensor networks as its name implies, but additional components such as microprocessors, communication devices, and power supply are also required to operate the system. The sensor nodes are featured to acquire physical data of the measured system with humidity, temperature, pressure and similar sensors or electrical data with current and voltage sensors. The microprocessor can be any platform including 16- or 32-bit cores. The WSN is an emerging application of sensor networks that is widely used in smart grid, M2M, and IoT applications. The essential idea behind WSN is its wireless communication infrastructure transmitting the acquired data in short or long distances corresponding to transmitter specifications. In the WSN infrastructure, each sensor is supplied by a portable power source as battery and utilizes wireless communication interface. The sensor node can be stable at any measurement section or it can be mobile. Although preliminary WSN applications are implemented for military and defense usages, now it has a widespread use in health, industrial, residential, and several engineering applications. Some of WSN usage areas can be summarized as environmental issues as agriculture and wildlife researches, energy harvesting applications, smart metering, industrial measurement, and remote monitoring applications, tracking and localization systems.

WSNs are mostly exposed to intrusion and threats since they transmit valuable data on communication medium. Therefore, security is a crucial topic in WSN researches that aim to prevent DoS attacks and intrusion attempts. The passive attacks try eavesdropping to obtain and analysis the traffic data as introduced in the previous subsection. On the other hand, active attack types as interruption, modification and fabrication are widely met in WSN as seen in other communication infrastructures. There are several methods such as node protection, cryptography, key management, secure routing and data aggregation have been proposed to increase WSN security. The security issues of WSN are analyzed in two groups as protocol security and trust and privacy concepts. The protocol security means entire layer protection instead of layer-by-layer security. The layer protection includes authentication, encryption, secure wakeup nodes, and tamper proofing approaches. Some combination security models have been implemented to fulfill the requirement on layer security. These security approaches are identified as low-level mechanisms while high level mechanisms include secure group management, intrusion detection, privacy-awareness, and secure data aggregation.

Security requirements play a vital role in the operation of WSN, but it should not impair bandwidth, availability, and data transmission reliability, which are essential QoS indicators. The QoS is a remarkable research area in the context of IoT and WSN studies. The QoS researches have defined two fundamental perspectives as individual QoS and collective QoS in recent years. In individual QoS, applications require particular features for sensor types and numbers. On the other hand, the communication and delivery parameters of each sensor by oneself are not cared but the corresponding requirements are considered.

7.5 IoT applications in smart grid

IoT is a widespread and complex communication infrastructure itself. The system is based on acquisition, processing, transmission, and storage of data in secure, reliable and efficient way. On the other hand, smart grid presents many novel technologies at the generation, transmission, distribution, and consumption levels as well as IoT. In one hand, the knowledge society demands to know the recent situation of any technology related to their life instantly, and to manage their own microgrid sources, home appliances, EVs, to monitor the daily statistics on services such as electricity, gas, and water. On the other hand, service providers require ICT technologies for management, monitoring, detection of the faults and troubleshoot aims. The most recent communication structure is being improved by using IoT technology in smart grid applications. The infrastructure is comprised by low-power and low-cost microprocessors, digital communication systems including security and reliability supports, and adaptable communication layer and protocol structures to interact among numerous different hardware and software technologies. The improvements seen on IoT have attracted attention of smart grid operators and developers. The preliminary communication methods and network infrastructures that have been widely used in smart grid are being updated to meet the requirements of IoT to benefit from cost and power efficient communication technologies. The easy access opportunity of IoT with a widespread mobile device support enables to implement infrastructures for monitoring sensors, actuators, surveillance cameras, smart cities, smart home applications, medical applications, smart microgrid structures, and security issues. Minoli et al. define IoT applications for smart city with five key areas as energy, water, mobility, buildings, and governments from a system perspective [6]. The IoT applications that are improved in the context of smart grid have been categorized into five prominent titles as smart cities, smart home applications, energy harvesting issues, smart sensor networks, and monitoring and metering topics.

In the complex and heterogeneous ICT environment, IoT provides several communication technologies and network types. Some of them has been presented in the previous sections as communication infrastructures, services, protocols, security and privacy, and sensor networks. Furthermore, we have comprehensively presented IoT communications and networks improving the conventional and degraded ICT infrastructures by transforming to IoT-based smart grid applications. Many novel communication technologies including LPWAN, UMTS, LTE, LTE-A, and narrow band IoT (NB-IoT) have been emerged in addition to BLE in Personal Area Networks and Zigbee in the use of HEMS. The improved technologies provided long-range communication in the unlicensed bands for IoT and smart grid applications. The most prominent LPWAN technologies are known with the names of Ultra Narrow Band (UNB) by SIGFOX, LoRa by Semtech, LTE machine-type communications (LTE-M), and NB-IoT. The comparisons of well-known communication methods have been presented in Fig. 7.8 and Fig. 7.9. The first one illustrates communication range and data rate while the second depicts comparison of LPWAN, cellular and IEEE 802.15.4 technologies in terms of range, power consumption, bandwidth, microprocessor/microcontroller unit (MCU) cost, communication channel license cost, number of transmission station, latency, and coverage specifications. It visualizes advantages of LPWAN on range, low cost communication, low number station requirement, wide coverage up to tens of km, and additionally long battery life up to 10 years with a single battery.

The presented features of emerging LPWAN enable end-user IoT applications that require low cost devices, long power source and battery life, small amount of data transmission, and widespread areas where the cellular technologies are not appropriate. LoRa and Sigfox UNB have been extensively emerged among other LPWAN technologies that both are proprietary and they use unlicensed frequency spectrum. Semtech has presented LoRa in 2015. The LoRaWAN is arranged in star topology or cellular communication architecture where an end-device is connected to central server over gateways. The end-devices are operated in device originated call mode to decrease power consumption and wake up at predefined intervals for paging data. Thus, LoRa has found a wide usage area in many applications such as smart health monitoring, smart

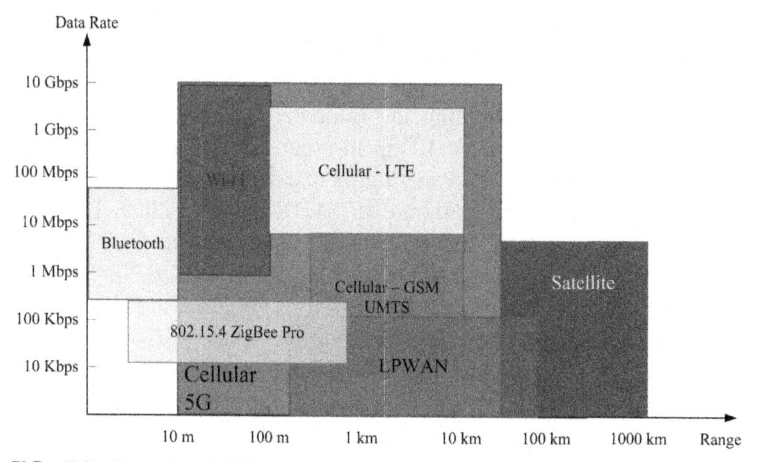

FIG. 7.8 Range-bandwidth spread of wireless communication technologies used in IoT applications.

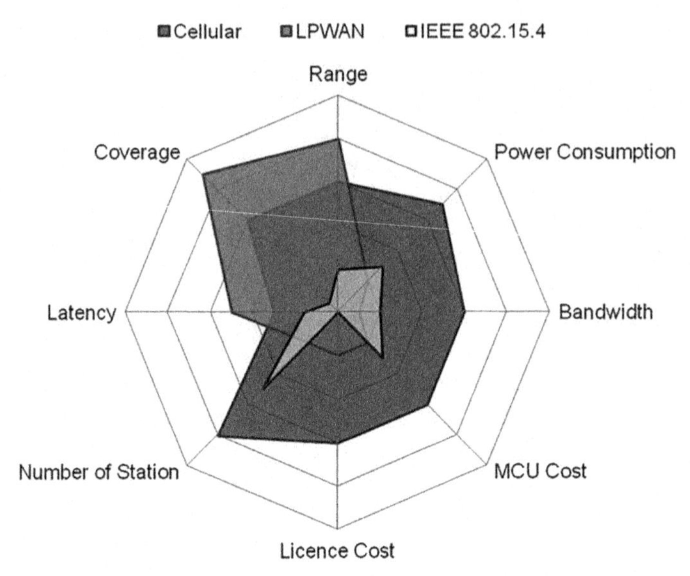

FIG. 7.9 Comparison of LPWAN, cellular and IEEE 802.15.4 technologies.

cities, industrial monitoring and metering applications. The improvement of LoRa is being maintained in two essential layers with particular techniques that Chirp Spread Spectrum (CSS) modulation technique for PHY and LoRaWAN protocol for MAC layer. In its current state, minimum transmission bandwidth of LoRa is 125 kHz and that supports 157 dB coupling loss at its maximum. The CSS is being improved to increase immunity to interference [54, 55].

The key features and comparisons of leading LPWAN technologies; LoRa, Sigfox, LTE-M, and NB-IoT, have been listed in Table 7.3. Sigfox UNB is another emerging technology supporting low-power end devices used in smart meters and home appliances. It provides a cellular technology approach operating in sub-GHz band with Binary Phase Shift Keying (BPSK) and differential BPSK (D-BPSK) modulations at 868 and 902 MHz frequencies. Sigfox provides up to 400 channels by dividing the frequency spectrum into 100 Hz, and coverage area is extended up to 10 km in urban area while it reaches up to 50 km in rural areas. Besides LoRa and Sigfox that both operate in unlicensed spectrums and sub-GHz, 3GPP has started IoT MTC study in 2009. The first user equipment (UE) of IoT technology has been introduced by 3GPP as LTE-M in Release 12. The proposed technology is later defined as NB-IoT that operates 200 kHz carrier in Release 13. NB-IoT uses OFDMA modulation for downlink, single-carrier frequency division multiple-access (SC-FDMA) for uplink, rate matching, interleaving, and enhanced channel coding methods. Despite of LTE-M, NB-IoT only supports tail-biting convolution coding in the downlink, eliminating turbo-decoding requirement at the UE [54, 55].

The IoE infrastructure associating IoT and smart grid demands is shown in Fig. 7.10. The applications include smart home automation system based on

TABLE 7.3 Summary of leading LPWAN technologies

Technology	LoRa	Sigfox UNB	LTE-M	NB-IoT
Standard	LoRaWAN	N/A	LTE (Release 12)	LTE (Release 13)
Modulation method	GFSK, SS Chirp	D-BPSK	BPSK, QPSK, OFDMA	π/2 BPSK, π/4 QPSK
Data rate	0.3–38.4 Kbps	100 bps	200 Kbps–1 Mbps	Up to 100 Kbps
Receiver sensitivity	−137 dBm	−147 dBm	−132 dBm	−137 dBm
Max coupling loss	157 dB	162 dB	155 dB	160 dB
Interference immunity	Very high	Low	Medium	Low
Bidirectional	Yes	No	Yes	Yes
Minimum transmission bandwith	125 kHz	100, 600 Hz	180 kHz	3.75 kHz
Frequency band	Sub-GHz	Sub-GHz (868, 902 MHz)	Licensed cellular	Licensed cellular
Security	32-bit	16-bit	32-bit	N/A
Range	2.5–15 km urban, up to 50 km rural	3–10 km urban, 30–50 km rural	35 km-GSM 200 km-UMTS, LTE	2.5–15 km urban, up to 50 km rural
Power efficiency	Very high	Very high	Medium	Very high
Transmitter power	20 dBm	15 dBm	23 dBm	23 dBm
Battery lifetime	8–10 years	7–8 years	7–8 years	1–2 years

appliance monitoring and operating interfaces, smart city operations in a wide variety of WSN enabled systems, energy-harvesting systems comprised by microsources and DERs, remote monitoring and metering systems. The figure summarizes hardware, software, and communication infrastructures in the context of layer structure of IoT that has been presented in Fig. 7.5.

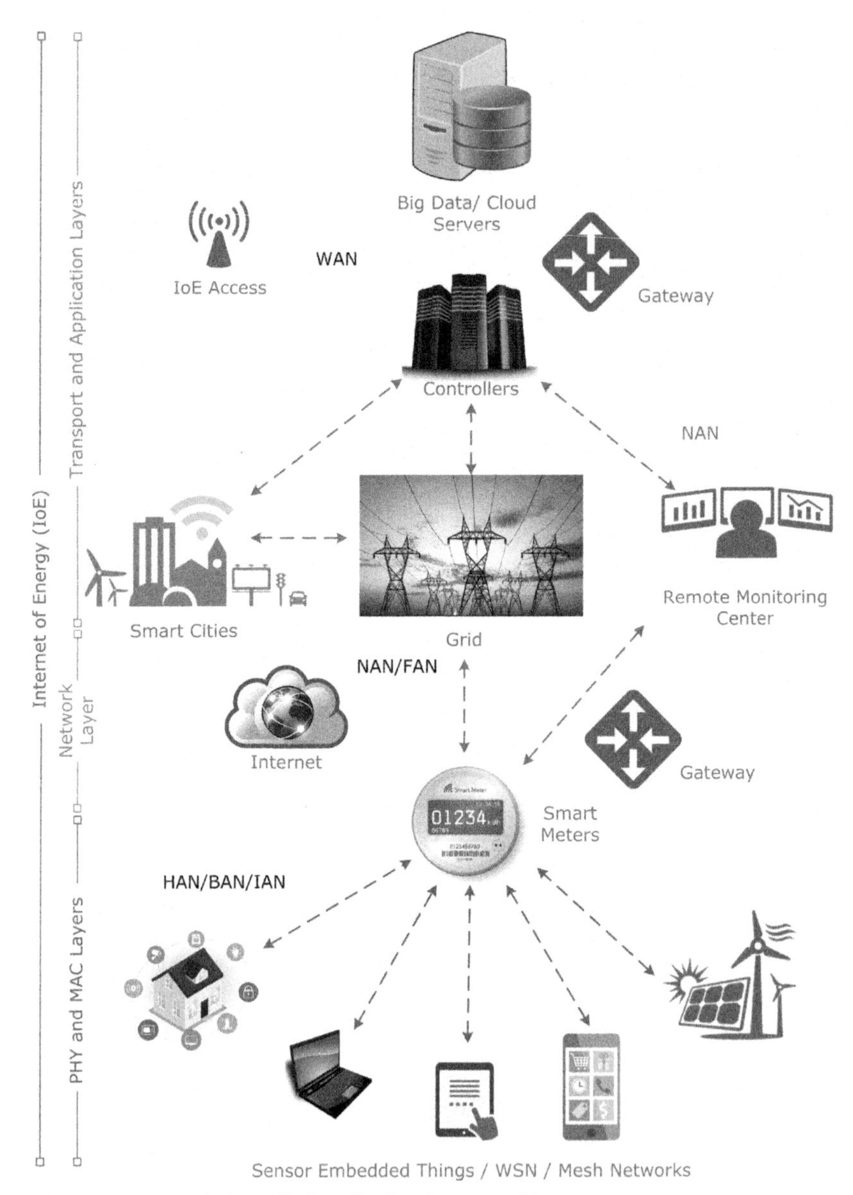

FIG. 7.10 A general view of IoT applications in smart grid.

7.5.1 IoT smart city applications

Although the smart city concept has not been officially accepted by any organization, it has been widely accepted in application level where the fundamental goals are to increase quality of public life, reducing the public costs, utilizing the public resources in a secure and sustainable way. The urban IoT term is used to imply this infrastructure used in public services in some occasions. The IoT, IoE and urban IoT applications include conventional public services such as intelligent transport, parking, street and traffic lighting, control and management of energy systems, waste management, environmental monitoring, healthcare services, medical help, structural health monitoring (SHM) systems, and many other [28, 29].

Environmental concerns such as climate change, carbon emission, greenhouse gas effects have attracted much attention for monitoring, predicting, and management requirements. The urban IoT has contributed several capabilities to legacy monitoring and management systems including management information system (MIS), and environmental information system (EIS), geographical information system (GIS), global positioning system (GPS) used for environmental factors. Fang et al. surveyed EISs and proposed an integrated information system (IIS) for environmental monitoring [56]. An environmental condition monitoring system that is based on similar strategy with ZigBee and WSN infrastructure has been proposed for home appliances [57]. Both studies have been implemented considering a scheme as shown in Fig. 7.10 where data acquisition processes are performed by using RFID, WSN, and other sensor-based interfaces, a metering and communication system, and a remote monitoring interface comprised by an application or device.

WSN is obviously one of the key interfaces of IoT applications that are also widely used in urban IoT for smart city applications. SHM is an emerging application that is integrated to the existing structures to detect risk levels where five topics are followed as detection, localization, classification, assessment, and prediction. Industrial plants, residences, and social places are monitored and are characterized in terms of energy and grid operations in urban IoT applications. Building Automation and Control Networks (BACnet) is a web-based and proprietary solution that is used in aforementioned automation systems in addition to ZigBee or WSN based monitoring and management applications. It is remarked in [58] that BACnet employs a set of alternatives to ZigBee in terms of IP technology by employing a virtual link layer (VLL) by adapting the underlying network and transport protocols.

As discussed earlier, smart city concept should deal with several technical challenges. A hierarchical layer structure that is called smart city infrastructure (SCI) has been proposed in [59]. Four layers have comprised the SCI where the first one is comprised by IoT to perform context management, ICT infrastructure, and computational methods. The second layer includes smart home that integrates smart services and user interfaces via a server. The third layer has been defined as Cloud of Things (CoT) that associate smart home and services of smart city that are located in the last layer. Layer 4 includes additional service infrastructures, resource management systems, and integration to IoT communication systems. It has been noted that well designed CPS systems in an IoT environment can increase energy efficiency, enhance the safety in city, and can increase the comfort at smart homes. The proposed middleware approaches provide interoperability and correlation of heterogeneous systems including building information models (BIMs), SIMs, and GIS.

Some featured urban IoT applications have been improved in several cities. May IoT projects on public spaces, lighting, and mobility to improve sustainable living and working were get started in the city of Amsterdam, Netherlands in 2006. Cisco and Philips have developed new concepts and innovations around network-enabled LED street lighting in the context of smart city applications. The City 24/7 smart screen project that provides an information platform has been improved in New York, USA. The project delivers smart screens to present required information incorporating audio, voice, and touch screen technologies. The smart screens can also be accessed via Wi-Fi at any time and on any device. Another smart city framework has been implemented in the city of Nice, France. The citywide project includes four main services of smart transport, smart lighting, smart waste management, and smart environment monitoring [60].

7.5.2 IoT applications in smart home environment

Several smart devices and things communicating with a central system comprise a featured smart home management system. The central system is defined with several names as distributed services middleware, home gateway, gateway and integrator, ZigBee based intelligent self-adjusting sensor (ZiSAS), and so on. Regardless of its name, smart home management systems (SHMS) receive monitoring data from sensors, controls smart environment by transmitting commands, and informing the user in case of critical changes occurred. The several SHMS works that are composed by hardware and software have been surveyed in [61]. Some of the presented studies are based on mobile applications to control the SHMS while high share of remainder is based on IoT middleware including smart metering applications.

The US government has been forced to pursue a considerable persistence on the DR Management due to the energy consumption of residential users. The HEMS have been improved regarding to the insistence of government. Thus, it enabled consumers to get benefit from DR programs while decreasing waste consumptions. HEMS modules integrated to the customer smart meters operate as a smart and autonomous agent of the infrastructure. The SHMS and HEMS tools enable consumers to increase the efficiency and to get involved to service provider programs such as time-of-day pricing for decreasing energy costs. Service providers and governments also take advantage of DR programs due to SHMS and HEMSs. SHMS include a specific management type that is sometimes named as home energy management system that particularly targets DSM and DR control. Although both are assumed as same concept, there is a nuance between them. DSM targets to increase the efficiency of electricity consumption in a general infrastructure while DR aims to change user habits on electricity consumption and control the consumption in an indirect way. The DR programs are classified into two groups as incentive-based programs (IBP) and price-based programs (PBP). IBP includes widely known direct load control and capacity market categories while PBP are TOU, real time pricing (RTP), and critical peak pricing (CPP) programs that indirectly affect the electricity consumption habits.

The network communication is performed by a series of area networks such as NAN. The neighborhood area includes smart home groups interacting via HANs and data aggregators in a specific geographic area. The NAN is comprised by a series of smart devices such as smart meters, DERs, energy storage devices, and loads. The aggregated data is transmitted over gateways to monitoring and control centers that are located at top of the schematic as shown in Fig. 7.10. The SHMS is also configured in a layer structure as done in other IoT applications. The PHY and MAC layers have been presented as device layer that includes sensors, actuator, and gateways. Several wired and wireless communication networks and HAN, BAN, NAN structures are listed in network layer. The protocols and services surveyed above are used to perform data acquisition and aggregation processes as shown in Fig. 7.11. The cloud management layer has been proposed as a featured layer holding cloud and management

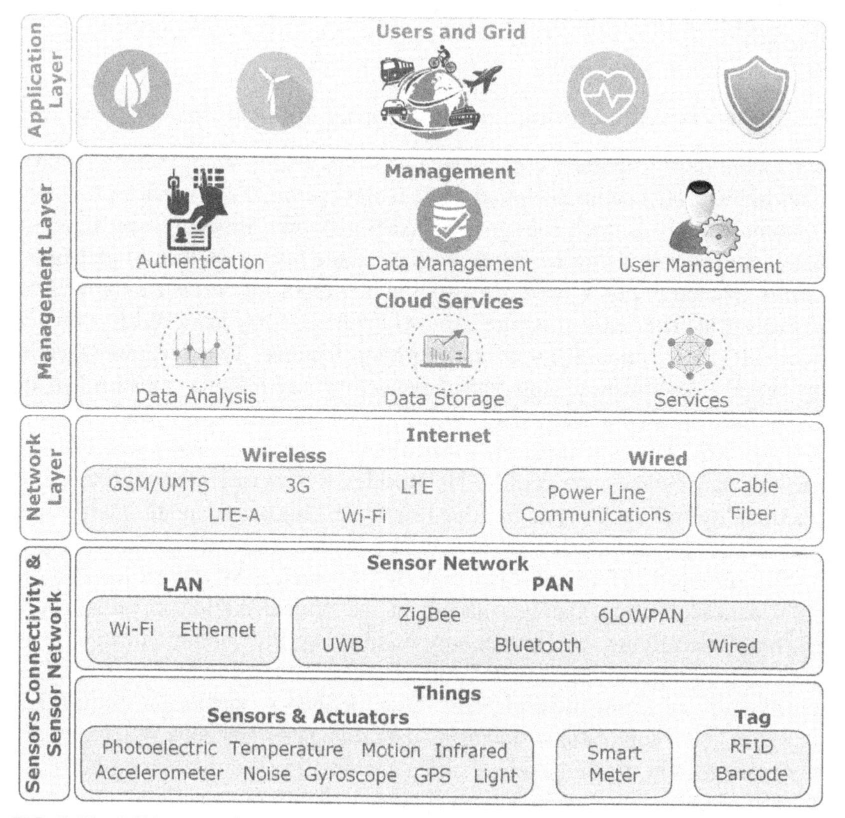

FIG. 7.11 IoT layers and applications in smart home environment.

services [62]. The application layer is responsible to provide required services and interfaces to users. It includes energy-centric management applications for DSM, DR, dynamic pricing, and featured applications in addition to SHMS and HEMS applications. All devices and applications can use any secure IoT protocols such as HTTPS, MQTT, CoAP, XMPP.

It is noted that smart grid is nominee to be one of the first and largest IoT infrastructure with its CPS and ICT background. The smart grid promises to be more efficient, more secure, and cost-effective system by integrating novel IoT network. The energy generation, transmission, and distribution studies present a wide literature on integrating smart grid power network and IoT communication networks.

The discussed works show that smart grid applications at generation, transmission, and distribution networks are still based on conventional, verified, and rugged communication infrastructures that are assumed to be converted to emerging IoT system. However, unlicensed communication bands comprising a huge share of IoT networks are not considered due to security and reliability issues. The bulk generation and transmission levels of smart grid are operated at WAN backbones. The AMI and last-meter that is located at the consumption side are being paid much attention considering LTE based technologies [63, 64].

7.5.3 IoT-based metering and monitoring applications

One of the most important components of smart grid is smart meter that provides bidirectional communication and enables customer and service providers to monitor the consumed energy rates. AMI networks are composed of smart meters and certain gateways configured in single hop or multi hop networks. AMI is located in NAN structure as being a component on the customer side of smart grid. The NAN structure is mostly comprised by using WMNs owing to its self-organization and self-configuration features. These features enable any node to establish an automatic connection and reliable transmission in NAN. As discussed earlier, IEEE 802.11s that improve single hop function of IEEE 802.11a/b/g/n standards to multihop is one of the most widely used open standards in the context of NANs. Besides, it increases internet connection functionality and MAC capacities that enable the standard to create smart meter mesh network including several meters in a NAN area.

The surveyed IoT based smart metering researches have been focused on WSN, data acquisition, gateway placement and implementation, automatic billing, real-time pricing, wireless energy monitoring, PLC communication, privacy, and computational methods to improve SMs [20, 63, 65–68]. The contributions of smart metering have been defined as supporting smart grid and smart home applications, acquiring data from heterogeneous WSNs, secure data management, tracking sensor and actuator data. The interaction between DSM and AMI environment improves the capability of smart grid system in terms of load control and energy efficiency management. The bidirectional data transmission ability of AMI enables service providers to develop automatic billing and real-time pricing programs to direct DSM. The most important contribution of AMI and DSM integration is to determine the instant load demand that enables service providers to manage energy price and tariff considering peak demand and real time load analyses. The dynamic price management is inherited from DR concept where it is classified within three kinds as ToU, RTP, and CPP as discussed earlier. The RTP that is the most effective method since it decreases the peak consumption demand among them can be achieved by using a complete deployed AMI into distribution network. The communication methods that are used for AMI applications are similar to any other IoT application as being PLC in wired medium or Wi-Fi, cellular technologies, Bluetooth, WiMAX, and other wireless technologies. Although PLC has been mentioned in some applications, a high share of the communication technologies is comprised by wireless systems owing to their coverage range, privacy and security precautions, efficient bandwidth, and low error rate.

PLC technology achieved great interest since it utilizes the existing power networks as transmission medium. Several researches and surveys on PLC communication of smart grid have been proposed in literature. Although it is an advantage that PLC does not require a transmission channel installation, technical features of PLC causes a number of noises and attenuation on the transmitted signal. In order to tackle these drawbacks of PLC, associating the PLC with wireless communication infrastructure may overcome these and limited bandwidth drawbacks. Mahmoud et al. proposed LTE cellular networks in AMI application due to reliability, availability, and coverage area issues that an AMI LTE network has been implemented on 802.11s-based WMN [1, 64, 68, 69].

A heterogeneous AMI structure including several wireless communication systems modeled and analyzed in several researches where a number of heuristic models implemented to solve current AMI network drawbacks. It was remarked that the implemented model was capable to define available number of resources, routes among the SMs, the UDAPs, and the cellular base stations. The interference issues were particularly researched for network initialization, address distribution and routing control situations in HTAMI as an alternative to current studies considered for low traffic AMI applications.

7.6 Open issues and future research directions

Despite extensive researches for transforming conventional utility grid to smart grid have been performed up to date, there still major challenges exist to be solved for improving interoperability, connectivity, reliability and security of smart grid CPS. Since the smart grid infrastructure encompasses comprehensive interaction of power generation, transmission, distribution and consumption environments, it requires a widespread and reliable communication interface to control, manage and monitor this system. The complexity and heterogeneity of power and communication systems comprising the smart grid infrastructure poses challenges on interoperability of devices, cyber-physical security, resiliency, and data management issues. Although the widespread operation area of smart grid draws a heterogenous ecosystem, the major challenges can be categorized into threefold as control, communication and security of system. The challenges on control issues are met at generation, transmission, distribution, and consumption levels that include DER integration, remote monitoring, transmission line monitoring, PMU and PQ analysis, RES integration, DSM, DR, load management, interoperability, and emerging EV applications. It is widely accepted that most of these challenges relating to control issues have been solved, and recent studies have proposed various control methods to overcome power network deficiencies. However, use of communication systems based on vast amount of different technologies and protocols poses several challenges on interoperability and security issues.

The smart grid communication network is defined as system of systems due to massive diversity of communication technologies. The surveyed smart grid studies indicate that modeling, analysis, and implementation researches may meet many of the new challenges in terms of power grid and communication systems integration. The operating, control, and monitoring systems of power grid requires wired and wireless communication infrastructures, and thus a wide variety of ICT and CPS interaction researches are required to improve communication network challenges. While the communication networks are fundamental to improve development of smart grid, several challenges are addressed for a robust, secure, resilient, and operational smart grid communication network. The smart meters that are connected in HAN, BAN and NAN topologies are prominent data sources of smart grid since they generate massive data bases and require secure and efficient communication interfaces. Moreover, data concentrator configurations and MDMS are faced to several drawbacks in communication base. The two-way communication infrastructure of smart meters is essentially enabled by various wireless technologies such as IEEE 802.15.4, IEEE 802.11, cellular, and area networks. Therefore, the diversity of communication technologies operating in the same infrastructure requires standardized protocol and application interfaces to ensure interoperability of entire system. To this end, generic APIs and middleware developments are required to tackle major challenges of heterogenous infrastructure of communication technologies.

In addition to monitoring and management researches, industrial WSN network and IIoT applications as ZigBee, WirelessHART, ISA100.11a, WIA-PA and smart grid interaction researches have been determined in the literature. WirelessHART, ISA100.11a, and WIA-PA were particularly designed for automation applications, while ZigBee has been introduced for industrial automation applications. As discussed earlier, most recent and robust technology for these industrial standards is 6LoWPAN that is an IP based standard. On the other hand, it is obvious that gateway studies on IEEE 802.15.4 based standards pose challenges due to low-rate data transfer capacity. However, recently improved IEEE 802.15.4 g standard can solve data rate and interoperability issues by its efficiently operating capability in heterogenous networks and inference prevention. The specific protocol-based applications are another emerging research area in the context of smart grid and IoT integration where CoAP, AMQP, MQTT, and JMS are prominent ones. The security and privacy researches are also expected to emerge due to several different communication protocols and frameworks are utilized in IoT and smart grid interaction with particular coding and encryption infrastructures. Some researchers surveyed on this topic have remarked that public key infrastructure (PKI) usage may tackle the security issues. The security concerns have also increased the researches on PHY and MAC layer security, IEEE 802.15.4 end-to-end security, network layer, IP and 6LoWPAN security, and routing security.

The discussed issues show that IoT systems are still emerging in its current ICT technologies and many smart grid infrastructures are tranformed to IoT technologies. Most of the research papers indicates that the progress of IoT is not clearly predicted. Due to wider connectivity and improvement of cellular technologies, it is assumed that billions of devices are going to be connected to each other by WAN and LPWAN infrastructures. It is expected that predicted progress will require tremendous growth in WSN, WMN, data clouds, ICT systems, and CPS infrastructures. One of the most crucial components of IoT and smart grid integration is WSN that provides the main player of system data. It is obvious that WSNs will become prevailing technology owing to featured properties of wireless devices and plenty of individual applications are expected to be developed. QoS definition and support is one of the most important points to be researched for WSNs.

The communication technologies and protocols that have been discussed earlier are also expected to be researched and improved to meet the requirements of WSNs. In the researched literature, it is seen that layer structure of IoT is not clear, and there are several approaches proposed in the literature. Beyond the layer definitions, cross-layer operation will likely become more important since it facilitates the interaction between separate layers and provides to achieve the required QoS. The surveyed PHY and MAC layer studies have shown that WSN improvements should be met by appropriate services at these layers. The studies on these layers also focused on channel access management, data packet conversion, and optimization methods. Therefore, it can be said that cross-layer operation and layer based QoS researches can find place in literature.

It is apparent that integration of IoT and smart grid is not a simple task since many challenges on accessibility, reliability, availability, and management are also brought by IoT system. To improve efficient and reliable services, service providers and operators should address these challenges. A CPS including many sensors, transducers, actuators, devices and operators that are connected to web-based systems generate massive data stacks which is widely named as Big Data. The generated big data should be stored, processed and retrieved in an efficient and reliable way. Such an infrastructure requires high capacity hardware and software assisted operation centers. The big data analytics pose several challenges that are addressed by using several computational solutions such as cloud computing, fog computing, edge computing and so on. Although these computing models are expected to overcome big data analytics, they are required to meet synchronization, standardization, reliability, and management requirements. The heterogonous CPS structure of an IoT based smart grid poses to another challenge on node-to-node interoperability due to different hardware and software are used on the same platform. Thus, interoperability is an outstanding criterion on integrating IoT and smart grid infrastructures.

Security and privacy of data acquisition, storage, and retrieving are crucial for IoT as well as CPS security. Security poses a serious challenge for IoT integration to smart grid infrastructure due to inadequacy of commonly embraced security architecture in IoT. The literature surveys show that attacks on IoT systems targets the impersonation, interception, data manipulation, access threats, and authorization issues. On the other hand, user and consumer privacy is another significant priority in IoT based smart grid infrastructure. These vulnerabilities of smart grid and IoT to cyber-attacks should be coped with novel protection algorithms. Fig. 7.12 illustrates the relationship of IoT applications along smart grid segments and layers where the applications have been grouped into six classes as generation, transmission, distribution, consumption, big data analyses, and security considering the surveyed papers. Moreover, open issues and future research directions have been listed into five subsections that have been foreseen to be extensively researched and studied in near future. The significant topics are discussed in the context of IoT architectures, IoT based smart building management systems, cyber security and privacy, big data and cloud management, and computational methods that are used in IoT-based smart grid for data processing and data fusion. The open research topics are summarized regarding to previously discussed challenges. To this end, IoT architectures and improvement subsection presents a summary on recent computing methods considering fog and edge types. The open issues of smart building management and smart grid application in daily life are listed in second subsection while future research directions on cyber security and privacy issue are given in third subsection. The prospectus research areas on big data and cloud computing, and computational methods for IoT-based smart grid applications are presented in the following subsections. Finally, a subsection is presented for standardization and interoperability of IoT-based smart grid.

7.6.1 IoT architectures and improvements

Most of the IoT applications are based on cloud computing methods that are supported by ICT companies hosting cloud services. The cloud services and applications provide essential efficiency, flexibility, and data management

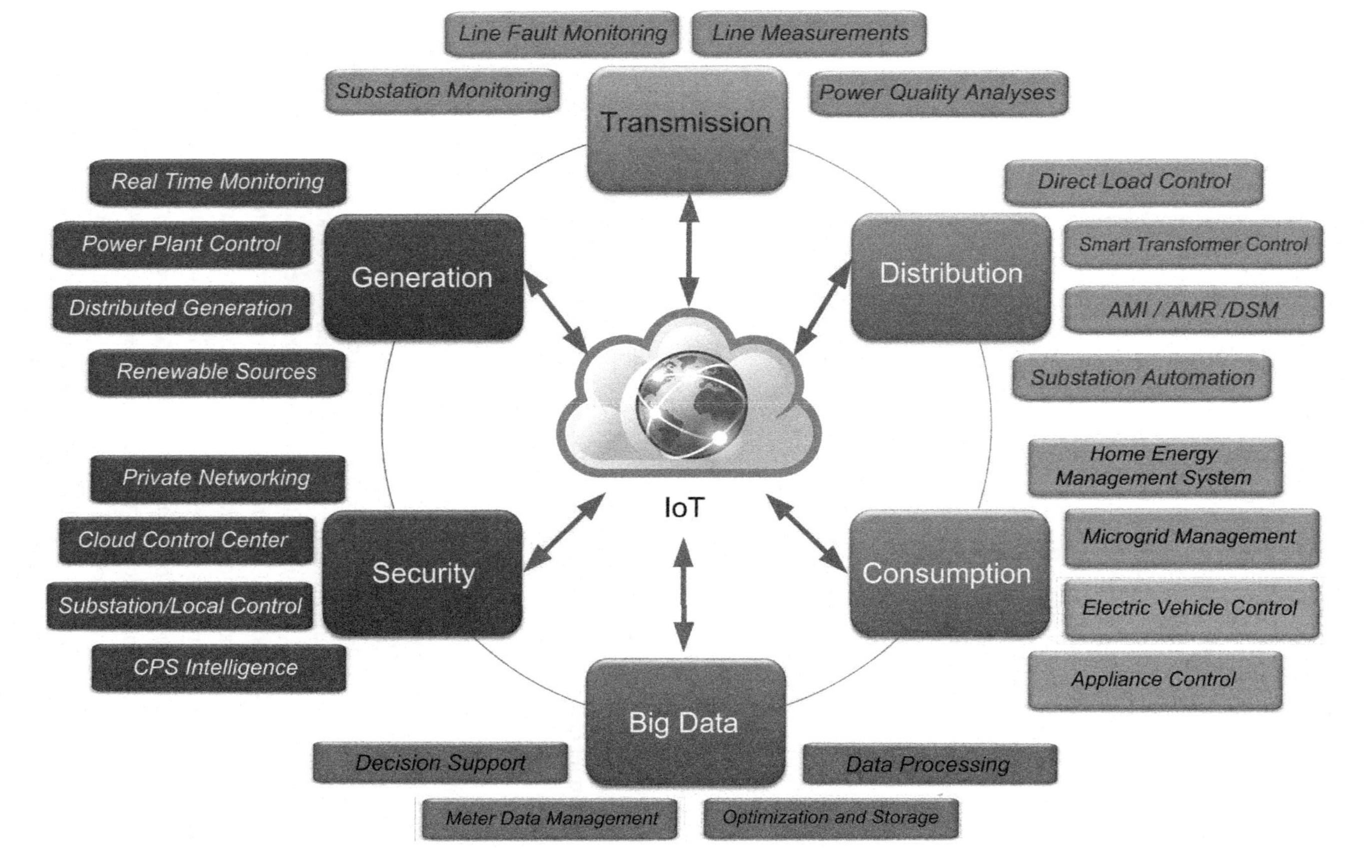

FIG. 7.12 The relationship of IoT applications along smart grid segments and layers.

capabilities to smart grid integration. It is seen that IoT-based applications are extensively operated on clouds, but colossal data sets require efficient storage, processing, and analyze features. Lin et al. express that cloud computing is a mature technology and fog/edge computing-based IoT is an emerging architectural approach in terms of improvement [70]. The fog/edge computing can install a connection between cloud and IoT applications. Thus, the applications and services can be done at nodes, things level, and distributed networks instead of centralized structure of cloud. The fog/edge architecture can be easily comprised by using any device and "thing" that is capable to provide storage, processing, communication, and management opportunities. The fog/edge approach can ensure the efficiency and privacy of smart meter deployment and smart city applications due to its distributed resource opportunities.

The researches on energy-efficient IoT architectures are state-of-the-art to manage resources and communication infrastructure with low-power consumption and highly efficient. The refined architectures that are composed by sensing and control layers, information processing layers, and application layers are being widely researched in the context of next generation IoT architectures. Therefore, it is predicted that there will be an intensive interest on architecture improvements of IoT and smart grid interaction.

7.6.2 IoT-based smart building management

It is noted that smart home management and smart building systems will be rapidly advanced with the aid of IoT in next 5 years [6]. In this concept, the most important contributions are expected to be provided by improvements of BMS, IPv6, 6LowPAN, cloud services, power over Ethernet, and smart meters. The home appliances and devices at household use comprise a WSN in the context of smart building management system. The internal WSN builds a CPS requiring smart grid and IoT applications to be managed by any device. Moreover, the regulations are increasingly focused on smart building management due to energy efficiency, DSM, and environmental concerns in recent years.

The control requirements on temperature, use of electricity and gas, surveillance systems, smart metering applications, and cost-efficiency have been demanded by users and authorities. Besides the home management systems, industrial plants and governmental buildings are also taken into account as smart building management concerns. The increased communication opportunities and cloud support facilitates integration of building management systems to existing infrastructures by using network devices. The smart meters are obviously most important device in this evolving management system. The interoperability and security are crucial topics in smart building management systems where the authentication and privacy-aware architectures are required. The surveyed studies express that microgrid management, appliance control, EV control and monitoring researches are also ever-evolving systems in the context of

IoT-based smart grid applications. The intelligent control features are defined as one of the research directions in the smart building management topics.

7.6.3 Cyber security and privacy in IoT architecture

The authorization, authentication, encryption, PKI, trust and identity control are key topics that have been determined according to researches in the context of cyber security and privacy issues of IoT. The security requirements of an IoT-based system are comprised by identifying the IoT devices and things, protecting the data transmitted along communication line, and preventing intrusions to the system. The authorization mechanism provides integrity and identification of users and things in the IoT based smart grid ecosystem. The encryption mechanism and PKI are essential on confidentiality requirements. On the other hand, the trust and identity control mechanism meet the integrity requirements either data or user is trusted [6].

The conducted researches also revealed that a secure infrastructure is required for smart grid applications such as smart cities, smart homes, energy generation, monitoring, measurement, and management applications. Since any subnetwork within smart grid infrastructure presents a heterogeneous structure with its own tools, capabilities, and requirements, a secure communication framework interoperating all components is required. There a wide variety of researches that implemented ICT and CPS for heterogeneous smart grid networks are conducted in the literature. On the other hand, substation automation and AMI networks are expected to become much more challenging research areas in the context of smart grid application with IoT. The IEC61850 standard is assumed to interact with recent ICT technologies in terms of TCP/IP and PHY-MAC layer protocols. We believe in that the assumed challenges and improvements will extremely facilitate security, connectivity, and interoperability of smart grid and IoT applications.

It is obvious that new technologies are required to manage colossal and large-scale systems comprised by IoT appliances in industrial, residential, and commercial environments. The CPS in industrial environments that are assumed as IIoT infrastructures require additional security and privacy due to critical loads are accommodated.

7.6.4 Big data and cloud computing

The data acquired from millions of devices should be stored and processed in an intelligently installed system. The colossal amount of data should be converted to meaningful information by using several data filtering, elimination, and aggregation processes. The data filtering operations require a specifically designed interface like middleware designs. The big data operations are based on several proactive and reactive events in the data clouds of IoT systems. Therefore, big data operations require featured and tailored middleware to interconnect IoT and cloud servers. The agent-based, virtual-machine based, and special application middleware provide several benefits on big data management and processing.

The data acquisition which is crucial for of big data and cloud operation is initial step. Due to complexity of infrastructures and interoperability requirements, it is noted that interdisciplinary researches are requested in this context. These issues are presented in the next chapter with details.

7.6.5 Computational methods for IoT-based smart grid

The huge amount of IoT data acquisition and management require high level processing capabilities for data filtering, predicting, decision making mechanisms. The recent term for this mechanism is known as data fusion. The computational methods such as fuzzy logic controllers (FLCs), artificial neural networks (ANNs), machine learning systems, and data mining middleware are used to derive meaningful results from crowded and imprecise data stacks. The computational methods provide adaptive learning and self-organization features to prevent faults. There have been several studies surveyed on IoT-based smart grid applications including wind speed forecasting, WSN data fusion operations, and intelligent methods using supervised machine learning approach for increasing reliability and accuracy of data processing.

FLC of artificial intelligence has been proposed for data fusion of WSN. It proposes an essential infrastructure to process data acquired from a wide variety of sensors and performs successful data fusion. The data mining methods provides benefits of data awareness, data reliability, qualified results, and trusted data filtering.

7.6.6 Standardization and interoperability of IoT-based smart grid

IoT is being used in a broad area including industry, power networks, and daily life. Moreover, it is forecasted improvements of IoT technologies will continue for many years. The literature surveys we conducted have shown that predictions indicate that IoT will encompass more than 50 billion devices all over the world by 2020. Many IoT applications and services have been adopted several smart grid applications including smart city, smart power networks, smart building management systems, and ICT infrastructures. IoT is an enabling technology connecting several other technologies such as communication networks, ICT systems, data acquisition and sensor technologies, control systems, application and services. Moreover, cellular technologies and big data play crucial roles in development of IoT and smart grid integration.

Standardization is a crucial issue for such a dynamic and rapidly evolving technology. The prominent standards on IoT reference architecture and conceptual models are described by several organizations as IEEE, ISO, IEC, ITU, ETSI, 3GPP, W3C, and so on. The ISO/IEC 30141 standard defines IoT system characteristics in eight issues as auto-configuration, function and management capabilities separation, highly distributed systems, network communication, network management and operation, real-time capability, self-description, and service subscription. On the other hand, IoT service characteristics are described as content-awareness, context-awareness, and timeliness that content-awareness describes the property of being aware of information in an IoT component and its associated metadata. The content-awareness property enables devices and services to adapt interfaces, abstract application data, improve information retrieval, discover services, and appropriate user interactions. The context-awareness defines the property of an IoT device, service or system being able to monitor its own operating environment. The determined information on events within that environment are provided in when (time awareness), where (location awareness), or in what order (awareness of sequence of events). The context-awareness enables flexible, user-customized and autonomic services based on the related context of IoT components and/or users. Context information is used as the basis for taking actions in response to observations, possibly through the use of sensor information and actuators. To fully utilize an observation and effect an action, the understanding of context is often critical.

IoT component characteristics are described as composability, discoverability, modularity, network connectivity, shareability, unique identification. Besides these characteristics. On the other hand, IoT system requirements are listed as compatibility, usability, flexibility manageability robustness security and protection of personally identifiable information that all defined property are described to ensure interoperability of heterogenous technologies within a unique infrastructure. To address the standardization and interoperability challenges, ISO/IEC/IEEE 42010:2011 standard has been described that is also used by Industrial Internet Consortium to define Industrial Internet Architecture Framework (IIAF) [32, 70, 71]. Although there are many researches have presented descriptions on interoperability, the improved applications and services should comply with aforementioned standards.

References

[1] Y. Kabalci, A survey on smart metering and smart grid communication. Renew. Sust. Energ. Rev. 57 (2016) 302–318, https://doi.org/10.1016/j.rser.2015.12.114.

[2] X. Fang, S. Misra, G. Xue, D. Yang, Smart grid-the new and improved power grid: a survey. IEEE Commun. Surv. Tutorials 14 (2012) 944–980, https://doi.org/10.1109/SURV.2011.101911.00087.

[3] L.M. Camarinha-Matos, Collaborative smart grids—a survey on trends. Renew. Sust. Energ. Rev. 65 (2016) 283–294, https://doi.org/10.1016/j.rser.2016.06.093.

[4] IEEE Standards Committee, IEEE guide for smart grid interoperability of energy technology and information technology operation with the electric power system (EPS), end-use applications and loads, Institute of Electrical and Electronics Engineers, New York, NY, 2011.http://ieeexplore.ieee.org/servlet/opac?punumber=6018237. Accessed 5 September 2017.

[5] IEEE Standards Association—Working Group Site & Liaison Index. Standard for an Architectural Framework for the Internet of Things IEEE P2413, n.d. http://grouper.ieee.org/groups/2413/ (Accessed 6 September 2017).

[6] D. Minoli, K. Sohraby, B. Occhiogrosso, IoT considerations, requirements, and architectures for smart buildings—energy optimization and next generation building management systems. IEEE Internet Things J. (2017) 269–283, https://doi.org/10.1109/JIOT.2017.2647881.

[7] C. Arcadius Tokognon, B. Gao, G.Y. Tian, Y. Yan, Structural health monitoring framework based on internet of things: a survey. IEEE Internet Things J. 4 (2017) 619–635, https://doi.org/10.1109/JIOT.2017.2664072.

[8] M.A. Razzaque, M. Milojevic-Jevric, A. Palade, S. Clarke, Middleware for internet of things: a survey. IEEE Internet Things J. 3 (2016) 70–95, https://doi.org/10.1109/JIOT.2015.2498900.

[9] M. Rana, Architecture of the internet of energy network: an application to smart grid communications. IEEE Access 5 (2017) 4704–4710, https://doi.org/10.1109/ACCESS.2017.2683503.

[10] A. Keyhani, A. Chatterjee, Automatic generation control structure for smart power grids, IEEE Trans. Smart Grid 3 (2012) 1310–1316.

[11] H. Lu, L. Zhan, Y. Liu, W. Gao, A microgrid monitoring system over mobile platforms. IEEE Trans. Smart Grid (2016) 749–758, https://doi.org/10.1109/TSG.2015.2510974.

[12] P. Garcia, P. Arboleya, B. Mohamed, A.A.C. Vega, Implementation of a hybrid distributed/centralized real-time monitoring system for a DC/AC microgrid with energy storage capabilities. IEEE Trans. Ind. Inf. 12 (2016) 1900–1909, https://doi.org/10.1109/TII.2016.2574999.

[13] M. Firouzi, G.B. Gharehpetian, B. Mozafari, Power-flow control and short-circuit current limitation of wind farms using unified interphase power controller. IEEE Trans. Power Delivery 32 (2017) 62–71, https://doi.org/10.1109/TPWRD.2016.2585578.

[14] S. Howell, Y. Rezgui, J.-L. Hippolyte, B. Jayan, H. Li, Towards the next generation of smart grids: semantic and holonic multi-agent management of distributed energy resources. Renew. Sust. Energ. Rev. 77 (2017) 193–214, https://doi.org/10.1016/j.rser.2017.03.107.

[15] A. Cataliotti, V. Cosentino, D. Di Cara, P. Russotto, E. Telaretti, G. Tine, An innovative measurement approach for load flow analysis in MV smart grids. IEEE Trans. Smart Grid 7 (2016) 889–896, https://doi.org/10.1109/TSG.2015.2430891.

[16] N. Moreira, E. Molina, J. Lázaro, E. Jacob, A. Astarloa, Cyber-security in substation automation systems. Renew. Sust. Energ. Rev. 54 (2016) 1552–1562, https://doi.org/10.1016/j.rser.2015.10.124.

[17] S.S. Mousavi-Seyedi, F. Aminifar, S. Afsharnia, Application of WAMS and SCADA data to online modeling of series-compensated transmission lines. IEEE Trans. Smart Grid 8 (2017) 1968–1976, https://doi.org/10.1109/TSG.2015.2513378.

[18] M. Liserre, G. Buticchi, M. Andresen, G. De Carne, L.F. Costa, Z.-X. Zou, The smart transformer: impact on the electric grid and technology challenges. IEEE Ind. Electron. Mag. 10 (2016) 46–58, https://doi.org/10.1109/MIE.2016.2551418.

[19] L. Ferreira Costa, G. De Carne, G. Buticchi, M. Liserre, The smart transformer: a solid-state transformer tailored to provide ancillary services to the distribution grid. IEEE Power Electr. Mag. 4 (2017) 56–67, https://doi.org/10.1109/MPEL.2017.2692381.

[20] E. Inga, S. Cespedes, R. Hincapie, C.A. Cardenas, Scalable route map for advanced metering infrastructure based on optimal routing of wireless heterogeneous networks. IEEE Wirel. Commun. 24 (2017) 26–33, https://doi.org/10.1109/MWC.2017.1600255.

[21] E. Avelar, L. Marques, D. dos Passos, R. Macedo, K. Dias, M. Nogueira, Interoperability issues on heterogeneous wireless communication for smart cities. Comput. Commun. 58 (2015) 4–15, https://doi.org/10.1016/j.comcom.2014.07.005.

[22] N. Javaid, I. Ullah, M. Akbar, Z. Iqbal, F.A. Khan, N. Alrajeh, M.S. Alabed, An intelligent load management system with renewable energy integration for smart homes. IEEE Access 5 (2017) 13587–13600, https://doi.org/10.1109/ACCESS.2017.2715225.

[23] A. Kaur, J. Kaushal, P. Basak, A review on microgrid central controller. Renew. Sust. Energ. Rev. 55 (2016) 338–345, https://doi.org/10.1016/j.rser.2015.10.141.

[24] N. Mahmud, A. Zahedi, Review of control strategies for voltage regulation of the smart distribution network with high penetration of renewable distributed generation. Renew. Sust. Energ. Rev. 64 (2016) 582–595, https://doi.org/10.1016/j.rser.2016.06.030.

[25] Y. Kabalci, E. Kabalci, Modeling and analysis of a smart grid monitoring system for renewable energy sources. Sol. Energy 153 (2017) 262–275, https://doi.org/10.1016/j.solener.2017.05.063.

[26] E. Kabalci, A smart monitoring infrastructure design for distributed renewable energy systems. Energy Convers. Manag. 90 (2015) 336–346, https://doi.org/10.1016/j.enconman.2014.10.062.

[27] C. Gamarra, J.M. Guerrero, Computational optimization techniques applied to microgrids planning: a review. Renew. Sust. Energ. Rev. 48 (2015) 413–424, https://doi.org/10.1016/j.rser.2015.04.025.

[28] S.E. Bibri, The IoT for smart sustainable cities of the future: an analytical framework for sensor-based big data applications for environmental sustainability. Sustain. Cities Soc. 38 (2018) 230–253, https://doi.org/10.1016/j.scs.2017.12.034.

[29] A. Zanella, N. Bui, A. Castellani, L. Vangelista, M. Zorzi, Internet of things for smart cities. IEEE Internet Things J. 1 (2014) 22–32, https://doi.org/10.1109/JIOT.2014.2306328.

[30] Q. Ou, Y. Zhen, X. Li, Y. Zhang, L. Zeng, Application of internet of things in smart grid power transmission. in: IEEE, Vancouver, BC, 2012, pp. 96–100, https://doi.org/10.1109/MUSIC.2012.24.

[31] F. Mattern, C. Floerkemeier, From the Internet of Computers to the Internet of Things, Act. Data Manag. Event-Based Syst. More.(2010) pp. 242–259.

[32] A. Al-Fuqaha, M. Guizani, M. Mohammadi, M. Aledhari, M. Ayyash, Internet of things: a survey on enabling technologies, protocols, and applications. IEEE Commun. Surv. Tutorial 17 (2015) 2347–2376, https://doi.org/10.1109/COMST.2015.2444095.

[33] P.P. Ray, A survey on internet of things architectures. J. King Saud. Univ. Comput. Inf. Sci. (2016)https://doi.org/10.1016/j.jksuci.2016.10.003.

[34] X. Xiaojiang, W. Jianli, L. Mingdong, Services and key technologies of the internet of things, ZTE Commun. 8 (2010) 26–29.

[35] M. Gigli, S. Koo, Internet of things: services and applications categorization. Adv. Internet Things 01 (2011) 27–31, https://doi.org/10.4236/ait.2011.12004.

[36] M. Jaradat, M. Jarrah, A. Bousselham, Y. Jararweh, M. Al-Ayyoub, The internet of energy: smart sensor networks and big data management for smart grid. Procedia Comput. Sci. 56 (2015) 592–597, https://doi.org/10.1016/j.procs.2015.07.250.

[37] Z. Guan, J. Li, L. Wu, Y. Zhang, J. Wu, X. Du, Achieving efficient and secure data acquisition for cloud-supported internet of things in smart grid. IEEE Internet Things J. 4 (2017) 1934–1944, https://doi.org/10.1109/JIOT.2017.2690522.

[38] N. Modadugu, E. Rescorla, The Design and Implementation of Datagram TLS, NDSS, San Diego, CA, 2004. http://www.isoc.org/isoc/conferences/ndss/04/proceedings/Papers/Modadugu.pdf. Accessed 7 September 2017.

[39] R. Kopmeiners, P. King, J. Fry, J. Lilleyman, S. Lancashire, D. Liu, A Standardized and Flexible IPv6 Architecture for Field Area Networks: Smart-Grid Last-Mile Infrastructure, Cisco Systems Incorporation, 2014.http://www.cs.cmu.edu/afs/.cs.cmu.edu/Web/People/jorjeta/Papers/ip_arch_sg_wp.pdf. Accessed 14 July 2017.

[40] Z. Zhou, J. Gong, Y. He, Y. Zhang, Software defined machine-to-machine communication for smart energy management. IEEE Commun. Mag. 55 (2017) 52–60, https://doi.org/10.1109/MCOM.2017.1700169.

[41] N. Chen, M. Wang, N. Zhang, X.S. Shen, D. Zhao, SDN-based framework for the PEV integrated smart grid. IEEE Netw. 31 (2017) 14–21, https://doi.org/10.1109/MNET.2017.1600212NM.

[42] Q. Wang, J. Jiang, Comparative examination on architecture and protocol of industrial wireless sensor network standards. IEEE Commun. Surv. Tutorial 18 (2016) 2197–2219, https://doi.org/10.1109/COMST.2016.2548360.

[43] T. Kim, I.H. Kim, Y. Sun, Z. Jin, Physical layer and medium access control design in energy efficient sensor networks: an overview. IEEE Trans. Ind. Inf. 11 (2015) 2–15, https://doi.org/10.1109/TII.2014.2379511.

[44] I.-G. Lee, M. Kim, Interference-aware self-optimizing Wi-Fi for high efficiency internet of things in dense networks. Comput. Commun. 89–90 (2016) 60–74, https://doi.org/10.1016/j.comcom.2016.03.008.

[45] U. Raza, P. Kulkarni, M. Sooriyabandara, Low power wide area networks: an overview. IEEE Commun. Surv. Tutorial 19 (2017) 855–873, https://doi.org/10.1109/COMST.2017.2652320.

[46] Y. Trivedi, Ethernet, the networking standard: more mature, more powerful where the whole world is going with ethernet, IEEE Commun. Mag. 54 (2016) 5–11.

[47] P. Gandotra, R.K. Jha, Device-to-device communication in cellular networks: a survey. J. Netw. Comput. Appl. 71 (2016) 99–117, https://doi.org/10.1016/j.jnca.2016.06.004.

[48] Á. Asensio, Á. Marco, R. Blasco, R. Casas, Protocol and architecture to bring things into internet of things. Int. J. Distrib. Sens. Netw. 10 (2014)https://doi.org/10.1155/2014/158252.

[49] O. Vermesan, P. Friess, Internet of Things—From Research and Innovation to Market Deployment, River Publishers, Denmark, 2013.

[50] O. ITU, Overview of Ubiquitous Networking and of Its Support in NGN, Geneva, Switzerlandhttp://www.cttl.cn/itu/itubz/itut/201203/P020120319759043977765.doc, 2010. Accessed 11 September 2017.

[51] J. Granjal, E. Monteiro, J. Sa Silva, Security for the internet of things: a survey of existing protocols and open research issues. IEEE Commun. Surv. Tutorial 17 (2015) 1294–1312, https://doi.org/10.1109/COMST.2015.2388550.

[52] S. Raza, S. Duquennoy, J. Höglund, U. Roedig, T. Voigt, Secure communication for the internet of things-a comparison of link-layer security and IPsec for 6LoWPAN: secure communication for the internet of things. Secur. Commun. Netw. 7 (2014) 2654–2668, https://doi.org/10.1002/sec.406.

[53] M. Zheng, W. Liang, H. Yu, Y. Xiao, Performance analysis of the industrial wireless networks standard: WIA-PA. Mob. Netw. Appl. 22 (2017) 139–150, https://doi.org/10.1007/s11036-015-0647-7.

[54] W. Yang, M. Wang, J. Zhang, J. Zou, M. Hua, T. Xia, X. You, Narrowband wireless access for low-power massive internet of things: a bandwidth perspective. IEEE Wirel. Commun. 24 (2017) 138–145, https://doi.org/10.1109/MWC.2017.1600298.

[55] Y.D. Beyene, R. Jantti, O. Tirkkonen, K. Ruttik, S. Iraji, A. Larmo, T. Tirronen, J. Torsner, NB-IoT technology overview and experience from cloud-RAN implementation. IEEE Wirel. Commun. 24 (2017) 26–32, https://doi.org/10.1109/MWC.2017.1600418.

[56] S. Fang, X. Li Da, Y. Zhu, J. Ahati, H. Pei, J. Yan, Z. Liu, An integrated system for regional environmental monitoring and management based on internet of things. IEEE Trans. Ind. Inf. 10 (2014) 1596–1605, https://doi.org/10.1109/TII.2014.2302638.

[57] S.D.T. Kelly, N.K. Suryadevara, S.C. Mukhopadhyay, Towards the implementation of IoT for environmental condition monitoring in homes. IEEE Sensors J. 13 (2013) 3846–3853, https://doi.org/10.1109/JSEN.2013.2263379.

[58] N. Bui, A.P. Castellani, P. Casari, M. Zorzi, The internet of energy: a web-enabled smart grid system, IEEE Netw. 26 (2012) 39–45.

[59] P. Lynggaard, K. Skouby, Complex IoT systems as enablers for smart homes in a smart city vision. Sensors 16 (2016) 1840, https://doi.org/10.3390/s16111840.

[60] S. Mitchell, N. Villa, M. Stewart-Weeks, A. Lange, The Internet of Everything for Cities, Cisco, California, 2013.http://pie.pascalobservatory.org/sites/default/files/ioe-smart-city_pov.pdf. Accessed 17 September 2017.

[61] B. Hafidh, H. Al Osman, J.S. Arteaga-Falconi, H. Dong, A. El Saddik, SITE: The simple internet of things enabler for smart homes. IEEE Access 5 (2017) 2034–2049, https://doi.org/10.1109/ACCESS.2017.2653079.

[62] S.K. Viswanath, C. Yuen, W. Tushar, W.-T. Li, C.-K. Wen, K. Hu, C. Chen, X. Liu, System design of the internet of things for residential smart grid, IEEE Wirel. Commun. 23 (2016) 90–98.

[63] E. Spano, L. Niccolini, S.D. Pascoli, G. Iannaccone, Last-meter smart grid embedded in an internet-of-things platform. IEEE Trans. Smart Grid 6 (2015) 468–476, https://doi.org/10.1109/TSG.2014.2342796.

[64] M.M.E.A. Mahmoud, N. Saputro, P.K. Akula, K. Akkaya, Privacy-preserving power injection over a hybrid AMI/LTE smart grid network. IEEE Internet Things J. 4 (2017) 870–880, https://doi.org/10.1109/JIOT.2016.2593453.

[65] Q. Sun, H. Li, Z. Ma, C. Wang, J. Campillo, Q. Zhang, F. Wallin, J. Guo, A comprehensive review of smart energy meters in intelligent energy networks. IEEE Internet Things J. 3 (2016) 464–479, https://doi.org/10.1109/JIOT.2015.2512325.

[66] A.F.A. Aziz, S.N. Khalid, M.W. Mustafa, H. Shareef, G. Aliyu, Artificial intelligent meter development based on advanced metering infrastructure technology. Renew. Sust. Energ. Rev. 27 (2013) 191–197, https://doi.org/10.1016/j.rser.2013.06.051.

[67] Y. Kabalcı, E. Kabalcı, Design and implementation of wireless energy monitoring system for smart grids, Gazi Univ. J. Sci. Part C 5 (2017) 137–145.

[68] E. Kabalci, Y. Kabalci, A measurement and power line communication system design for renewable smart grids. Meas. Sci. Rev. 13 (2013)https://doi.org/10.2478/msr-2013-0037.

[69] M. Armendariz, C. Johansson, L. Nordström, A. Yunta Huete, M. García Lobo, Method to design optimal communication architectures in advanced metering infrastructures. IET Gener. Transm. Distrib. 11 (2017) 339–346, https://doi.org/10.1049/iet-gtd.2016.0481.

[70] J. Lin, W. Yu, N. Zhang, X. Yang, H. Zhang, W. Zhao, A survey on internet of things: architecture, enabling technologies, security and privacy, and applications. IEEE Internet Things J. 4 (2017) 1125–1142, https://doi.org/10.1109/JIOT.2017.2683200.

[71] A. Rayes, S. Salam, The internet in IoT—OSI, TCP/IP, IPv4, IPv6 and internet routing. in: Internet Things Hype Real, Springer International Publishing, Cham, 2017, pp. 35–56, https://doi.org/10.1007/978-3-319-44860-2_2.

Chapter 8

Big data, privacy and security in smart grids

Chapter outline

8.1 Introduction

The big data is a recent and trend term referring to data mass obtained from several digital sources such as sensors, transducers, mobile devices and computers, internet, and so on. The rapid improvement of sensor technologies, wireless sensor networks and digital media, huge amounts of dataset have been generated by any communication platform. In addition to data types and high volumes, the raw data collection produces enormous data sizes that are required to be analyzed and meaningful outcomes to be generated. The conventional data processing methods such as model based analysis and decoupling systems are based on assumptions and summarizing approaches. The developments of digital technologies and artificial intelligence have leveraged data processing procedures from conventional approaches to much more accelerated and sophisticated processing systems. The data processing algorithms of recent technologies are based on 4Vs or 5Vs features of data that are volume, velocity, variety, veracity and value in big data processing.

The volume of data which is first V in this approach grows exponentially due to massive data sources and generates excessive databases that are hard to be processed by using traditional methods. The second V, velocity, implies generation and data transmission speed along internet-based sources. Variety denotes diversity of data types and forms while veracity refers to quality, accuracy, and

From Smart Grid to Internet of Energy. https://doi.org/10.1016/B978-0-12-819710-3.00008-9

reliability of the data sets. All of these Vs are assumed as crucial parameters to achieve the last V which is value providing decision making and success of big data evaluation [1, 2].

It is noted in [3] that International Data Corporation has reported that 1.8ZB data was generated and reproduced in 2011 that is expected to be increased around 50 times up to 2020. The generated data size is also increased by the improvement of smart grid technologies since they are based on widespread data acquisition applications inherited over wireless sensor networks (WSNs). The entire power network is equipped with several types of sensor for instant detection of generation, transmission, distribution, and consumption data. The vital services and components such as demand side management (DSM), distributed energy resources (DERs), renewable energy sources (RESs) and others of power networks require some detailed and rapid monitoring infrastructures. The digitalization of power networks causes several challenges as economic operation, control issues, stability and reliability of system.

The big data applications of networks bring new opportunities in terms of smart energy management and incorporates several operational capabilities. The smart grid is defined as system of systems integrating two-way flow of communication and power. Therefore, large amount of measurement, monitoring and control signals are required to be transmitted over smart grid infrastructure. The generated and transmitted data sets are utilized to perform monitoring, optimization, forecasting, planning and management issues. Thereby, big data analysis methods play vital role on management and processing of large amount of data generated by smart grid infrastructure. It provides rapid detection against failures, dynamic system restoration, rapid response to load and source fluctuations, and increasing the reliability and flexibility of entire grid [3].

It is obvious that smart grid is advancing on green and environmentally friendly generation of energy. It can be roughly said that the most important components differing smart grid from conventional one is measurement and monitoring devices. It is reported in [4] that the phasor measurement units (PMUs) which are crucial on detecting magnitude and phase of a power network generate around 900 TB data per year. On the other hand, smart meters, WSNs that are used for estimation and predictions, intelligent electronic devices (IEDs), smart transformers, smart reclosers and circuit breakers generate large amounts of data. Therefore, big data and data analytic methods can leverage advancements of smart grid by providing various opportunities on load planning, demand management, forecasting, and data analytics. The applications and data analytics of smart grid that are based on big data processing operations are illustrated in Fig. 8.1. The big data analytics are listed into two main categories as grid operation and customer operations. The storage cycle of big data servers is provided by several data suppliers from generation, transmission, distribution and consumption systems. Moreover, the DER systems are

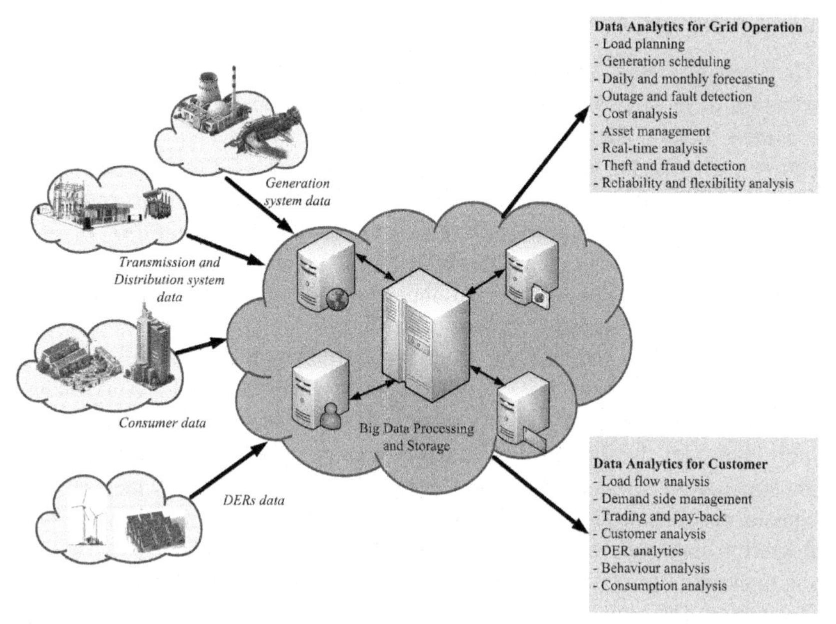

FIG. 8.1 Big data analytics in the smart grid.

also integrated to big data system in order to sustain data acquisition procedures. The data analytics of smart grid which are listed into two fundamental aspects as grid operation and customer operations. The data analytics required for grid operation include load planning, generation scheduling, daily and monthly forecasting for energy demand, outage and fault detection, cost analysis, asset management, real-time analysis, theft and fraud detection, reliability and flexibility measurements. The data analytics of customer side applications are required to detect load flow, demand side management, trading and pay-back operations, customer behavior analysis, customer DER monitoring, and consumption analytics [4].

The big data analytics provide more precise results for management and decision-making issues in smart grid applications. Although it enables to reach several opportunities, big data brings several challenges on smart grid plane.

The main concerns are related with efficient data acquisition, storage and management, analyzing and mining the collected data, producing meaningful outcomes, and protection against vulnerabilities and intrusions for ensuring the privacy. This chapter deals with big data properties, data analysis methods, and their application in smart gird infrastructure. The big data privacy and smart grid security are presented in the following titles where particular threats, challenges, and privacy preserving applications are discussed in detail.

8.2 Overview of big data

It is noted that Big data has been proposed by McKinsey in a report with definition of its size, storage, and management methods. It has been accepted as a new method facilitating technological innovations and leveraging the economic growth. On the other hand, big data does not refer to just data masses of several TBs. The fundamental approach behind the data processing is generation and distribution of smart intelligence among huge numbers of computers, server, and service providers. The big data analysis cycle is comprised by data acquisition, data storage and management, data modeling, data mining and data processing steps [4, 5]. One of the most widely used open-source data analysis tool is known as Hadoop which is benefited by Siemens, ABB, General Electric, IBM, Facebook, Microsoft, Google, HP, Yahoo, Netflix, Amazon and many more [4]. The popular application of Hadoop is MapReduce technology and it facilitates the data analysis among large amounts of databases. The big data applications of smart grid are defined by seven fundamental tasks as data acquisition, transmission and storage; data cleaning and preprocessing; data integration and selection; data mining and discovery; representation, visualization, application; decision making and real-time processing; and smart energy management. The quality of different smart grid applications such as management, power generation and transmission control, and DSM are certainly improved by big data analytics [3]. The technological infrastructure of big data processing stages is illustrated in Fig. 8.2 starting from data acquisition to security end node. The inherited data are processed at each feedforward stage and fundamental technologies are used to ensure reliability and security of process.

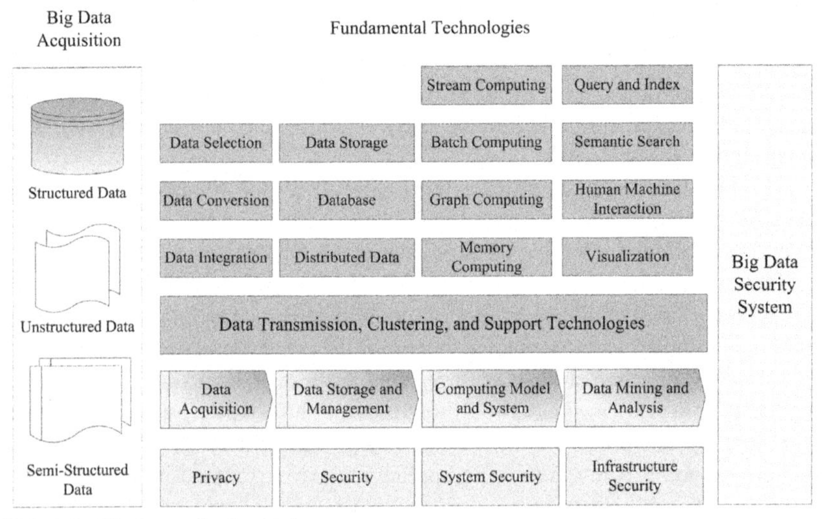

FIG. 8.2 Big data technology infrastructure.

Although the challenges and technical deficiencies, decision-making methods based on data inheritance are widely accepted by authorities. The data is a strategic component generated by technical and natural resources. The collected data are not usually ready for processing and defined as raw-data that are required to be located, identified, understood, and prepared for effective processing. At first step, data integration and cleaning are required to convert inherited raw data for storage. The big data differs from conventional data management systems due to their heterogenous formats such as structured, unstructured and semi-structured data sets as seen in Fig. 8.2. It is noted by reports that nearly 85% of inherited data are semi-structured or unstructured that are treated by nonrelational analytic technologies such as MapReduce or Hadoop. The three Vs which are volume, velocity, and variety are very important among others for data analytics in big data. The data volume is enormously growing year by year and it is expected to reach up to 40 Zeta bytes (ZB) until 2020. Therefore, the velocity of data acquisition and processing should be as fast as volume of growing data size. On the other hand, the variety of data is another interest of big data researches since the data types and databases are differing in terms of structured or unstructured, public or private, shared or confidential types [6, 7].

In addition to challenges in data acquisition, big data applications bring several problems on generating correct metadata which is related with processing the acquired and stored data. The data analysis challenges are tackled by using sophisticated data mining techniques that provide to discover integrated, meaningful, clear and accessible data stacks. The gradually increased data sizes and volumes force researchers to improve computational methods for efficient data processing processes. The big data analytics require some efforts such as integration of massive data types with data correlation procedures, reliable and rapid processing models, real time processing and sampling capabilities of processors, and interactive user interfaces for managing the data processing ecosystem. The data processing operations are based on utilization of linear equation solvers, optimization algorithms, linear and nonlinear prediction procedures such as Wiener and Kalman filters, canonical correlation analysis, linear discriminant analysis, and adaptive sampling processes such as belief propagation, sensing, and k-nearest neighbor algorithms [6]. The stages of big data processes are presented in the following sections according to data generation, data acquisition and storage, machine learning methods, and Internet of Things (IoT) applications in big data ecosystem.

8.2.1 Big data generation

The data generation is preliminary step of big data operations. The critical applications, measurement and control devices, ICT interfaces used in smart grid and smart sensors generate the highest share of big data in smart grid applications. The big data of smart grid is a combination of all the inherited data from smart infrastructures. It is also related with operation and trading requirements of companies, enterprises, and consumer interactions. The source system-based data are generated as raw-data and it is required to handle these data stacks with extraction. Most of the challenges met in big data processing stages start with extraction, transformation, and loading processes. Once the data is extracted, it is filtered and normalized in transformation step, and then it is loaded to system for processing [8–10].

The big data analytics stages are listed as acquisition, extracting/cleaning, integration/aggregation, and interpretation. The big data generation sources can be listed as enterprise level, IoT, internet data, biomedical sources, and other generation sources as communication, experimental fields, and multimedia fields. The reliable and efficient result generation is depended to cleaning of big data. Another important issue is synchronizing the data sources and big data processing platforms with internal organization [7].

8.2.2 Data acquisition and storage

The second stage of big data infrastructure is big data acquisition and storage applications which include data capture, transmission, and pre-processing. When the data is acquired from its source, it is required to be transmitted to data storage centers to ensure operation of analytical methods. The acquired datasets may sometimes include unnecessary and unrelated data, and this causes increments on storage volumes. Therefore, data compressing methods are convenient to be used for decreasing database sizes. On the other hand, pre-processing methods are used to improve data storage efficiency. The storage challenges require innovative solution against volume, velocity, and variety of acquired big data stacks. The conventional hard disks are not proper solution for these requirements. Thus, cloud computing and cloud storage opportunities are robust solutions to meet requirements of large amount of data systems.

The data transmission can be progressed in various stages of big data processing as: (i) transmission of acquired data to storage, (ii) integration of data acquired from multiple sources, (iii) management of integrated data, and (iv) transferring the stored data from storage to analysis server. The main data acquisition methods are defined as log files, sensors, network data. The log files are used to store regularly generated data, and they are integrated with all digital devices. The data acquisition is performed by using almost all digital devices such as internet servers, log files, sensors, and network devices. Sensors are widely preferred in big data transmission since they convert the physical amounts to readable and storage data. The WSNs provide widespread use of applications and are convenient devices for data transmission by using their multi-hop infrastructures. The raw data acquisition is followed by data transmission process where transferred and stored data are prepared for processing procedures [2, 8]. Machine learning, clustering and some other artificial intelligence methods are widely used for data analysis and processing procedures of big data stacks.

8.2.3 IoT and big data

The IoT is a comprehensive definition including all type of devices such as sensors, detectors, transducers, measurement devices, and interactive actuators which are connected to internet over similar or different network types. All such devices comprise intelligent or smart device interface on data acquisition and data processing planes. The "*things*" are convenient to inherit, transmit, share, and deploy the data thanks to internet infrastructure. It is known that IoT concept has been derived from preliminary applications of radio frequency identification (RFID) technology that has been firstly proposed by Ashton [11]. Nevertheless, the definition and conceptual environment of IoT have been gradually evolved day by day. The IoT defines interactive devices that are sensitive to their environment to detect changes, conditions, and operation circumstances, and they are able to interact with each other to inform entire network. IoT have found widespread usage areas including transportation, energy, health, security, public and social life, education, and communication applications. It is expected almost all electronic devices and components will be connected to internet and comply with IoT environment. There are several leading technologies such as human to machine (H2M) and machine to machine (M2M) communication, semantic sensor networks (SSNs), big data analytics, machine learning, and ubiquitous computing (UbiComp) providing progress of IoT [11, 12].

In addition to wireless connections, the wired communication methods are also used in IoT applications to solve daily life problems and to improve connectivity. The components of an IoT infrastructure comprise cyber-physical system (CPS) that includes hardware for data acquisition, data processing and software for transmission and data analytics. The information technologies are adopted to smart grid environment as well as other application areas of IoT due to large number of device usage. The heterogeneous device variety and generated massive data sets are handled by big data applications to produce meaningful data on demand, and to facilitate the selection of precise data among huge stacks. The big data is one of the state-of-the-art research areas on IoT aspects that includes volume, variety, velocity, and value of data. The generated massive datasets are handled by using sophisticated big data analysis methods such as data classification, regression, and clustering solutions to decrease communication and processing costs [13]. The required efficient data processing algorithms are presented in the following section in detail.

Smart grid integration is one of the recently widespread application of IoT. The large number of sensing nodes and measurement devices generate heterogenous data stacks for storing and processing at each monitoring interval. The inherited data are comprised by measurements of generated power rates, consumer energy demand data, network control data, DSM commands and programs, smart meter data, blackout and fault monitoring data, and many more related measurement and sensing data. The wide variety of application types and data acquisition frequencies force service providers to install high capability servers and workstations that are equipped with storage, processing, and management capacities. The supervisory control and data acquisition (SCADA) is one of the legacy communication and metering system which is used for long years in power networks. The SCADA systems inherit monitoring and measurement data over sensor nodes and transmit control signals from operational centers to transmission and distribution controllers. Installed infrastructures provide convenient control and monitoring capability for system operators to achieve DSM and distributed generation management. However, the increased data volumes and growing structure of power network require much more sophisticated and modernized information and communication technologies (ICTs) integrating legacy systems and cloud computing infrastructures [14].

The general architecture of IoT Big Data infrastructure is illustrated in Fig. 8.3. The fundamental services and functions of IoT bring several challenges for Big Data analytics. The prominent challenges are related with number and variety of IoT based devices, operational malfunction drawbacks of devices, data storage frequency, storage period, and privacy and security issues.

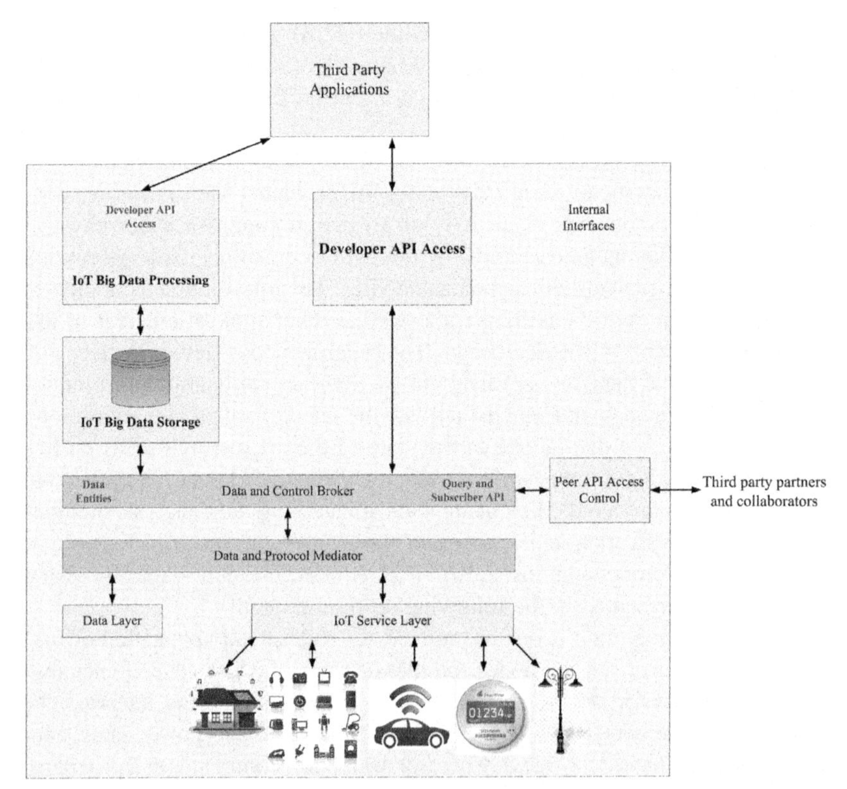

FIG. 8.3 IoT Big data infrastructure and architecture.

The IoT based device number is expected to reach up to billions that can transmit instant data. In addition to data volumes, variety and structures of transmitted data will be heterogeneous depending IoT device types, structures, and vendors. The IoT devices are foreseen to be converted from simple sensors to much more sophisticated systems which have their own processors and storage devices. Therefore, malfunction of such systems may cause to severe failures which cannot be handled with basic troubleshooting methods. On the other hand, the updating frequency of data acquisition is another challenging topic in IoT Big Data operations. The substantial part of IoT devices are assumed to transmit measurement and control signals with higher frequencies while some others such as cars and electric vehicles will generate data stacks at lower frequencies [15].

The IoT devices and sensors comprise entity definitions as shown in Fig. 8.3. The sensor-based communication infrastructure provides connection between devices and applications at IoT service layer.

The third-party applications are also integrated to IoT ecosystem over services interface. The interfaces are managed by application program interface (API) based software which provide authentication and security controls according to standards. The data layer applications are operated regarding to subscriber data and enable users to store data. The inherited data stacks come from several sources and devices and collected by Data and Protocol Mediator that is responsible for acquiring provided data of IoT based devices and other sources. The data and protocol mediator layer convert control demands and requests to low level IoT protocols. The IoT big data storage layer enables massive and high volume IoT data acquisition for storage processes, and it presents data for analytics and research purposes. The IoT big data processing section allows to use hardware and analytics tools to generate data reports and intelligence outputs.

The data storage capability of IoT devices will define flexibility of Big Data applications since it will allow or prevent data derivation from intelligent devices for smart monitoring and smart control operations. Another significant challenge is related with privacy issues against intrusions and violations. Once the data which is generated by IoT devices is stored in a Big Data analytic system, and presented to use for third parties, it forces operators and service providers to improve system security and authentication controls. The functional sections and layers of presented IoT Big Data architecture enable to obtain IoT and Big Data services in a wide range. These services include management functions, direct data control features, external data acquisition, establishing connection between IoT data platforms and devices, and delivery of IoT based data stacks. The Big Data analysis is based on data acquisition methods and data access opportunities. The function provides data storage feature such as storing the small scale data to hard disk drives or multiple disk usage depending the increased data volumes. The stored data stack should be planned by considering high volumes over than 50 TB in Big Data usage. On the other hand, database processing systems should consider to meet requirements of high volume of data, relations between data stacks, indexing methods such as column and table structures, and data transmission capability of devices. It is noted that high volume data of Big Data analytics are handled by using products such as Apache Cassandra, Hadoop, Mango DB and so on [15].

The integration of IoT technology with smart power network causes additional storage and processing cost due to massive volume of data inherited from every measurement and monitoring nodes. The generated data provides information about power demand of consumers, power line parameters, network situations, demand response programs, advanced metering data, DSM and outage control, and similar other information. Therefore, power system operators are faced to meet software and hardware requirements to handle storage, management and data processing duties. One of the most appropriate solution of this situation is merging Big Data analytics with IoT technologies in terms of power grid aspects. SCADA is one of the most widely used communication and measurement system in power networks due to its real time metering and control capabilities. The SCADA architecture is based on data acquisition over smart sensors and measurement nodes along transmission and distribution systems. The inherited data types are transmitted to control center by using highly reliable communication lines that are also used to transmit control commands to nodes of smart or controllable devices on the same basis. The increasing amount of power networks and distributed generation sources require improvements on installed and aging power grid architectures. Therefore, SCADA based legacy measurement and monitoring systems are being converted to IoT and emerging communication system by the help of Big Data analytics. The cloud computing methods provide appropriate solutions to IoT based measurement and management applications due to its storage, API, service, and network supports. It is noted in [14] that cloud computing operators provide three types of services named as Infrastructure as a Service (IaaS), Platform as a Service (PaaS), and Software as a Service (SaaS). The IaaS defines the environment that is comprised by operating systems, storage and network infrastructures, and database services through the cloud as its name implies. On the other hand, the PaaS accommodates programming and software development interfaces for end users, and provides several libraries in the cloud ecosystems, while SaaS presents APIs for end users. The storage support of IaaS services provides benefits to system operators in addition to excessive processing capabilities with low cost investments. On the other hand, security and reliability of measured data are ensured by IaaS services supported by cloud operators. The vulnerabilities of cloud computing which are caused by data share with other parties can be prevented by fog computing approaches. The fog computing increases system security by the devices located at the edges of network, and data transferring nodes are decreased by this way [14].

8.3 Big data analysis methods

The big data analytics define all processes and procedures for discovering the required data from databases by using particular methods, tools, and analyses. The amount of acquired data is gradually increasing due to spreading use of big data infrastructures, and improves the requirements of obtaining true data among stacks. One of the most important challenge in Big Data analytics is transforming relevant data to predicted outcomes and making the possible most appropriate decision. The fundamental hindrances are related by noisy, incorrect, and biased structure of obtained data. Therefore, the quality of data analytics is directly related with quality of the data. The reliability can be classified regarding to used data quality metrics that can provide information on how user data are precise. The evaluation of acquired data and data quality is the first step of detecting data extraction from massive stacks. The evaluation criteria should include accuracy, reliability, completeness, and firmness since each of these provide assessment information about structured databases. Ardagna et al. proposed a data quality service architecture which has been shown in Fig. 8.4 [16]. The data quality profiling and assessment section of proposed architecture is comprised by data quality profiler and source analyzer that are fed by several data sources. The second part of this module is data quality assessment section that is used by data quality adapter and custom settings blocks. The data quality profiling block defines required measurement and monitoring parameters to determine overall quality of inherited datasets. On the other hand, the data quality assessment block calculates data quality dimensions and concepts that are computed regarding to selected data stacks and quality metrics.

Obtained calculation result are saved to quality metadata stack that is used by data quality assessment section to be achieved on demand. The data quality service interface which is another feeder of data quality profiling and

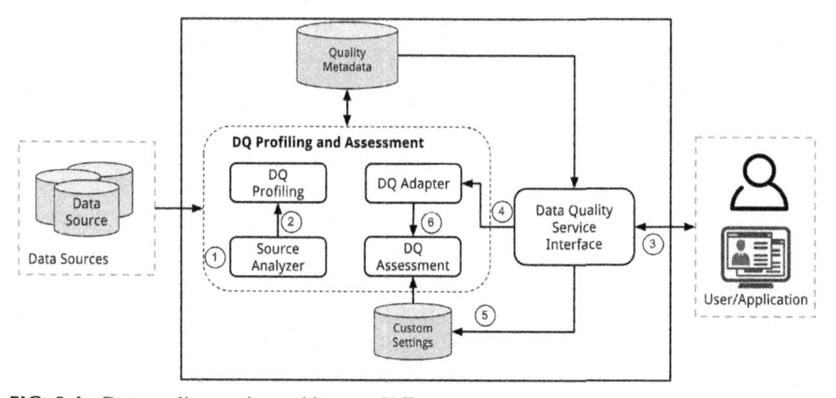

FIG. 8.4 Data quality service architecture [16].

assessment section is responsible for providing data quality service for users and applications, and supplies metadata on demand [16].

There are two scaling approaches that are vertical and horizontal are used to determine various structure of big data. The vertical scaling defines processing devices such as memory, processing units, and other computing devices that are based on single operating system. Therefore, entire workload can be shared with several processing units and parallel computing devices. The horizontal scaling is based on increased number of computing nodes that can be extended as required. The main drawback of horizontal scaling is its use of multiple operating systems due to parallel computing infrastructure. The two featured platforms that are used in vertical and horizontal scaling environments are high performance computing clusters and Apache Hadoop [17].

8.3.1 Data mining methods

It is clear that distributed computing methods can cope with massive databases and can handle data analytics on overloaded data stacks. The distributed computing architecture provides high capabilities in terms of storage and processing speed. In addition to resource contribution, distributed computing allows to use different machine learning algorithms and soft computing methods. Iosifidis et al. refers to kernel based learning algorithms for novel big data analytics approaches in [18]. On the other hand, the distributed analysis approaches are performed by several programming models such as MapReduce which is implemented by Google to cope with massive databases. Its name implies data processing steps as mapping at first step and reducing the computing requirements in the second step. The Hadoop based Apache Spark clustering is another approach to decrease complex structures and to increase the processing speed. The power and robust structure of Hadoop comes from two components that one is Hadoop Distributed File System (HDFS) and the other is MapReduce framework. In addition to these components, Directed Acyclic Graph (DAG) scheduling structure improves capability of Hadoop architecture [7, 18].

The data mining provides solutions to many difficulties of Big Data analytics such as data searching, capturing, management, and result generation. The conventional data management steps are listed as data cleaning, aggregation, encoding, data storage, and data access which are applicable in Big Data stacks. However, main challenge is how to manage the complexity of big data caused by 4Vs, and how to process them in a distributed processing infrastructure. The data scientists and researchers are focused on data acquisition, integration, storage, and processing massive data stacks with limited hardware and software requirements. The big data management refers obtaining clean data for reliable outcomes, aggregating various resource data and encoding capability to provide security and privacy on processing. Therefore, data management should be performed accessible, manageable, and secure ways.

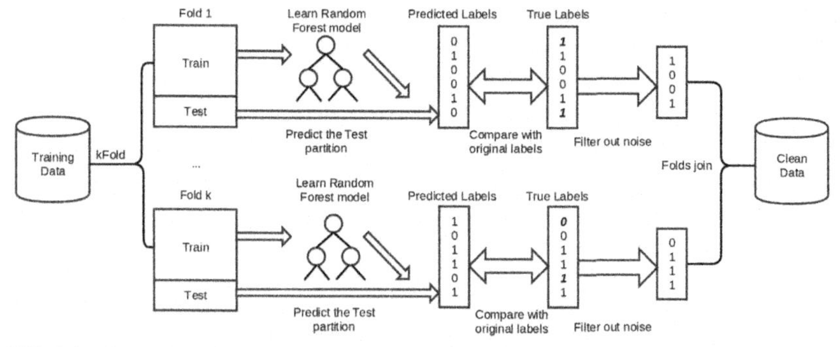

FIG. 8.5 A sample noise filtering for data cleaning application with MapReduce [19].

The data cleaning which is also known as noise filtering can be performed by using some algorithms based on ensembles. An example homogenous ensemble filtering approach is illustrated in Fig. 8.5 that is proposed in [19]. Garcia-Gil et al. have implemented the filtering algorithm by using Apache Spark expressions that provide extending the MapReduce operation as shown below. Apache Spark structure is located on top of resilient distributed dataset which is denoted as RDD in table. The input parameters are defined as data, partitions, and number of trees for random data forest.

Algorithm 8.1: HME-BD Algorithm

1: **Input:** *data* a RDD of tuples (label, features)
2: **Input:** *P* the number of partitions
3: **Input:** *nTrees* the number of trees for Random Forest
4: **Output:** the filtered RDD without noise
5: *partitions* ← *kFold(data, P)*
6: *filteredData* ← ∅
7: **for all** *train, test* in *partitions* **do**
8: *rfModel* ← *randomForest(train, nTrees)*
9: *rfPred* ← *predict(r fModel, test)*
10: *joinedData* ← *join(zipWithIndex(test), zipWithIndex(r f Pred))*
11: *markedData* ←
12: **map** *original, prediction* ∈ *joinedData*
13: **if** *label(original)* = *label(prediction)* **then**
14: *original*
15: **else**
16: *(label* = ∅, *features(original))*
17: **end if**
18: **end map**
19: *filteredData* ← *union(filteredData, markedData)*
20: **end for**
21: *return(filter(filteredData, label* ≠ ∅*)*

The *kFold* parameter denotes the list of RDD pairs for each *k* fold numbers and test data. The filter is result of each new RDD with predicted values while map applies transformation for each element along RDD sets. The *zipWithIndex* is used to zip RDD with its elements. It is ordered firstly on the partition index and then it is applied to each partition. Thus, the first partition is listed as Index 0 and the last partition gets the highest index number.

Random forests are a popular family of classification and regression methods in Spark APIs. They are used as ensembles of decision trees, and combine many decision trees for reducing the risk of overfitting. The block diagram shown in Fig. 8.5 is proposed by Garcia-Gil et al. where each partition is iterated for a random forest model, and predicting test data are used in learned model. Once the test data and predicted data are obtained to compare classes, *zipWithIndex* operation has been applied at each RDDs. Afterwards, the map function is operated for each RDD classes and for the predicted class. In case any difference detected at comparison of predicted and actual classes, this situation and differences are defined as noise that are removed by filter function [19].

Another sample noise filtering method proposed by [19] is heterogenous ensemble (HTE-BD) method which is based on three classifiers as random forest, logistic regression, and kNN. Unlike heterogeneous ensemble algorithm, the homogeneous one (HME-BD) was based on just random forest classification algorithms. However, the learning algorithms used in HTE-BD provides increased detection capability in noise filtering and data cleaning operations. It is mainly improved regarding to ensemble filter with increased classification algorithms. The decision tree classification algorithm is one of the most widely used machine learning method used in data mining. The decision tree is started with a single node and then each inherited outcome generates another node. The Apache Spark provides optimization in decision tree scalability since it is grown by several nodes. Random forest is the combination of decision trees and their nodes that are collected by algorithms called ensembles. Spark provides individual training capability to each tree along the random forest and thus, distributed operation feature is obtained for each random forest. The kNN which is another component of HTE-BD proposed by Garcia-Gil et al. is a supervised learning algorithm for classification. The algorithm presented below has been improved regarding to ensemble filter where algorithm benefits from learning algorithms such as decision tree, kNN, and linear machine learning. Authors suggest to use Spark random forest model instead of pure decision tree algorithm, and logistic regression has been used for linear machine learning. Each *train* and *test* operation of input data are run by three algorithms as seen in Fig. 8.6 for any fold from first one to *k*th one. The trained algorithms predict the test data and creates RDD triples as (*rf, lr, knn*) which are compared with original ones. The required input parameters are defined as database (*data*), partition number (*P*), number of trees in Random Forest (*nTrees*) and the voting strategy (*vote*).

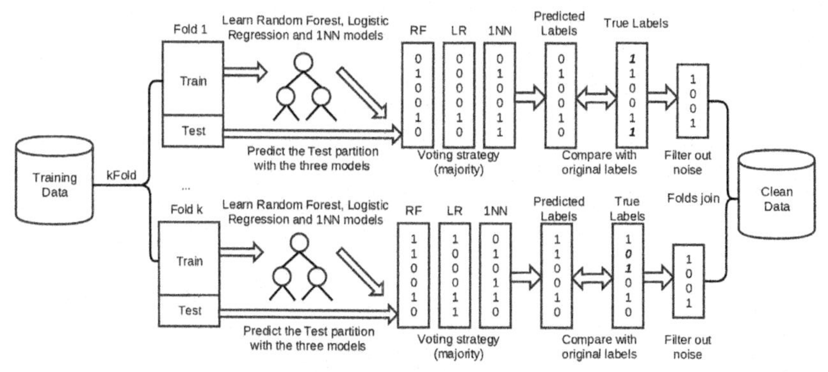

FIG. 8.6 A sample noise filtering with HTE-BD [19].

Algorithm 8.2: HTE-BD Algorithm

1: **Input:** *data* a RDD of tuples (label, features)
2: **Input:** *P* the number of partitions
3: **Input:** *nTrees* the number of trees for Random Forest
4: **Input:** *vote* the voting strategy (majority or consensus)
5: **Output:** the filtered RDD without noise
6: *partitions* $\leftarrow$ *kFold(data, P)*
7: *filteredData* $\leftarrow \varnothing$
8: **for all** *train, test* in *partitions* **do**
9:     *classifiersModel* $\leftarrow$ *learnClassi f iers(train, nTrees)*
10:     *predictions* $\leftarrow$ *predict(classifiersModel, test)*
11:     *joinedData* $\leftarrow$ *join(zipWithIndex(predictions), zipWithIndex(test))*
12:     *markedData* $\leftarrow$
13:     **map** *r f, lr, knn, orig* $\in$ *joinedData*
14:         *count* $\leftarrow 0$
15:         **if** *r f* $\neq$ *label(orig)* **then** *count* $\leftarrow$ *count* + 1 **end if**
16:         **if** *lr* $\neq$ *label(orig)* **then** *count* $\leftarrow$ *count* + 1 **end if**
17:         **if** *knn* $\neq$ *label(orig)* **then** *count* $\leftarrow$ *count* + 1 **end if**
18:         **if** *vote* = *ma jority* **then**
19:             **if** *count* ≥ 2 **then** *(label* $= \varnothing$*, features(orig))* **end if**
20:             **if** *count* < 2 **then** *orig* **end if**
21:         **else**
22:             **if** *count* $= 3$ **then** *(label* $= \varnothing$*, features(orig))* **end if**
23:             **if** *count* $\neq 3$ **then** *orig* **end if**
24:         **end if**
25:     **end map**
26: *filteredData* $\leftarrow$ *union(filteredData, markedData)*
27: **end for**
28: *return(f ilter(f ilteredData, label* $\neq \varnothing$*))*

The training data has been evaluated by three algorithms as voting strategies as seen in Fig. 8.6. The predicted tables and true labels are compared to determine noise levels that are filtered out to generate clean data. This filtering and cleaning procedures are repeated until last fold reached.

The recent progress has brought several different analysis approaches such as data mining, visualization, statistical methods, deep learning and machine learning. The machine learning methods facilitate to discover knowledge and intelligent decisions by using massive databases. It is analyzed in three categories according to its learning bases as supervised, unsupervised, and reinforcement learning. The conventional data mining methods such as association mining, clustering and classification lack in terms of efficiency, and they are not able to provide scalable and accurate outcomes when they are applied to Big Data stacks. The size, speed and variety of data streams prevent conventional data mining methods to analyze data stacks permanently. Therefore, researchers have improved new optimization methods and analytical approaches for improving processing capability with limited resources.

8.3.2 Machine learning in big data analytics

The machine learning is a research area of computing science and an application area of artificial intelligence that is based on processing inductive models trained by limited data input. It is improved regarding to pattern recognition and computational learning systems. The input data provide patterns for learning algorithm to define relationships among parameters of the database which is called as training set and samples. The learning categories of a machine learning system is comprised by three types of approaches as supervised, unsupervised, and reinforcement. The supervised learning taxonomy is based on predicting and output vector due to inherited knowledge from training set of input vectors and corresponding relations. The supervised learning methodology is based on classification and regression methods where classification denotes category variables while regression defines prediction of numerical variables. On the other hand, the unsupervised learning does not provide any training set and there is not any labeling required for predicting the variables. These learning structures are known as clustering algorithms or recommender systems. The reinforcement learning addresses learning problem for particular action or a set of actions to improve reliability of outcomes for a predefined situation. The most widely used machine learning algorithms and data processing methods are presented in Table 8.1 [20–22].

The machine learning process is performed at a few steps by following data acquisition, preprocessing, selection, extraction, model selection, and validation stages. Different datasets and inputs are combined at data acquisition and preprocessing steps while data cleaning is also performed at this stage. The predefined particular features are selected and extracted in the next step where it is followed by model selection step. All the selected and processed data

TABLE 8.1 Mostly used machine learning algorithms in big data analytics

Algorithm type	Data processing method
Naive bayes	Classification
K-nearest neighbor	Classification
Support vector machine (SVM)	Classification
Linear regression	Classification/regression
Support vector regression	Classification/regression
Classification and regression trees	Classification/regression
Random forest	Classification/regression
Bagging	Classification/regression
Artificial neural network	Clustering/classification/regression
Feed forward neural network	Clustering/classification/regression
K-means	Clustering
Density based spatial clustering	Clustering

are validated at the last step where classification, regression and evaluation of processed data are performed. The classification and prediction are the most important initial processing steps since they provide filtering, cleaning, validation, and model selection on input databases. The model selection of machine learning algorithm enables to use learning datasets. The models include various duties as classification, regression, detection, sampling, noise filtering, and other solutions. The support vector machines (SVMs) and artificial neural networks (ANNs) are most widely used models in machine learning systems. The conventional SVMs are binary classifiers which are used to find training sets with maximum benefit among others. The binary classifier feature of SVM is used to determine a hyperplane as a linear function of input data. Another important feature is related to training points requirement where SVM needs a few points that are called support vectors to classify next data points. SVMs are accepted as the best supervised learning models due to their efficiency dealing with high volume datasets by using limited memory resources. However, SVM causes to drawbacks since it is not capable to provide direct probability estimations [20, 22].

The ANN is based on processing of larger datasets, improved initialization algorithms, robust learning models, and multilayered structure which is called deep learning. The complex structure of ANNs that are formed by hidden layers and intermediate layers is simplified by feedforward architectures that are operated in single direction and eliminates extra cycles between neurons. The regular training time of ANNs may be long lasting, but the efficiency on classification and regression increases its operating speed [20].

The main objective of linear regression method is to learn a function that is denoted as $f(x,w)$ where $f{:}\varphi(x) \rightarrow y$ mapping is obtained and a linear combination of a linear or nonlinear input parameters of $\varphi_i(x)$. This fundamental function is formed as $f(x,w) = \varphi(x)^T w$ where w is the weight vector. The fundamental linear regression functions are composed of polynomial, radial, sigmoidal, and Gaussian functions. Widely used training models are known as Ordinary Least Square, Regularized Least Squares, Least-Mean-Squares (LMS) and Bayesian Linear Regression approaches. LMS is the most rapid one among others and it can be scaled for high volume of datasets. The combinational models can be formed as Classification/Regression and Clustering/Classification/Regression approaches. The linear regression, classification and regression trees, support vector regression, random forests, and bagging methods are some of widely used combinational machine learning algorithms [22].

In addition to aforementioned algorithms, there are sophisticated tools presented by corporates for efficient data analytics. Most of the widely used analytics tools are operated in Standard Query Language (SQL)-based structures. Moreover, dedicated data servers are also preferred by corporations to decrease data storage and data processing costs. The R and Weka are corporate level machine learning tools that are widely used in big data analytics by deploying machine learning algorithms and models. It is obvious that big data provides several benefits to corporates in any area including healthcare, smart environments, traffic managements, financial sectors, security, banking and many more. Therefore, another important issue to be considered is security and privacy for secure data processing [23].

8.4 Overview of smart grid privacy

The smart grid devices generate big datasets due to widely used SCADA, advanced metering infrastructure (AMI), and IEDs at each 2 or 4 s intervals. The PMUs provide much faster data stacks since they generate 30 or 60 samples per second, and attach time stamp to measurement data. The replacement of regular electricity meters with the advanced meter read (AMR) devices comprised another important component of smart grid big data. Each AMR device generates 96 metering data a day which comprises 2880 measurement data for each consumer in a month. The deployment of AMR, AMI, PMU and IEDs have accelerated generation of massive data sets. A sample calculation has been noted as 100 PMUs with 20 measurements a day generates over 100 GB data at standard sampling rates [24]. In some respects, smart grid systems can be assumed as integration of internet and numerous IEDs used in the power networks. Each section of power network including generation, transmission,

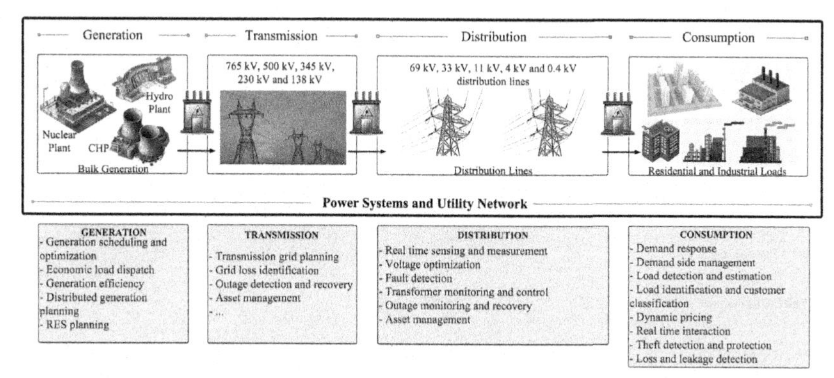

FIG. 8.7 Big data analytics requirement in smart grid applications.

distribution and consumption is equipped with IEDs and sensor networks in this convergence. The big data analysis and decision support tools operated on smart grid plane are illustrated in Fig. 8.7 where each requirement has been noted. The big data analytics are operated at seven steps in smart grid applications as data acquisition, data transmission, storage, cleaning, preprocessing, integration, feature and model selection stages [3, 25].

The overall schematic of smart grid infrastructure shown in Fig. 8.7 is completely depended on ICT operations since all sections should be instantly monitored and controlled. The two-way communication is indispensable for each application and Big Data analytic requirement depicted on the lower plane of figure. Each application presented in Fig. 8.7 should be dealt with seven steps of big data analysis framework. The generated data of thousands of smart devices should be inherited and sampled in a few seconds. The generated data could be brought by any section of generation, transmission, distribution or consumption levels of smart grid network. The generation level data sources can be distributed generation sources, forecasting measurements, power plants, RES plants and so on. The enormous number of data sources are located at consumer side since many nodes are comprised by each customer which have also different IEDs in their reserves. The residential data sources are mostly generated by AMIs, smart meters, home energy management systems, electric vehicles, micro sources, and by many more sensors used for surveillance, forecasting, and management demands.

The data acquisition process is performed in three steps as data capture, data transmission and data preprocessing in smart grid big data analysis. The generated data of sensors and nodes are captured by centralized of distributed agents. The acquired data are transmitted to a central Hadoop cluster for storing and local master nodes are comprised. The stored raw data are transferred to data storage system for initial processing. The data integration process is required at step since the inherited data are provided by a wide diverse of devices and different file formats or information types are generated. This stage is named

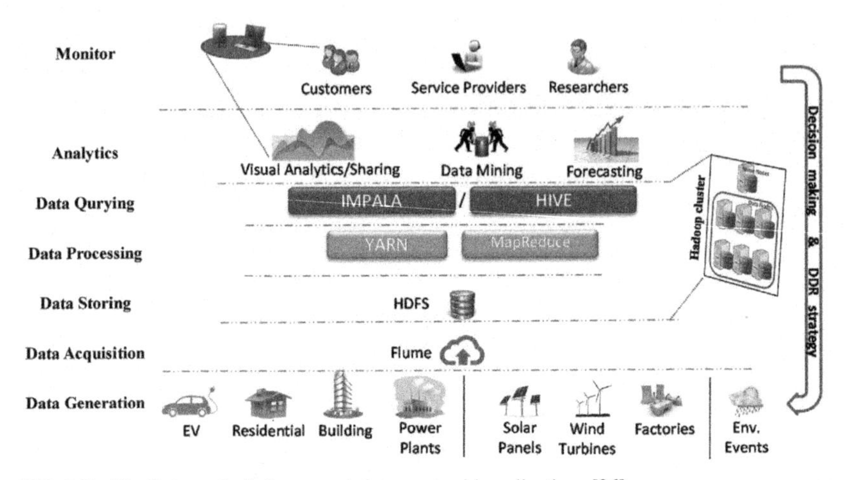

FIG. 8.8 Big Data analysis framework for smart grid applications [26].

as data preprocessing for integration of the inherited data stacks. The unified data stacks are stored in comma-separated value (csv) files with several identifying informations such as timestamp, device ID number, generated source and location data. Apache Flume is one of the most widely used distributed data collecting tool that collects, aggregates, and transfers huge amount of data to a Hadoop node. Once Flume server receives data stacks, it generates a few channels regarding to data sizes and transmits data to HDFS which authorizes Flume for data write process. The hierarchical organization of seven step processing is shown in Fig. 8.8 which has been originally depicted in [26]. HDFS stores the received smart grid data where it generates clusters comprised by NameNodes. The metadata are managed by DataNode and prepares datasets for computational processing to be handled by Hadoop Yarn that operates simultaneously with HDFS on same nodes. The MapReduce is also capable to operate with HDFS and other fundamental components at processing stage. The data querying stage is comprised by different tools that most widely used ones are Impala and Hive which are convenient for data selection from HDFS repository, analyzing the data and generating required data selections [26].

The acquired, processed and queried data stacks are prepared for data analytics in the next step. The next stage is comprised by several analysis methods including Big Data analytics algorithms that are presented above for visual analysis, data mining, prediction and forecasting purposes. The data sharing operation at this level requires sophisticated security and privacy protections.

The data analysis based on data mining is not new to power network systems, but the methods used over passed years are evolved from SQL based analysis to more sophisticated algorithms. It is obvious that smart grid applications require more efficient and effective methods and tools for dealing with rapidly changing and increasing data stacks. The solution to processing massive data stacks is based on multisource based mining mechanisms and data mining algorithms that have been presented in the previous section. The most widely used machine learning algorithms are noted as k-means, linear support vector machines (LSVM), logistic regression (LR), locally weighted linear regression (LWLR), Gaussian discriminant analysis (GDA), back-propagation neural network (BPNN), expectation maximization (EM), naive Bayes (NB), and the independent variable analysis (IVA) in smart energy applications [24]. Since the big data analysis is in its evolution stage, applications face with several challenges and threats in terms of data storage, data integration, instant data processing, data compression and security issues. The threats and challenges are presented in the following sections considering privacy, security, and related issues.

8.4.1 Threats and challenges in privacy

The privacy requirements may vary depending to countries, legal approaches, personal rights, and regulations. Privacy issues refer to protection of sensitive data from intrusions and unauthorized accesses. It is noted that privacy can be handled in four categories as physical, informational, decisional, and dispositional [5]. The deployment of big data analytics has transformed privacy issue to a core problem in data mining applications. Although the identification, encryption and several other methods are used to enforce data privacy, ICT based security risks may not completely eliminated [27].

The Big Data analytics can help to operation of distribution and transmission system operators in terms of deployment and calibration of smart grid applications with different tools such as simulation and modeling studies. However, the facilitated applications may cause to several challenges and systems can be posed to several threats. It should be taken into consideration that networks are not tolerant to threats and should be operated under heavy reliability and security standards. Moreover, consumer privacy plays vital role in network operation [25]. One of the prominent challenges of big data analytics in smart grid is related with multisource data integration and storage issues. Despite the conventional data analyses dealing with single planes, big data is a fusion of multiple source-based data stacks with different formats and presentations. Therefore, data storage and multiple source structure should be proven by HDFS and similar systems. The rapid reactions such as fault detection and transient protection plays crucial role in utility networks. The cloud-based data storage and processing systems can cause to latencies due to complicated and heavy analysis algorithms. Thus, real time data processing is important requirement to prevent latency threats against rapid reactions. These threats are mostly tackled by using storing databases on rapid and local memories. Data compression is another crucial solution for massive databases such as wide area monitoring system (WAMS) [24].

8.4.2 Privacy preserving methods

The protection effect of data privacy is defined as disclosure risk which represents the probability of any intruder may disclose private information from deployed data. The privacy preserving methods are listed as data perturbation, secure multiple computing, storage encryption, identity authentication, access control and so on. Data perturbation, data encryption, and data anonymization methods are the most widely used one among others. The data perturbation method is a complex operation comprised by a few steps that replaces original data with anonymous perturbation and random variables, generating random offset values or fuzzy sets, and adding perturbation information for computing. Nevertheless, it cannot be said that data perturbation can completely prevent intrusions to private data. The data encryption method which uses several encryptions to hide original data during data mining is widely used in distributed applications, and it ensures authenticity, reversibility, and robustness of data against intrusions. The data encryption method uses several technologies such as secure multiparty computation, symmetric encryption, public key encryption, differential privacy protection, authentication and access control techniques. The data anonymization method is based on hiding the identities of users and databases to improve privacy. It uses some techniques such as anonymous protection technology, digital signature, secret sharing technology, *k*-anonymity, *l*-diversity, *t*-closeness, anonymized publication, anonymization with high utility and so on to achieve privacy preserving operation conditions. The privacy preserving technologies are considered in degree of privacy preserving, missing data amount, and performance of run algorithm. The degree of privacy preserving value of an algorithm is evaluated with its disclosure risk. The missing data amount is the indicator of privacy preserving method that lower missing data measure implies higher success. It is the difference between recovered data and original transmitted data [5].

Although the human machine interaction brought by big data analytics provide many opportunities and progress, it also causes to several challenges for current ICT systems. The security and privacy are one of the crucial challenges among others in big data processing systems. The complex and dynamic structure of big data stream force operators and users to face several unpredicted threats in data storage, analysis, and management issues. The challenges of big data privacy and security are classified into four categories as infrastructure security, data privacy, data management, integrity and reactive security. The infrastructure security is related with secure computing in distributed programming frameworks and security schemes for data storage. The distributed computing infrastructure requires multiple nodes and devices which cause increased number of nodes. In Hadoop systems, a mapper can expose privacy of customers by analyzing a special data as personalized data or commercial reports in databases. The data privacy can be exposed to scalable privacy preserving data mining and analysis challenges. The malicious or violating users can abuse the datasets to acquire other users' data and track the private data. The most recent AI systems improve the reliability of big data analytics to tackle this challenge.

In big data applications, encryption mandatory is a convenient method to provide end-to-end protection for the security of the data centers. The integrity of data also can be used to prevent data falsification for diverse end-point data. The underlying attacks may bypass the control algorithms and provide direct access to databases. Therefore, it is required to decrease visibility of any data at the source databases. The big data analytics require a fine-grained and scalable database access control that is provided by granular access control protocols. It enforces the system by preventing unauthorized access and maintaining privacy. The privacy preserving methods used in data management are related with secure data storage, granular audits, and data provenance [5].

The data source and users are separated in big data infrastructure where the user have not complete control of data and is not able to know the exact location of stored data. The transaction logs are generated to locate stored data information that reach to higher volumes with the increment of stored data size. Therefore, security of the stored data and transaction logs should be ensured. Another important privacy challenge is seen in real time security monitoring applications since it is based on tracking the dynamic analytics. It is required to prevent intrusions and unauthorized accesses to big data infrastructure. The monitoring applications are sensitive to denial of service (DoS) attacks, and advanced intrusion attacks. Therefore, instant and persistent real-time monitoring applications are convenient to predict and prevent intrusions [5].

8.4.3 Privacy enhancing applications

The sophisticated features of smart grids such as self-healing, remote control, self-monitoring, and distributed control properties have increased the attention to this new kind of power network. The widespread use of services and opportunities of smart grid have brought concerns on protecting the acquired data which include personal information of consumers. The collected data could be used to define personal behaviors, lifestyle, and habits. The demand response programs which are tailored solutions for each consumer is one of the smart grid application that privacy and security should be ensured in big data analytics. Other smart grid applications provide personal data about consumers as well as demand response or smart meter data. Nowadays, several privacy preserving and secrecy techniques have been taken into consideration to protect such critical data. The prominent applications are listed as anonymization, access control, encryption, differential privacy protecting methods and so on [5, 28].

The anonymization is one of the most widespread protection method that targets to hide user ID and sensitive personal data. The raw data is anonymized during processing and before deploying to nodes by using generalization, decomposition, replacement and interference operations. The generalization and compression processes hide the attribution identities while decomposition and permutation methods change sensitive identifiers by grouping and decoupling information between operators and sensitive properties. Four privacy preserving methods that are expressed above are *k*-anonymity, *l*-diversity, *t*-closeness, and differential privacy processes. It is noted that the key methods and basic means on anonymity protection are still in their development stage for structured data stacks in big data [5]. The access control is proposed as an efficient process for data sharing applications. The difficulties are related to prediction of authorization of each user in big data, variety of access requests, and providing privilege in massive databases. The complexity and diversity of big data prevents efficient calculation for privacy preserving algorithms. Thus, encryption and decryption methods are proposed as alternative control operations in big data. The privacy preserving methods can be achieved by ensuring communication security and using encryption algorithms in complex big data contents. Therefore, the encryption based methods are mostly used in distributed application as data mining, distributed queries and distributed data deployment applications [5].

References

[1] X. He, Q. Ai, R.C. Qiu, W. Huang, L. Piao, H. Liu, A big data architecture design for smart grids based on random matrix theory. IEEE Trans. Smart Grid (2015) 1, https://doi.org/10.1109/TSG.2015.2445828.

[2] C. Yang, Q. Huang, Z. Li, K. Liu, F. Hu, Big data and cloud computing: innovation opportunities and challenges. Int. J. Digital Earth 10 (2017) 13–53, https://doi.org/10.1080/17538947.2016.1239771.

[3] K. Zhou, C. Fu, S. Yang, Big data driven smart energy management: from big data to big insights. Renew. Sust. Energ. Rev. 56 (2016) 215–225, https://doi.org/10.1016/j.rser.2015.11.050.

[4] Z. Asad, M.A. Rehman Chaudhry, A two-way street: green big data processing for a greener smart grid. IEEE Syst. J. 11 (2017) 784–795, https://doi.org/10.1109/JSYST.2015.2498639.

[5] W. Fang, X.Z. Wen, Y. Zheng, M. Zhou, A survey of big data security and privacy preserving. IETE Tech. Rev. 34 (2017) 544–560, https://doi.org/10.1080/02564602.2016.1215269.

[6] R.C. Qiu, P. Antonik, Smart Grid Using Big Data Analytics, John Wiley & Sons, Inc, Chichester, West Sussex, 2017.

[7] A. Oussous, F.-Z. Benjelloun, A. Ait Lahcen, S. Belfkih, Big data technologies: a survey. J. King Saud. Univ. Comput. Inf. Sci. 30 (2018) 431–448, https://doi.org/10.1016/j.jksuci.2017.06.001.

[8] M. Chen, S. Mao, Y. Zhang, V.C.M. Leung, Big data generation and acquisition. in: Big Data, Springer International Publishing, Cham, 2014, pp. 19–32, https://doi.org/10.1007/978-3-319-06245-7_3.

[9] I. Yaqoob, I.A.T. Hashem, A. Gani, S. Mokhtar, E. Ahmed, N.B. Anuar, A.V. Vasilakos, Big data: from beginning to future. Int. J. Inf. Manag. 36 (2016) 1231–1247, https://doi.org/10.1016/j.ijinfomgt.2016.07.009.

[10] T. Rabl, H.-A. Jacobsen, Big data generation. in: T. Rabl, M. Poess, C. Baru, H.-A. Jacobsen (Eds.), Specifying Big Data Benchmarks, Springer, Berlin, Heidelberg, 2014, pp. 20–27, https://doi.org/10.1007/978-3-642-53974-9_3.

[11] O.B. Sezer, E. Dogdu, A.M. Ozbayoglu, Context-aware computing, learning, and big data in internet of things: a survey. IEEE Internet Things J. 5 (2018) 1–27, https://doi.org/10.1109/JIOT.2017.2773600.

[12] P.P. Ray, A survey on internet of things architectures. J. King Saud Univ. Comput. Inf. Sci. (2016)https://doi.org/10.1016/j.jksuci.2016.10.003.

[13] W. Ejaz, A. Anpalagan, Internet of Things for Smart Cities: Technologies, Big Data and Security, Springer Nature Switzerland AG, New York, NY, 2018.

[14] M. Jaradat, M. Jarrah, A. Bousselham, Y. Jararweh, M. Al-Ayyoub, The internet of energy: smart sensor networks and big data management for smart grid. Proc. Comput. Sci. 56 (2015) 592–597, https://doi.org/10.1016/j.procs.2015.07.250.

[15] GSMA Report, IoT Big Data Framework Architecture, GSM Association, 2018.

[16] D. Ardagna, C. Cappiello, W. Samá, M. Vitali, Context-aware data quality assessment for big data. Futur. Gener. Comput. Syst. 89 (2018) 548–562, https://doi.org/10.1016/j.future.2018.07.014.

[17] A.M.S. Osman, A novel big data analytics framework for smart cities. Futur. Gener. Comput. Syst. 91 (2019) 620–633, https://doi.org/10.1016/j.future.2018.06.046.

[18] A. Iosifidis, A. Tefas, I. Pitas, M. Gabbouj, Big media data analysis. Signal Process. Image Commun. 59 (2017) 105–108, https://doi.org/10.1016/j.image.2017.10.004.

[19] D. García-Gil, J. Luengo, S. García, F. Herrera, Enabling smart data: noise filtering in big data classification. Inf. Sci. 479 (2019) 135–152, https://doi.org/10.1016/j.ins.2018.12.002.

[20] A. Stetco, F. Dinmohammadi, X. Zhao, V. Robu, D. Flynn, M. Barnes, J. Keane, G. Nenadic, Machine learning methods for wind turbine condition monitoring: a review. Renew. Energy 133 (2019) 620–635, https://doi.org/10.1016/j.renene.2018.10.047.

[21] I. Portugal, P. Alencar, D. Cowan, The use of machine learning algorithms in recommender systems: a systematic review. Expert Syst. Appl. 97 (2018) 205–227, https://doi.org/10.1016/j.eswa.2017.12.020.

[22] M.S. Mahdavinejad, M. Rezvan, M. Barekatain, P. Adibi, P. Barnaghi, A.P. Sheth, Machine learning for internet of things data analysis: a survey. Digital Commun. Netw. 4 (2018) 161–175, https://doi.org/10.1016/j.dcan.2017.10.002.

[23] N. Dey, A.E. Hassanien, C. Bhatt, A.S. Ashour, S.C. Satapathy (Eds.), Internet of Things and Big Data Analytics Toward Next-Generation Intelligence, Springer International Publishing, Cham, 2018https://doi.org/10.1007/978-3-319-60435-0.

[24] C. Tu, X. He, Z. Shuai, F. Jiang, Big data issues in smart grid—a review. Renew. Sust. Energ. Rev. 79 (2017) 1099–1107, https://doi.org/10.1016/j.rser.2017.05.134.

[25] B.-A. Schuelke-Leech, B. Barry, M. Muratori, B.J. Yurkovich, Big data issues and opportunities for electric utilities. Renew. Sust. Energ. Rev. 52 (2015) 937–947, https://doi.org/10.1016/j.rser.2015.07.128.

[26] A.A. Munshi, Y.A.-R.I. Mohamed, Big data framework for analytics in smart grids. Electr. Power Syst. Res. 151 (2017) 369–380, https://doi.org/10.1016/j.epsr.2017.06.006.

[27] M. Marjani, et al., Big IoT data analytics: architecture, opportunities, and open research challenges. IEEE Access 5 (2017) 5247–5261, https://doi.org/10.1109/ACCESS.2017.2689040.

[28] H. Li, Enabling Secure and Privacy Preserving Communications in Smart Grids, Springer, Cham, 2014.

Chapter 9

Roadmap from smart grid to internet of energy concept

Chapter outline

9.1 Introduction

Recently, several concerns have appeared based on the climate change, scarcity of fossil resources and greenhouse gases caused by conventional energy generation systems. Energy generation systems based on renewable energy sources (RESs) have gained a great attention depending on these issues, and the use of RESs such as solar and wind energy systems has become widespread nowadays. The most important advantages of these sources are both generating energy with low cost and being eco-friendly. The use of RESs needs to become more widespread in order to handle these significant issues. However, these green systems come with new challenges such as power grid integration issues of RES systems. In addition, real-time monitoring of these microgrids is another important challenge for these systems to ensure sustainability of energy generation systems [1–5]. The widespread use and extensive studies on smart grid (SG) is improving knowledge based on it. However, it is not still adequate to overcome of the complexity, diversity, variety and massive data sizes of entire system. Although the SG is interested in single type of energy as electricity, there others such as chemical, thermal and electromagnetic energy provide wide usage areas. On the other hand, electricity flow is managed by distribution system operators (DSOs) even in SG applications where two-way power feed exist between consumer and DSOs. The internet of energy (IoE) or energy internet (EI) is assumed as the next revolutionary phase of SG since it brings flexibility of using any type of energy besides electricity, and enables consumers and prosumers to route energy in any direction. The EI is integration of

From Smart Grid to Internet of Energy. https://doi.org/10.1016/B978-0-12-819710-3.00009-0

information and communication technologies (ICTs), energy distribution systems and operators as well as distributed generation systems [1].

On the other hand, the SGs which have been introduced as the new generation power grid system have attracted great interest from both academia and industry [5]. The SGs target a modern and exclusive vision integrating power systems, communication networks and smart metering systems [6]. Unlike conventional power grid structures, the SGs allow bidirectional flows of energy and communication signals to present more reliable, secure and efficient grid structure. Even though this modern concept offers remarkable advantages, current researches indicate that the concept is insufficient against some issues such as increasing complexity and instability of power grids. Despite the SGs are able to manage only one type of energy (only electricity), there exist different types of energy such as electromagnetic, thermal and chemical. Another important issue for the SGs is the use of current power distribution grids that limit the two-way electricity flow possibility of SG systems. In addition, this situation leads to some problems on routing, scheduling and allocation of energy flow processes [1, 2]. In order to overcome these issues, a new concept called "Energy Internet (EI)" has been introduced which intends to improve SG systems by including various energy forms and new capabilities. The differences between EI and SG concepts can be sorted out as follows [1, 2].

- While SGs perform regional system control constantly, EI allows accessing distributed generation (DG) and distributed energy storage systems.
- SGs mostly point out centralized power and integrate limited type energies whereas EI supports different energy types involving nuclear, solar, tidal, wind and so on.
- Communication systems and conventional power systems are effective in the SGs while Internet and additional information networks are in the EI.
- Whereas SGs contain communication, information and control technologies, EI concept combines smart metering, smart monitoring, energy distribution and auto adjust controlling technologies.
- Furthermore, EI supports plug-and-play interfaces to enable energy and information sharing in various ways. SGs deliver energy only over current power grid systems.

This chapter provides an outlook for the EI systems. First of all, fundamental targets and requirements of this vision are analyzed by taking into account communication technologies and other components. Then, the most important components of the EI systems are introduced. Recent developments about energy routers, information systems and network systems of EI systems are also investigated. In addition, potential communication technologies for EI systems are analyzed and comparison of EI communication systems with other existing communication systems such as Internet, SGs and so on is provided. Furthermore, the essential parts of the EI systems are examined in a detail. Besides, several issues affecting EI systems are introduced and potential solutions for these issues are also discussed.

9.2 Vision and motivation of IoE

Energy generation, transmission, distribution and storage systems can be managed more smartly in the EI concept since all of the energy sources are connected to each other thanks to the Internet. The EI intends to augment the use rate of energy and substantially to encourage the utilization of RESs. The EI also promotes various distributed energy resources (DERs) to meet the requests of entire energy market. The information of clients and energy providers can be acquired promptly and accurately to set energy allocation due to advances on Internet of Things (IoT) technologies. The structure of EI develops in order that the energy market progressively depend on the renewable and distributed energy generation systems [3, 7].

The conventional power network has been transformed in the late 20th century by increasing use and combination of RESs to generation and distribution infrastructures. The decreasing amount of fossil-fuels and degrading hydroelectric plants have accelerated this transformation. The economic and environmental concerns were the trigger factor of researchers that are performed on increasing source variety, efficiency, and decreasing the generation, transmission and operating costs. The energy storage systems (ESSs) become critical players in the smart grid transformation to improve reliability and flexibility of this new power network structure. In addition to two-way power flow characteristic of new grid concept, the more important contribution has been provided by communication technologies which were crucial to increase management and monitoring capability of power network. Therefore, two-way transmission is not only achieved for power but also for data and control commands. The intelligent data processing and control capability that is brought by energy management system (EMS) facilitates real time monitoring, remote measurement, and control of numerous generation and transmission substations in a centralized infrastructure. The rapid innovations seen in generation and transmission networks have enabled consumers to have their own generation sources that transforms consumers to prosumers. The load types are also changed and evolved in the transformation era of smart grid. In addition to conventional static loads, consumers were encouraged to use intelligent and controllable loads in order to facilitate remote monitoring and control requirements of smart grid. The deployment of electric vehicles (EVs) provided to achieve mobile ESSs due to charging stations, and sophisticated wired and wireless communication architectures have been integrated to each load sites [8, 9].

The EI concept that is assumed as transformation of SG is established depending on the energy routers to allow energy conversion, transmission and information network with plug- and-play feature. In addition, this new concept purposes to present sustainable computing by combining several energy types to greatly flexible grid analogous Internet [6]. Moreover, the EI supports comprehensive packing and routing services. Massive amounts of energy data arise, namely big data, while performing energy generation, transmission, distribution, storage and energy management processes in the EI [4, 9–11]. There is a great need for multi-advanced communication technologies in order to expand coverage range of electrical power grids. Effective, secure and cooperative communication systems that are utilized for controlling energy management among device and systems connected each other are fundamental part of the EI systems. Even though the communication structure of EI systems is established depending on the Internet and SG communication systems, the communication infrastructures of EI systems are quite different from the SG communication systems. The communication infrastructures of EI systems are compared with Internet, wireless sensor network (WSN) and SG communication systems, which are listed in Table 9.1. As can be seen from the table, communication infrastructures of EI systems are completely based on energy routers that manage energy and information data. On the other hand, Internet and SG communication systems are only responsible for conveying information data. In other words, the most important difference between EI communication systems with other ones is energy transmission via Internet [1].

The integration of energy and communication architectures comprises fundamental base of EI concept which is assumed as the characteristic infrastructure of novel grid network. The contributions of new grid has been listed into four categories in [8] as follows;

- ***Controllable intelligent loads:*** The controllable loads include residential heaters, air-conditioning devices, smart appliances and intelligent electronic devices that are capable to communicate with EMS,
- ***Generation and load demand control:*** The intermittent structure of RESs that are depended to solar and wind variations, varying state of charge value of ESSs, and EV technologies require energy management infrastructures,
- ***Source and communication deployment:*** Widespread DER amounts and types, smart meters, advanced metering infrastructures (AMIs) and remote monitoring nodes contribute to the excessive manner of source and communication variety on the smart grid plane. Therefore, a generic and convenient communication infrastructure should be installed.
- ***Distribution grid transformation:*** The diversity of sources and distributed generation capability which is enabled with development of smart grid will improve the amount and volume of distribution network. The two-way power flow will improve control and operation requirements of existing DSOs.

In order to overcome communication and power flow problems, data transmission and processing requirements and improve the operational capabilities of DSOs, a wide variety of decentralized communication architectures and emerging infrastructures are required. In addition to communication technologies, data mining and big data analytics are also required to deal with huge amount of data that have been generated and transmitted from numerous sensor nodes.

TABLE 9.1 Comparison of EI communication with other communication methods

Communication categories	Internet	WSN	SG	EI
Transmission type	Data	Data	Information (measurements and control commands)	Energy and information
Supply type	Heterogeneous data	Sensor data	Heterogeneous data	Multiple energy sources and heterogeneous data
Routing system	Switches and routers	Sensor and/or sensor nodes	SMs	SMs and energy routers
Service types	Network connection, network management	Network connection, data management	Grid connection, grid management	Grid connection, grid management
Standard	Numerous international standards	Numerous international standards	Several international standards	–

The big data analytics, security and privacy issues have been presented in the recent chapter. Architecture of EI system and components used to comprise an EI infrastructure are presented in the following sections.

9.3 Architecture of EI systems

A typical EI communication structure including DESs, energy management center (data, control center and smart energy), electric vehicles, residential, commercial and industrial customers are depicted in Fig. 9.1. The DESs are composed of several types of energy sources such as hydroelectric, wind, solar and energy storage systems [12]. EI management center includes data center, smart energy management center and control center. Energy routers are established to ensure energy and information services where a wide area is considered to allow energy and communication transfer between various areas [11, 13, 14]. Therefore, energy routers have very critical mission to accomplish energy management transactions in the EI communication [8]. The energy routers located in the wide area network (WAN) perform overall energy allocation and these energy routers are able to manage two-way energy flows. Data center and control center that are located in management areas assistance EI system for controlling energy management in both power and user side. Smart meters

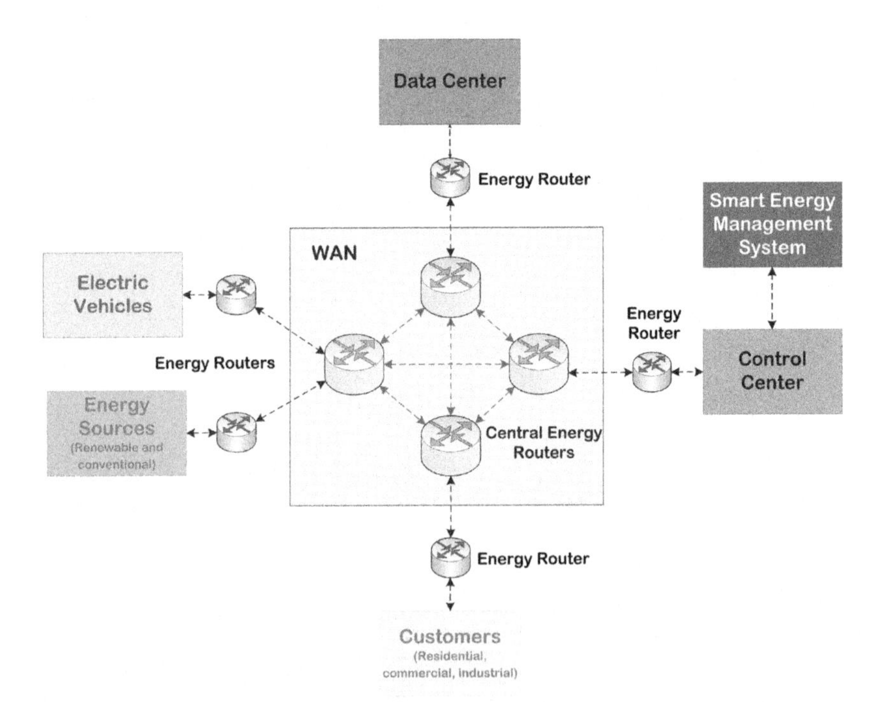

FIG. 9.1 General block diagram of EI concept.

(SMs) perform metering of electricity and energy quality at the same time. Several efficient communication technologies such as cognitive radio (CR), power line communication (PLC), fiber optic and cellular are recommended to improve communication performance of EI systems. The essential intention of the EI communication is to allow data exchange to accomplish real-time balance among energy generation rates and consumption rates.

The general EI architecture is comprised to accomplish energy routing applications that energy routers are located at any node of EI infrastructure as seen in Fig. 9.1. The fundamental requirement of energy router is to deploy, manage and control the energy flow in planned and programmed operation since it is controlled by ICT interfaces. The centralized organization of this power network provide interconnection of several intelligent loads, DERs, storage systems, customers, control and management centers. It is noted in [13] that energy routers are proposed as a convenient interface in RES integration to active distribution network and multi-agent systems comprise the core controller of energy routers for accomplishing the control requirements. On the other hand, the solid-state transformers (SSTs) are proposed as dynamic energy routers due to their power handling and coordinating capabilities. The SSTs are comprised by power converters in DC and AC operations where dynamic energy management and power flow controls are available in ICT based converter architecture. A novel energy router has been proposed in [13] that is designed in modular structure and provides multiple operation modes including single-phase and three-phase power interfaces, stability control features, and distribution network capabilities.

A general EI system is composed of three main structures called energy subsystem, network subsystem and information subsystem. These subsystems are connected together via energy routers. Fig. 9.2. shows the connection of energy

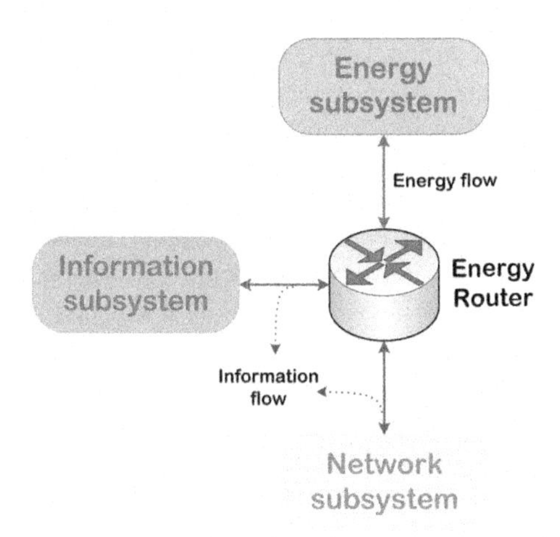

FIG. 9.2 Interaction of energy router in EI system.

routers in EI that is considered as the most important component of the EI systems due to managing the energy and data flow transactions. The first research about energy routers are reported in [15] where energy routers are responsible for two important task as efficient scheduling for energy flow processes and managing communications among devices in real-time. Then, different architectures of energy routers such as *E*-energy, energy hub and digital grid router (DGR) have been introduced by combining information and modern communication systems [1, 16, 17].

The communication systems which are spread into information and network systems in order to allow monitoring, controlling and management in real-time are very important in EI concept to carry out goals of this new vision. The information systems of EI are composed of smart sensing and computing structures and are very critical for data collecting analyzing and managing gathered data. The network systems of EI are responsible for connection entire devices existing in the EI system by utilizing modern wired and wireless communication technologies such as CR, PLC, cellular communications, ZigBee, software-defined network (SDN) and Worldwide Interoperability for Microwave Access (WiMAX). It is important to note that the combinations of these technologies are exploited in the EI concept in order to establish a flexible and efficient structure. Even though several technologies have been adopted to SG systems, there exist still many challenges to be solved regarding combining these communication technologies in EI concept. For instance, in order to ensure a real-time communication infrastructure among users, there are critical requirements such as high data rate, efficient two-way communication structures and lower latency that is not accomplished by the SGs [1, 5]. The following subsections present explanations for substructures of the EI concept.

9.3.1 Energy routers

The EI system presents an innovative outlook for managing both energy and information flow together. The most important component of the EI is certainly energy router that is taken into account as a smart management and energy transformer device. The authors recommended in [6] that this architecture need to firmly integrate energy flows with information transmissions. Unlike the traditional electrical equipments, this structure presents several important characteristics as follows. It supports two-way energy conversion with high quality between various types of terminals. In addition, it supports plug-and-play features for enabling a perfect connection among electrical systems and consumers. Furthermore, energy routers accomplish optimum energy management for entire EI system [1, 18]. Researches reported in [19, 20] propose energy router systems as a bridge structure between several energy types and loads. Different types of energy routers are introduced as energy hub and energy interconnector as mentioned before. An energy interconnector can manage combinations of electricity, gas and heat transmissions whereas an energy hub can administer processes regarding monitoring, storage and conversion of several energy sources for increasing stability of power systems. Besides, energy routers should have capability to collaborate between each other.

The EI architecture is improved in an internet-like structure that is mimicking routing data where power routing is also included in addition to communication signals. The energy router concept is firstly proposed with the concept of the future renewable electric energy delivery and management (FREEDM) system in [21]. The proposed energy router which is the core of EI architecture with generation, distribution, storage systems and loads could be used to control flow and amount of power and energy. Actually, the FREEDM system is designed as a distribution system which is based on SST configurations interfacing residential and industrial consumers. It was presenting three key technologies as plug-and-play feature for any source, the second is providing an energy router namely intelligent energy management device connecting 12 kV ac distribution network to 120 V ac residential and 400 V dc buses, and the third feature is to bring an open standard operating system which is called distributed grid intelligence (DGI) embedded into energy routers that are comprised by SSTs. An example energy router that has been proposed as the first-generation SST of FREEDM is shown in Fig. 9.3 where the distribution bus is comprised by

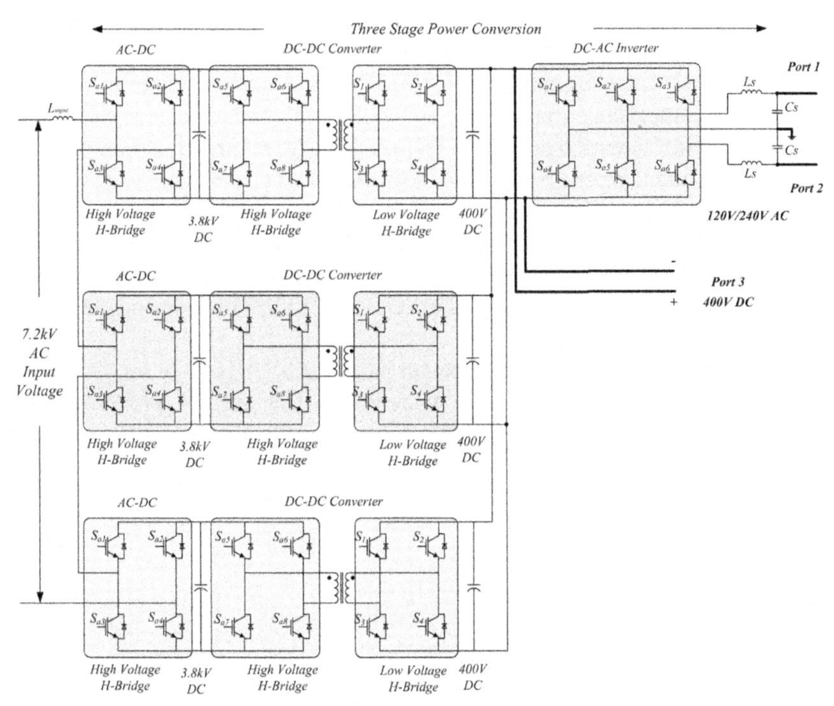

FIG. 9.3 Energy router comprised by first generation SST of FREEDM with 6.5 kV-25A Si IGBTs.

7.2 kV ac input voltage. The first-generation energy router is built with 6.5 kV/25A Si insulated gate bipolar transistors (IGBT) for ac-dc and dc-dc converters as seen in the blocks of figure. The input converters which are used to rectify and regulate dc bus voltages are cascaded to increase rated power of SST. Each dc-dc converter is followed by a full-bridge inverter that is fed by a single dc intermediate bus at 400 V. The output converter provides three ports for 120 V/240 V ac output voltages and a 400 V dc output voltage.

The improved versions of FREEDM architectures are classified and Gen-II and Gen-III due to accomplished developments in solid state technologies and device topologies. The Gen-II SST which is also proposed as energy router is comprised by three-stage architecture similarly to first generation, but it was differing with single level ac input voltage, dual half-bridge dc converter topology and SiC semiconductors operating at 15 kV/10A voltage and current ratings. Moreover, the SiC mosfets of Gen-II SST was capable to interface distribution bus voltage up to 20 kV with decreased number of active switching devices. The third-generation SST of FREEDM is improved by replacing dual active bridges with series resonant converter and by increasing 97% overall efficiency of Gen-II to 98.2% in Gen-III architecture [22].

The SSTs that are proposed as energy routers are essentially based on power switches. The semiconductors used in SSTs are selected from high switching frequency devices which are not suitable to be operated at high voltages. Therefore, serial or cascaded connections of IGBTs and mosfets are used to comprise energy routers. However, conventional Si semiconductors are high in volume and weight comparing to novel SiC and GaN semiconductor devices. 15 kV SiC IGBTs which are used in Gen-III SST seen in Fig. 9.4 are capable to operate at 10 kHz for reducing the transformer size and volume, and decreasing overall size of energy router. On the other hand, SiC mosfets provide higher switching frequencies up to 50 kHz comparing to SiC IGBTs. The Gen-III energy router SST has three connection terminals at different voltage levels as 7.2 kV input bus that is connected to 12 kV distribution line, 120 V/240 V ac and 400 V dc output ports [21, 22].

The control operation of energy router based on SSTs is more sophisticated comparing to conventional power converters. The control processes are

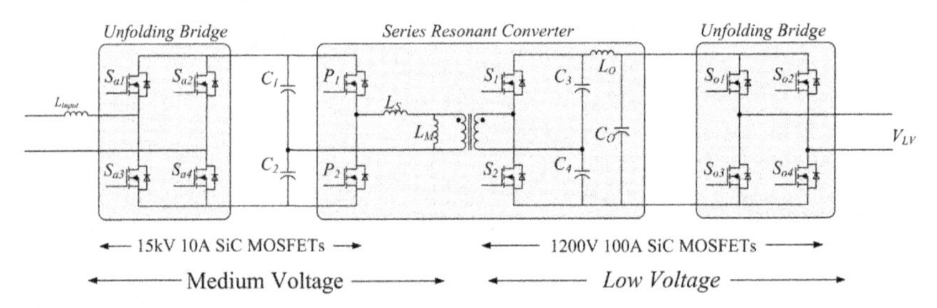

FIG. 9.4 Gen-III energy router of FREEDM with 15 kV-10A SiC Mosfets.

classified into two categories as intelligent energy management and intelligent fault management. The management nodes comprise the distributed grid intelligence which is cyber-physical system of FREEDM.

9.3.2 Information sensing and processing

The main target of EI concept is providing more flexible energy systems where EI concept is capable of better managing entire energy systems with combining several important structures such as data sensing, metering, collecting, processing, transmission and services. In this context, advanced metering infrastructure (AMI) undertakes a great importance and is a promising technology for enabling metering and monitoring processes among users and utilities [23–25]. On the other hand, the most important component of the AMI systems is the SMs that are modern electricity meters with two-way energy supporting and advanced features when compared conventional electricity meters. These modern meters are established at the user side for acquiring users' electricity information. A SM is responsible for collecting real-time energy consumption rates accurately to transmit energy management center and analyzing these measurement data to present information to users about various information regarding measurements [5]. In the EI concept, it is possible to adapt cloud computing technologies to handle big data.

The control interface of EI plane is software based which is organized in a centralized logic. The data and energy controllers are designed in software defined (SD) structure for controlling data and energy flow. The data flow control is closely related with cloud computing in EI. The SD data controller allows to configure the network devices and optimizing the data management infrastructure. The reliability and efficiency of energy transmission is also another important aspect of SD controllers. The centralized energy controllers are required to maintain reliability of EI during generation or consumption operations [26].

9.3.3 Network topologies for EI

The EI system can be taken into account as a massive network which associate several equipment and users in an effective and secure way. Generally, the network topologies of EI systems are classified in three main categories as SDN, energy-efficient routing in CR and cellular network structures. The SDN architecture is considered as a central control network scheme. In this structure, relaying and managing stages are individually fulfilled via user-defined features that present several advantages for network management in terms of flexibility and credibility. A typical SDN scheme is composed of a switch, terminal equipment, a controller and an application layer [27]. In the SDN, it is also possible to set apart managing functions from network switches that involve separation of routing and managing components, advanced control characteristics and improved functions. Several wireless communication technologies can be implemented in the EI communications. Because of its effective spectrum resource management features, the CR systems are widely utilized in communication networks of SG systems [28, 29]. The idea of employing CR technologies has been arisen in various application areas such as device-to-device (D2D), WSNs and SGs. The authors researched architecture and hardware applications of CR technologies for SG systems in [30]. Network topology of current cellular systems contains many cell structures that enable uninterrupted data flows between cells [4]. These networks can be expressed analytically depending on the load areas, DESs and management areas. Therefore, the cellular networks can be efficiently utilized to accomplish advanced communications in the EI systems.

9.4 Current challenges of EI systems

This section presents various current challenges of EI concept that can be mainly sorted out as complexity, efficiency, reliability and security as can be seen from Fig. 9.5. Since the EI system is a novel technology aiming to integrate information technology, energy systems and network systems to present more reliable and flexible power grid, this new concept can be considered as system of systems that leads to massive and complex systems [1, 2]. Hence, communication systems to be developed for EI concept will certainly experience important problems. Researchers currently try to develop various models in order to overcome the complexity issue. This concept is created to provide a balance between energy sources and energy consumption. Thus, establishing a reliable information infrastructure among each components is a crucial issue to ensure efficiency of the EI system. The EI exhibits several behaviors in terms of scalability, stability and security depending on the system characteristics such as network scheme, management methods and so on. The selection of network topology may be also varied based on energy type and its operation procedures.

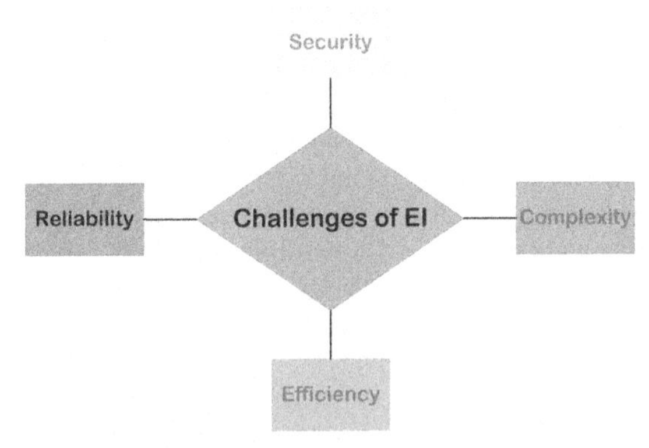

FIG. 9.5 Current important problems of EI system.

One of the critical challenges is the reliability issue, which is correlated with physical architecture of network and transmission techniques. The carrier of transmission method is realized by energy transmission network. In addition, important components of EI system such as energy routers, energy resources and control systems need to operate in cooperation to enhance reliability of the systems. Several recent challenges of EI concept have emerged depending on the developments of utilization communication technologies in power systems and security concerns. Security threats of EI systems can be classified into two categories as natural threats and artificial (man-made) threats. When the threats are considered for information systems of EI concept, the system should be taking measures for preventing malicious attacks, improving security for information transmission and privacy protection. When the issues are considered in terms of energy, the system should be taking measures for ensuring stability of energy transmission network and robustness of energy sources. It is also important to note that interference problems of communication and energy transfer systems should be taken into account.

The EI vision that aims to integrate energy, information and network systems is introduced as an innovative power grid. One of the most important components of EI systems is communication technologies that are very crucial systems to accomplish energy management processes between components of EI system efficiently. This chapter presents a general outlook for EI systems. Firstly, we introduce motivations of EI vision and explain superiorities of these systems comparing with SGs. Then, technical background and communication system requirements of these systems are analyzed. In addition, EI architecture is explained in three subcategories as energy routers, information sensing and processing and network topologies where it is clearly seen that energy routers are most important component of EI systems. Furthermore, different from the conventional and SG systems, the energy routers have very critical tasks in the EI systems such as controlling both of energy and information exchanges between all entire components and networks and achieving harmony between energy supplies and demands.

References

[1] K. Wang, X. Hu, H. Li, P. Li, D. Zeng, S. Guo, A survey on energy internet communications for sustainability. IEEE Trans. Sustain. Comput. 2 (2017) 231–254, https://doi.org/10.1109/TSUSC.2017.2707122.

[2] K. Wang, J. Yu, Y. Yu, Y. Qian, D. Zeng, S. Guo, Y. Xiang, J. Wu, A survey on energy internet: architecture, approach, and emerging technologies. IEEE Syst. J. (2017) 1–14, https://doi.org/10.1109/JSYST.2016.2639820.

[3] Q. Sun, Y. Zhang, H. He, D. Ma, H. Zhang, A novel energy function-based stability evaluation and nonlinear control approach for energy internet. IEEE Trans. Smart Grid 8 (2017) 1195–1210, https://doi.org/10.1109/TSG.2015.2497691.

[4] K. Wang, H. Li, Y. Feng, G. Tian, Big data analytics for system stability evaluation strategy in the energy internet. IEEE Trans. Ind. Informat. 13 (2017) 1969–1978, https://doi.org/10.1109/TII.2017.2692775.

[5] Y. Kabalci, A survey on smart metering and smart grid communication. Renew. Sust. Energ. Rev. 57 (2016) 302–318, https://doi.org/10.1016/j.rser.2015.12.114.

[6] L.H. Tsoukalas, R. Gao, From smart grids to an energy internet: assumptions, architectures and requirements. in: 2008 Third International Conference on Electric Utility Deregulation and Restructuring and Power Technologies, 2008, pp. 94–98, https://doi.org/10.1109/DRPT.2008.4523385.

[7] C.-C. Lin, D.-J. Deng, W.-Y. Liu, L. Chen, Peak load shifting in the internet of energy with energy trading among end-users. IEEE Access 5 (2017) 1967–1976, https://doi.org/10.1109/ACCESS.2017.2668143.

[8] F.F. Wu, P.P. Varaiya, R.S.Y. Hui, Smart grids with intelligent periphery: an architecture for the energy internet. Engineering 1 (2015) 436–446, https://doi.org/10.15302/J-ENG-2015111.

[9] H. Jiang, K. Wang, Y. Wang, M. Gao, Y. Zhang, Energy big data: a survey. IEEE Access 4 (2016) 3844–3861, https://doi.org/10.1109/ACCESS.2016.2580581.

[10] K. Wang, Y. Wang, X. Hu, Y. Sun, D.-J. Deng, A. Vinel, Y. Zhang, Wireless big data computing in smart grid. IEEE Wirel. Commun. 24 (2017) 58–64, https://doi.org/10.1109/MWC.2017.1600256WC.

[11] K. Wang, L. Gu, X. He, S. Guo, Y. Sun, A. Vinel, J. Shen, Distributed energy management for vehicle-to-grid networks. IEEE Netw. 31 (2017) 22–28, https://doi.org/10.1109/MNET.2017.1600205NM.

[12] W. Su, A. Huang, The energy internet and electricity market in the United States. Chin. Sci. Bull. (2016)https://doi.org/10.1360/N972015-00761.

[13] L. Chen, Q. Sun, L. Zhao, Q. Cheng, Design of a novel energy router and its application in energy internet. in: 2015 Chinese Automation Congress (CAC), 2015, pp. 1462–1467, https://doi.org/10.1109/CAC.2015.7382730.

[14] C. Wu, T. Yoshinaga, Y. Ji, T. Murase, Y. Zhang, A reinforcement learning-based data storage scheme for vehicular ad hoc networks. IEEE Trans. Veh. Technol. 66 (2017) 6336–6348, https://doi.org/10.1109/TVT.2016.2643665.

[15] Y. Xu, J. Zhang, W. Wang, A. Juneja, S. Bhattacharya, Energy router: architectures and functionalities toward energy internet, in: Smart Grid Communications (SmartGridComm), 2011 IEEE International Conference on, IEEE, 2011, pp. 31–36.

[16] M. Geidl, G. Koeppel, P. Favre-Perrod, B. Klockl, G. Andersson, K. Frohlich, Energy hubs for the future, IEEE Power Energy Mag. 5 (2007) 24–30.

[17] H. Guo, F. Wang, J. Luo, L. Zhang, Review of energy routers applied for the energy internet integrating renewable energy, in: Power Electronics and Motion Control Conference (IPEMC-ECCE Asia), 2016 IEEE 8th International, IEEE, 2016, pp. 1997–2003.

[18] J. Zhang, W. Wang, S. Bhattacharya, Architecture of solid state transformer-based energy router and models of energy traffic. in: 2012 IEEE PES Innovative Smart Grid Technologies (ISGT), 2012, pp. 1–8, https://doi.org/10.1109/ISGT.2012.6175637.

[19] P. Favre-Perrod, A vision of future energy networks. in: 2005 IEEE Power Engineering Society Inaugural Conference and Exposition in Africa, 2005, pp. 13–17, https://doi.org/10.1109/PESAFR.2005.1611778.

[20] M. Geidl, G. Koeppel, P. Favre-Perrod, B. Klöckl, G. Andersson, K. Fröhlich, The energy hub–a powerful concept for future energy systems, in: Third Annual Carnegie Mellon Conference on the Electricity Industry, 2007, p. 14.

[21] A.Q. Huang, M.L. Crow, G.T. Heydt, J.P. Zheng, S.J. Dale, The future renewable electric energy delivery and management (FREEDM) system: the energy internet. Proc. IEEE 99 (2011) 133–148, https://doi.org/10.1109/JPROC.2010.2081330.

[22] E. Kabalci, Y. Kabalci, Smart Grids and Their Communication Systems, Springer, Berlin, Heidelberg/New York, NY, 2018.

[23] Z.s. Zheng, W.d. Deng, Z.g. Zhang, W. Fan, Z.m. Xia, X.j. Qi, J.H. Zhang, Q.c. Qu, Study on establishing the power grid advanced metering infrastructure by applying on-site working standard. in: 2012 International Conference on Systems and Informatics (ICSAI2012), 2012, pp. 550–555, https://doi.org/10.1109/ICSAI.2012.6223057.

[24] D. Li, B. Hu, Advanced metering standard infrastructure for smart grid. in: 2012 China International Conference on Electricity Distribution, 2012, pp. 1–4, https://doi.org/10.1109/CICED.2012.6508429.

[25] S. Chen, K. Xu, Z. Li, F. Yin, H. Wang, A privacy-aware communication scheme in advanced metering infrastructure (AMI) systems. in: 2013 IEEE Wireless Communications and Networking Conference (WCNC), 2013, pp. 1860–1863, https://doi.org/10.1109/WCNC.2013.6554847.

[26] W. Zhong, R. Yu, S. Xie, Y. Zhang, D.H.K. Tsang, Software defined networking for flexible and green energy internet. IEEE Commun. Mag. 54 (2016) 68–75, https://doi.org/10.1109/MCOM.2016.1600352CM.

[27] D. Kreutz, F.M.V. Ramos, P.E. Veríssimo, C.E. Rothenberg, S. Azodolmolky, S. Uhlig, Software-defined networking: a comprehensive survey. Proc. IEEE 103 (2015) 14–76, https://doi.org/10.1109/JPROC.2014.2371999.

[28] A.S. Cacciapuoti, M. Caleffi, F. Marino, L. Paura, Sensing-time optimization in cognitive radio enabling smart grid. in: 2014 Euro Med Telco Conference (EMTC), 2014, pp. 1–6, https://doi.org/10.1109/EMTC.2014.6996657.

[29] F. Akhtar, M.H. Rehmani, M. Reisslein, White space: definitional perspectives and their role in exploiting spectrum opportunities. Telecommun. Policy 40 (2016) 319–331, https://doi.org/10.1016/j.telpol.2016.01.003.

[30] R.C. Qiu, Z. Hu, Z. Chen, N. Guo, R. Ranganathan, S. Hou, G. Zheng, Cognitive radio network for the smart grid: experimental system architecture, control algorithms, security, and microgrid testbed. IEEE Trans. Smart Grid 2 (2011) 724–740, https://doi.org/10.1109/TSG.2011.2160101.

Index

Note: Page numbers followed by *f* indicate figures, *t* indicate tables, and *b* indicate boxes.

C

D

F

G

H

I

Q

R

S

T

U

V

W

Z

Printed in the United States
By Bookmasters